D0845311

LAUGHING GAS,

VIAGRA, AND LIPITOR

Laughing Gas, Viagra, and Lipitor

The Human Stories Behind the Drugs We Use

Jie Jack Li

2006

OXFORD
UNIVERSITY PRESS

Oxford University Press, Inc., publishes works that further
Oxford University's objective of excellence
in research, scholarship, and education.

Oxford New York
Auckland Cape Town Dar es Salaam Hong Kong Karachi
Kuala Lumpur Madrid Melbourne Mexico City Nairobi
New Delhi Shanghai Taipei Toronto

With offices in
Argentina Austria Brazil Chile Czech Republic France Greece
Guatemala Hungary Italy Japan Poland Portugal Singapore
South Korea Switzerland Thailand Turkey Ukraine Vietnam

Published by Oxford University Press, Inc.
198 Madison Avenue, New York, New York 10016

www.oup.com

Library of Congress Cataloging-in-Publication Data

Li, Jie Jack.
Laughing gas, viagra, and lipitor : the human stories behind the drugs we use / Jie Jack Li.
p. cm.
Includes bibliographical references and index.
ISBN-13 978-0-19-530099-4
ISBN 0-19-530099-8
1. Drugs—Miscellanea. 2. Nitrous oxide. 3. Sildenafil. I. Title.
RM301.15.L5 2006
362.29'9—dc22 2006002586

9 8 7 6 5 4 3 2 1
Printed in the United States of America
on acid-free paper

To Sherry

O, mickle is the powerful grace place that lies
In herb, plants, stones, and their true qualities:
For nought so vile that on earth doth live,
But to the earth some special good doth give;
Nor aught so good but, strain'd from that fair use,
Revolt from true birth, stumbling on abuse:
Virtue itself turns vice, being misapplied;
And vice sometime's by action dignified.
Within the infant rind of this weak flower
Poison hath residence, and medicine power:
For this, being smelt, with that part cheers each part;
Being tasted, slays all senses with the heart.
Two such opposed kings encamp them still
In man as well as herbs, grace and rude will;
And where the worser is predominant,
Full moon the canker death eats up that plant.

—William Shakespeare, *Romeo and Juliet*
Act II, Scene iii

FOREWORD

Any person interested in understanding the history of medical progress and the way in which many of the most important medicines came into existence will be well rewarded by reading this latest book by Jie Jack Li, one of the most prolific and interesting authors in the field of molecular medicine and chemistry. And if that person happened to be aiming for or beginning a career in medicinal discovery, he or she would gain even more by so doing. The many case histories within this book contain a large number of valuable take-home lessons and extraordinary insights that have emerged from a spectrum of research over many years in medicinal science, or what can now be called molecular medicine. These stories reveal not only the complexity, unpredictability, and challenges of the discovery process but also the intellectual and human qualities upon which success depended. I found this to be a most inspiring, engaging, and broadening reading experience, and I cannot imagine that this sense of reward would not be shared by most individuals who want to know more about health sciences, creativity, discovery, and just plain luck (good and bad).

It is unimaginably difficult (and now very costly) to discover an important new therapeutic agent. Most people who have engaged in this effort for a lifetime have not been so fortunate. It is not easy to select a research objective that will turn out to be a winner. The enormous complexity of the human body (especially in molecular terms), the depth of our ignorance in the molecular and biological sciences, and the lack of more powerful and yet to be discovered scientific research tools all conspire, like the fog of war, to obscure our vision in the search for a great new discovery. So one needs not only keen intellect and extraordinary personal qualities to be successful but also as much perspective, perception, and intuition as one can muster. This book provides a clear picture of these human elements behind success. As a bonus, it is also entertaining and enjoyable to read.

It will be gratifying if the future audience for this fine work includes a wide swath of the general public, as they too will be fascinated by the inner workings of research at the frontier.

E. J. Corey
June 2005
Cambridge, Massachusetts

❧ ❧ ❧

Over the course of the past century, science has made such tremendous progress in the battle against human disease that many of the ailments that were lethal to our ancestors no longer pose a threat to us. Yet, for most people living in modern society, the process by which such a tremendous revolution has taken place remains a mystery, with drugs merely constituting a pill or liquid prescribed by a doctor and bought at a pharmacy. *Laughing Gas, Viagra, and Lipitor: The Human Stories Behind the Drugs We Use* fills this gap beautifully.

Through a series of enlightening and highly entertaining chapters, Jie Jack Li takes the reader on an unparalleled journey through the history of drugs, a trip that conveys both the science underlying the discovery and the personalities and drama that accompany them. Each section glows with behind-the-scenes glimpses at the brilliance, passion, and frailty of the people involved in the discoveries, from early pioneers to the heroic and landmark team efforts of today's pharmaceutical companies that amalgamate the expertise of scientists in such diverse fields as synthetic chemistry, biology, pharmacokinetics, drug metabolism, and toxicology. What is particularly striking in chapter after chapter is the astonishing and sometimes chilling role that serendipity plays in the drug discovery process. For instance, how many people would have died from bacterial infection had Fleming covered his petri dish while on vacation, or had the first in vivo studies of penicillin been performed on guinea pigs (as penicillin is toxic to them)? How many people know that the discovery of digitalis, used for the treatment of congenital heart failure, arose from a young lady's love of flowers?

These questions offer just a glimpse of the many hidden treasures that lie within this text. I offer my heartiest congratulations to Dr. Li for writing a uniquely original book that is a joy to read and that will undoubtedly serve as an inspiration for scientists who are tirelessly at work developing the cures of the future and for students who aspire to such a noble cause.

Phil S. Baran
April 2005
La Jolla, California

PREFACE

THIS IS A book on the history of drug discovery. More important, this is a book about drug discoverers that highlights their intellectual splendor, as well as their human frailty. Emil Fischer, perhaps the greatest organic chemist ever, once stated: "Science is not an abstraction; but as a product of human endeavor it is inseparably bound up in its development with the personalities and fortunes of those who dedicate themselves to it." In his book *Ascent of Man,* Jacob Bronowski wrote: "Discoveries are made by men, not merely by minds, so that they are alive and charged with individuality."[1] When I set out to take on this project, I did not intend to write another scholarly tome—various topics have been published extensively in specialized areas of medical history. In contrast, I wanted to place emphasis on the human aspect of drug discovery by telling fascinating stories about the discoverers: their aspirations and activities, their hard work and frustrations, their hopes and fears, their setbacks and triumphs. History is replete with examples of breakthrough medicines that have saved millions of lives. The discovery of ether as an anesthetic by Morton, of penicillin as an antibiotic by Fleming, and of insulin for diabetes by Banting are just a few examples. The discoverers of these medicines are certainly benefactors to mankind—for instance, without penicillin, 75% of us probably would not be alive and reading this book, because some of our parents or grandparents would have succumbed to infections. On the other hand, as a testimony to human frailty, some of the scientists suffered misunderstanding, antipathy, and heartbreak. Some of them turned from collaborators into acrimonious scientific rivals. As is often the case, reality is more dramatic than fiction.

I strongly believe that man, if given enough time, can overcome any medical problem. The future seemed seriously grim when AIDS became an epidemic 20 years ago. Now the advent of HIV protease inhibitors and reverse transcriptase inhibitors as antiviral agents have transformed AIDS from a death sentence to a disease that can be controlled.

In preparing this manuscript, I have had the great fortune to communicate with some scientists who played pivotal roles in discovering the drugs we use. They are Steven J. Brickner and Michael R. Barbachyn (Zyvox); Carl Djerassi (norethindrone); V. Craig Jordan (Tamoxifen); Bruce D. Roth (Lipitor); Nicholas K. Terrett (Viagra); Joseph Vacca (Crixivan); Robert Vince (Ziagen), and Jürg Zimmermann (Gleevec). There have been many myths and inaccuracies associated with these well-known drugs. The inventors' perspectives afford this book invaluable accuracy and insight, because history is not history unless it is true.

Drug discovery often involves many people across many disciplines—teamwork is of paramount importance. In this manuscript, I have often focused on the key players in the discovery of drugs. This is in no way meant to marginalize collaborations essential to the success of the venture.

In preparing this manuscript, I have also incurred many debts of gratitude to my academic friends, E. J. Corey at Harvard University and Phil S. Baran at Scripps Research Institute, who have provided much enthusiastic encouragement. Corey, a Nobel laureate in chemistry, read the entire manuscript and provided countless invaluable comments and suggestions. My friends in industry, Danielle Mills, Lorna H. Mitchell, Bruce D. Roth, and Dennis Vargo, also proofread the entire manuscript and offered numerous suggestions for improvements in terms of both science and English. Roth and Vargo, both veterans in the pharmaceutical industry, afforded precious insight into drug discovery. Friends and colleagues who also proofread portions of the manuscript are: Narendra Ambhaikar, Emily Andrews, Noah Burns, Michael DeMartino, Ben Hafensteiner, Douglas S. Johnson, Marc Klinger, David Lin, Carlo Lanza, Dan O'Malley, Thomas Peterson, Derek A. Pflum, Jeremy Richter, and Jacob Schwarz. A big thank-you goes to Jeremy A. Lewis at Oxford University Press, who has been instrumental to the success of every stage of this project. Pfizer kindly granted us permission to use Robert Thom's paintings from the book *Pictorial Annals of Medicine and Pharmacy* © 1999 by Warner-Lambert Company, which was acquired by Pfizer in 2000. Finally, I am grateful for permission to use the postage stamps from respective postal authorities, who still retain the copyrights for those stamps.

Disclaimer

No word in this book is to be regarded as affecting the validity of any trademark. Mention of specific companies, organizations, or drugs does not imply endorsement of the author and publisher, nor does mention of companies, organizations, or authorities in the book imply that they endorse the book. The author's statements are merely his own personal opinions, not those of his employer or the companies concerned. This book, an account of history in drug discovery, is not intended to dispense medical advice for individual health and drug-related issues. Please consult your physician and/or your pharmacist with regard to your medical needs.

Viagra and Lipitor are the trademarks of Pfizer, Inc. The trademarks of the drugs mentioned in this book are listed in Appendix B, "Trademarks of the Drugs," on page 277.

CONTENTS

LAUGHING GAS,
VIAGRA, AND LIPITOR

CHAPTER 1

Cancer Drugs

From Nitrogen Mustards to Gleevec

Discovery needs luck, invention, and intellect—
neither can do without the other.

—Johann Wolfgang von Goethe

How Does Cancer Begin?

Whereas infectious diseases are the scourge of developing countries, cancer is the most significant affliction in developed countries. At the beginning of 2005, the American Cancer Society announced that, for the first time, cancer had surpassed heart disease as the number one killer of Americans.

Cancer, which is the uncontrollable multiplication of cells, has been in existence as long as animals have; evidence of cancers has been found in dinosaur bones. Cancers have also been found on mummies dating back 2,500 years. An operation to remove cancer was documented in the Ebers papyrus found in Egypt. In ancient times cancer was a relatively rare disease, because infectious diseases often made the life span so short that cancer had little chance to proliferate. Hippocrates (460–370 B.C.) coined the word *cancer,* which means "crab" in Greek. There are over 110 types of cancer, which can be divided into four categories depending on the tissue involved: *carcinoma, lymphoma, leukemia,* and *sarcoma. Carcinomas* are the most common, with 85–90% of all cancers falling into this

Fighting Cancer © Belgium Post

category. Carcinomas are tumors that originate in epithelial tissue such as skin, breast, lung, prostate, stomach, colon, ovary, and so forth. *Lymphomas* are cancers of the lymphatic system. *Leukemia* is the cancer of the blood, bone marrow, and liver. *Sarcomas,* the rarest of all four types, are tumors arising from cells in connective tissue, bone, or muscle.

It seems inconceivable that we had almost no clue about the origin of cancers up until the mid-1970s, despite the existence of cancer that predates human life. The debate raged on as to what caused cancer, with one camp believing that carcinogens (cancer-causing agents such as chemicals, X-ray, and ultraviolet light) were to blame, whereas the other thought that viruses were the culprits.

The carcinogen theory took root first. As early as 1775, British doctor Percival Pott made the astute epidemiological observation that young English boys employed as chimney sweeps were more prone to develop scrotal skin cancers than their French counterparts.[1] Further scrutiny revealed that the continental sweeps bathed more frequently after work, which prompted Pott to speculate that long exposure to coal tar caused skin cancer. In 1915, 140 years later, Katsusaburo Yamagiwa and Koichi Ichikawa confirmed Pott's theory in an animal model.

Yamagiwa went to Germany in 1891 to study pathology with the legendary Rudolf Virchow. After returning to Japan in 1894, Yamagiwa spent more than 20 years studying cancer. Intrigued by Pott's hypothesis, Yamagiwa and his coworker, Ichikawa, applied coal tar to rabbit ears three times a day. After a year, they observed that 7 of the 137 rabbits developed grotesque skin cancer on their ears. Before this study, many scientists had tried the same approach, but none had run studies of sufficient duration to see the effects. Yamagiwa and Ichikawa presented their discovery at the 1915 Conference of the Tokyo Medical Society. This marked the first time that a chemical was shown to cause cancer in an animal model.[1] In 1926, the Nobel Prize in medicine was awarded for making great strides toward solving the mysteries of cancer. Unfortunately, Yamagiwa was not the recipient, which was considered by many to be a great injustice. The honor went to Johannes Fibiger from Denmark "for his discovery of *Spiroptera* carcinoma."[2] Fibiger's theory that a worm caused stomach cancer was later debunked because his experiments could not be reproduced.

The Germans were among the first to identify tobacco as a cause of lung cancer.[3,4] During World War II, the Nazis waged a campaign against smoking. Their propaganda machine constantly reminded citizens of the Third Reich that Hitler, Mussolini, and Togo were all nonsmokers, whereas Churchill, Roosevelt, and Stalin were all smokers. Some Nazi doctors even referred to smoking as "lung masturbation." Meanwhile, German scientists identified many carcinogens from tobacco tar. By 1964, the initial U.S. Surgeon General's warning began to appear on cigarette boxes, because the link between smoking and lung cancer was solidly established.

While the carcinogen theory was gaining momentum, the virus (meaning "poison" in Latin) theory emerged, proposing that viruses caused cancer. At least two events advanced the belief that viruses caused cancer: one involved the Rous sarcoma virus, the other the Abelson virus.

The Rous sarcoma virus was first isolated from a chicken. In 1909, a farmer from Long Island brought a Plymouth Rock hen to Peyton Rous at the Rockefeller Institute. The chicken had a large muscle tumor in her right breast, and the farmer wanted Rous to remove it. To the farmer's chagrin, Rous removed it all right—by slaughtering his prized hen! Rous then proceeded to grind up the tumor and inoculate another young chicken with the tumor extract. A few weeks later, the chicken developed the same type of cancer at the injection area. This was possibly the first time that a tumor was artificially produced in an animal using a tumor virus, which was later named the Rous sarcoma virus, or RSV (not to be confused with Respiratory Syncitial Virus).[1] Fifty-five years later, at the age of 89 and still working, Rous won the 1966 Nobel Prize in Medicine for "his discovery of tumor inducing viruses."[5]

The Abelson virus was isolated from mice. Herbert Abelson, a pediatrician, initially worked at the National Institutes of Health (NIH) for several years before moving to the Massachusetts Institute of Technology (MIT). During his research into childhood leukemia, he had injected leukemia viruses into mice, which then went on to develop severe bone marrow cancer. In effect, the mice mutated the leukemia virus to another form that caused bone marrow cancer. The new virus was named the Abelson virus, or *abl.*

In addition to RSV and *abl,* there are many virus genes, often represented by three-letter abbreviations such as *abl, src, ras, myb,* and *erb.* The aforementioned two stories may shed some light as to how these names originated.

Along the same lines, *src* (pronounced "sarc") is related to sarcoma virus; *ras*, rat sarcoma virus; *myb*, avian myeloblastosis virus; and *erb*, chicken erythroblastosis retrovirus. Other viruses implicated in oncogenesis are the Epstein-Barr virus (EBV) and the human T-cell leukemia virus (HTLV).

After all was said and done, the virus theory also fell short—like the carcinogen theory, the virus theory failed to explain the vast majority of human cancers that have no viral association at all. Thanks to the diligent work of numerous pioneers in cancer research, we have now amassed enough knowledge to appreciate that, although in some cases it may be caused by carcinogens or viruses (for instance, liver cancer, which is rampant in parts of the Third World, is caused in large part by hepatitis B and C viruses), cancer is ultimately a disease of genes.

Bruce Ames was a professor of genetics at the University of California, Berkeley. His research on bacterial genetics resulted in a powerful tool, known as the Ames test, to predict the ability of chemicals to damage human genes. In the Ames test, various putative mutagens or carcinogens are placed with a histidine-requiring bacterium (*Salmonella typhinurium*) that, because of its histidine-requiring state, is nonviable and will not produce bacterial colonies in the histidine-deficient medium in which it is placed. If the mutagen or carcinogen induces the bacteria to convert the available medium precursors into histidine, the bacteria will survive and grow in the medium. Colony growth, therefore, indicates mutagenesis and is thus considered a "positive" Ames test. Because the Ames test involves bacteria rather than mice or rats, it can be performed overnight, as opposed to requiring weeks or even months. The Ames test is now widely used to test the mutagenic potency of chemicals and drugs—their ability per unit weight to induce mutation in exposed DNA. In the 1970s, based on a vast amount of data, Ames proposed the "carcinogen = mutagen" theory: in essence, carcinogens such as X-rays and chemicals act by damaging DNA, thereby creating mutations in genes of targeted cells.

In 1975, J. Michael Bishop and his postdoctoral fellow Harold E. Varmus at the University of California, San Francisco, discovered the *src* viral gene that caused cancer in cells infected by the virus. They shared the 1989 Nobel Prize in Medicine or Physiology as a result of their work. In 1979, Robert A. Weinberg's group at MIT isolated a gene from a rat's brain tumor. Weinberg christened it the *neu* gene to signify its origin from a neurological tumor and moved on to other fields. He did not clone the protein of the neu gene, a relatively easy experiment to do. *Cloning* means to produce multiple, exact copies of a single gene. The monoclonal antibody earned its name because it contains a single antibody from the same clone and is thus identical in all aspects. Monoclonal antibodies also attack a single target selectively, thus offering a much better safety profile and less toxicity than chemotherapy. Had Weinberg carried out the cloning experiment, he could very well have been the third Nobel awardee a decade later, along with Bishop and Varmus.

Despite not having cloned the gene, Weinberg had his own share of good luck as well. In 1980, his group discovered the first human oncogene *ras,* isolated from rat sarcoma virus.[6] They successfully created tumor cells by adding genetic flaws to normal human cells. Gradually, evidence began to amass that pointed to the possibility that cancer was triggered by oncogenes. An oncogene—*onco* means "mass" or "tumor" in Latin—is a gene that causes normal cells to form tumors. There are about 50 oncogenes known thus far. In 1997, Weinberg's group cloned the first of some 20 now-known tumor-suppressor genes—the *Rb* (retinoblastoma) gene from a rare childhood eye cancer. Tumor-suppressor genes, as their names imply, suppress the development and growth of tumors.[7]

The findings of Weinberg and others revolutionized the way we think about how cancer arises. The current view is that cancer is a multistep process, characterized by mutations in several oncogenes and the loss of function of tumor-suppressing genes. Oncogenes are activated by either an inherited defect or exposure to an outside agent, a carcinogen. However, oncogenes are normally kept inactive by tumor-suppressing genes. In some cases those controlling genes may be mutated or removed, allowing oncogenes to run rampant. For a tumor to develop, one has to lose one tumor-suppressing gene in addition to having two or more oncogenes.

While Weinberg was working to prove his hypothesis that it took two or more oncogenes to cause cancer, a young scientist in Canada almost scooped him. In the late 1970s, Demetrios Spomdidos was a postdoctoral fellow in Louis Siminovitch's laboratories in Toronto. By using a new gene transfer technique, he claimed to have proved unequivocally that two cancer genes resided in a tumor cell, not just one. Moreover, Siminovitch achieved an extraordinary feat by converting cancer cells into normal noncancerous cells. When the work was submitted to the prestigious journal *Cell,* an expert reviewer gave it glowing praise. At the end of the review, the senior referee paternally pointed out that the paper contained so much experimental work that it would have taken his own laboratories 3–4 times longer to achieve the same results. Sadly, he was right! Out of curiosity, Siminovitch calculated that the number of petri dishes that Spomdidos reportedly used had vastly exceeded the total number of petri dishes that he purchased for the entire group. Siminovitch dismissed Spomdidos on the spot after he failed to provide a satisfactory explanation.[7,8] Even today, we do not know what truly transpired in the laboratory.

The understanding of how cancer arises now allows us to tackle the disease with an appreciation of its molecular basis. Recent targeted cancer drugs are exemplified by Herceptin, Erbitux, Avastin, Gleevec, and Tarceva, with more surely to follow.

Chemotherapy

Chemotherapy is the treatment of serious diseases such as cancer with strong cytotoxic drugs. *Cyto-* means "cell" in Latin; thus cytotoxic means that the drug destroys rapidly growing cancer cells. The principle of cytotoxic chemotherapy for cancer is based on the fact that cancer cells grow rapidly. Cytotoxic chemotherapy differentially affects cells that are rapidly growing, preferentially killing them. This is also the reason for the occurrence of the frequent side effects of cytotoxic chemotherapy—diarrhea, mucosal ulcers, hair loss, and so forth—that the organs and tissues involved are composed of rapidly growing cells.

Nitrogen Mustard

Mustard gas is a deadly chemical weapon developed at the turn of the twentieth century and first used during World War I.[9] Throughout the years, it has killed or maimed countless people. One famous victim was a 29-year-old German corporal, Adolf Hitler, who became temporarily blinded from the mustard gas assault launched by the Allies on the Western front in October of 1918. Ironically, chemotherapy using nitrogen mustard-like drugs was discovered as a result of mustard gas.

In 1943, during World War II, the Italian campaign was raging. Berthed at the Bari Harbor on the Adriatic Sea, along with another 30 munitions ships and tankers, was the *S. S. John E. Harvey.* The merchant ship was loaded with 100 tons of mustard gas contained in 2,000 M46A1

100-pound bomb casings. President Roosevelt, alarmed by the reports of imminent use of chemical weapons by the Axis, had ordered the United States to stockpile these weapons. The hope was that the Axis would be deterred by learning that the United States also possessed chemical weapons. On the evening of December 2, the *John Harvey,* along with 16 other Liberty merchant ships, was sunk by Luftwaffe Ju-88 bombers. Mustard gas spewed out over the harbor and the town of Bari. More than 1,000 Allied sailors and port personnel died immediately; more than 800 additional casualties were hospitalized, with 628 casualties sustaining different degrees of mustard burns. Sixty-nine more died within the ensuing weeks. Because everyone on the ship *John Harvey* who could identify the mustard gas cargo was killed in the raid, no one could attest to what had caused the deaths and the injuries. Lieutenant Colonel Stewart F. Alexander, a consultant with Chemical Warfare Medicine at the Allied Force Headquarters at Algiers, was dispatched by General Dwight Eisenhower to Bari to investigate. Alexander had spent a year at the beginning of the war in the Medical Research Division of the Edgewood Arsenal in Maryland, conducting research into various effects of mustard and nitrogen mustard agents. His experience proved invaluable in solving the Bari mystery, and he quickly determined that mustard was indeed the culprit in the catastrophe. Subsequent autopsies of 617 victims revealed that mustard gas destroyed most of their white blood cells. This observation suggested that the mustard gas attacked bone marrow preferentially. It occurred to Alexander that if mustard affected the rate of white blood cell division, it might also slow the rate of cancer cell division. Because fast division of cells is a hallmark of cancer, mustard gas–based drugs could be applicable in cancer treatment. Based on his studies of the victims, Alexander recommended the use of mustard compounds in the treatments of certain cancers. His observations specifically suggested the significance of compounds of this type for possible treatment of neoplastic disorders of the tissue that forms blood cells.

In 1942, a year before the Bari disaster, two pharmacologists at Yale University, Louis Goodman and Alfred Gilman, were studying the mechanism of action of nitrogen mustards under the mantle of military secrecy. To test it in an animal model, they treated mice with lymphoma (a cancer of the lymph nodes) using mechlorethamine (Mustargen). They confirmed that mechlorethamine indeed killed the faster growing abnormal cells, but not the normal cells. Alexander's landmark report on mustard poisoning infused more momentum into their approach. Subsequent clinical trials in cancer patients were rewarded with positive results. Even today, mechlorethamine is still a powerful weapon in the oncologists' arsenal of chemotherapeutics—it is used to treat Hodgkin's disease, leukemia, and brain tumors. In addition to having discovered one of the first nitrogen mustards in chemotherapy, Goodman and Gilman also coauthored one of the most popular textbooks on pharmacology, *Goodman & Gilman's The Pharmacological Basis of Therapeutics.* To date, 10 editions and thousands of copies have been printed and have educated generations of scientists and physicians.

After World War II, the U.S. government sponsored a large-scale screening program in search of anticancer drugs. In an army arsenal in Maryland, thousands of American troops were intentionally exposed to mustard gas. In all, more than 500,000 compounds were screened, and 45 anticancer drugs were discovered, a great boon to cancer treatment. Since then, scientists have prepared and experimented with thousands of mustard gas derivatives. To date, more than 100 nitrogen mustards have been used alone or in combination with other drugs to treat cancer, saving thousands of lives. Mechanistically, the antitumor effects of nitrogen mustards were exerted by simple alkylation reactions, which kill both resting and multiplying cancer cells. They work by preventing DNA from uncoiling, thereby blocking DNA replication and cell division. Unfortunately, chemotherapeutics such as nitrogen mustards annihilate cancer cells and normal healthy cells with equal ferocity. An analogy has been made of this chemotherapy to the "grenade approach" or a "carpet-bombing strategy." Common side effects of chemotherapy include vomiting, infections, profuse diarrhea, bouts of massive depression, and hair loss. Dozens of drugs have been developed just to combat the side effects of chemotherapy. Examples include Zofran for vomiting and Neupogen, a white-blood-cell booster.

Despite its terrible side effects, chemotherapy works wonders in several categories of cancer. For instance, Hodgkin's disease and childhood lymphoblastic leukemia, previously almost universally fatal, enjoy a more than 90% remission rate if the cancers are discovered and treated early.

Cisplatin

Barnett Rosenberg, a physics professor at Michigan State University, found Cisplatin to be an effective cancer chemotherapeutic.[10–13] He later reminisced: "The discovery of the biological action of the simple platinum coordination complex was a textbook case of serendipity. The motivation of the experiments came from a desire to study the effects of electric field on cells growing in culture."[10,13] In the early 1960s, Rosenberg and coworkers were experimenting to see whether alternating electromagnetic forces could have any effect on the division of cells. In 1964, they observed that electric current interfered with the cell divisions of *E. coli* bacteria in suspension. Curiosity piqued, they further investigated the source of electric current and found that it was generated by the platinum electrodes, which formed catatonic (cell killing) platinum. In 1967, Rosenberg tested Cisplatin as a drug against intestinal bacteria first and tumors later. Cisplatin demonstrated good activity in killing cancer cells despite some toxic effects. Cisplatin (trade name Platinol) is *cis*-dichloroamino-platinum (II), which was first synthesized in 1845. Alfred Werner, the 1913 Nobel laureate in chemistry, deduced the structure in 1893. Cisplatin is one of the most widely used chemotherapeutics in treating metastatic testicular cancer, ovarian tumors, and bladder cancer.

Another success story in the battle against cancers using platinum-based chemotherapy is the treatment of testicular cancer. Testicular cancer affects primarily young males. Prior to the use of platinum-based therapies and extensive surgical exploration, both pioneered at the Indiana University Medical Center by Lawrence Einhorn (oncologist) and John Donahue (surgical urologist), testicular cancer was generally fatal. When cyclist Lance Armstrong, the seven-time champion of the Tour de France, was diagnosed with testicular cancer, he was treated at Indiana University with a regimen of Cisplatin, bleomycin, and etoposide. Bleomycin, a macrolide antibiotic chemotherapy, interferes with cell division. Similarly, etoposide, a plant alkaloid, is a topoisomerase II inhibitor that stops cell division.

The royalties that Michigan State University receives from Cisplatin have risen steadily: from less than $1 million in 1969 to almost $20 million in 1998, plus several million dollars from Carboplatin, another platinum chemotherapy. The mechanism of action for the anticancer property of Cisplatin is similar to that of nitrogen mustards, namely rupturing DNA. It forms cross-links between and within the strands of DNA in the nucleus of dividing cells. The major side effects of Cisplatin include renal toxicity, vomiting, and nausea.

At the time of Rosenberg's discovery, platinum was hardly the only metal incorporated into drugs. Paracelsus's antisyphilis drug contained mercury; Ehrlich's 606 was an arsenic compound. The gold complex, RS-Au-PEt_3, is an antiarthritis drug; bis-(maltolato)oxovanadium (IV) is an insulin mimetic; and bismuth has long been used in antiulcer drugs such as Pepto-Bismol. However, few suspected that a platinum-containing compound would be active against cancer at the time. Indeed, the discovery of the remarkable biological activities of the anticancer platinum drug Cisplatin makes an impressive chapter in the history of drug discovery.

Vinca Alkaloids

In 1820, Pierre Joseph Pelletier and Jean Bienaime Caventou, two French researchers famous for their discovery of quinine from *cinchona* barks, isolated colchicine from the autumn crocus (*Colchicum autumnale L.*).[14] Colchicine was found to have cytotoxic activities in 1938. It was later found that colchicine served as a "spindle poison" to inhibit mitosis, the division of a cell's nucleus. Despite the insight that colchicine afforded into its mechanism of action, the drug itself has a narrow range of effectiveness against cancer. As a consequence, colchicine is mostly used for treating gout rather than as a chemotherapeutic.

Pierre Joseph Pelletier and Jean Bienaime Caventou © Warner-Lambert

Colchicine's cytotoxic activities nonetheless elicited much interest in searching for anticancer drugs derived from plants. A prominent example is the vinca alkaloids isolated from Madagascar periwinkle, which earned its name for its abundance on this African island. Its Latin name is *Catharanthus roseus,* which used to be called *Vinca rosea.*

Madagascar periwinkle. Photo by Alexandra H. Li

In 1958, during their pursuit of diabetes drugs, R. L. Noble and C. T. Beer at the University of Western Ontario in London, Ontario, Canada, isolated two important cancer drugs from Madagascar periwinkle: vincristine and vinblastine. As early as 1949, Noble had heard stories that Jamaicans were treating diabetes using *Vinca rosea* leaves. He and Beer procured some of the leaves and fed them to four rabbits but did not see any effects on the level of blood sugar. Having failed orally, they tried the injection route. To their dismay, after injection of the leaf extract, many rabbits died of an overwhelming infection induced by *Pseudomonas* bacterium. Most people would have given up on the project after this disastrous outcome, but Noble and Beer dug deeper. They found that rabbits treated with the plant extract developed critically low counts of white blood cells, leaving them with damaged bone marrow and defenseless against bacterial infections. Because one of the hallmarks of cancer is abnormal proliferation of white blood cells, they tried the plant extract on animals with transplanted tumors and saw tumor shrinkage. They then looked for the specific alkaloids in the extract that might slow or halt white blood cell production and identified two, vincristine and vinblastine, as the most potent. The Canadians presented their findings in 1958, and the chemotherapeutic effects of vinca alkaloids became known to the public. It was at that time that Noble learned that Gordon H. Svoboda at Eli Lilly had also made similar discoveries. In December 1957, Svoboda submitted extract of the whole vinca plant for testing. It showed a 60–80% prolongation of life for mice infected with P-1534 leukemia, an acute lymphoblastic leukemia.

The yield of vincristine was low (0.00025%) from the dry periwinkle plants, one of the lowest of any medically important alkaloids. American drug firms contracted farmers in the hill country of eastern India, near the Chinese border, and promised to buy all the vinca they could grow. The supply of vinca leaves imported to the United States was plentiful and largely steady, except a brief interruption during the China-India border war in October and November of 1962. Since 1964, Eli Lilly has marketed vincristine (trade name Oncovin) for acute childhood leukemia and vinblastine (trade name Velban) for lymphoma such as Hodgkin's disease, for advanced testicular cancer, and for advanced breast cancer. Similar to colchicine, the vinca alkaloids work by serving as a "spindle poison." They bind to tubulin, one of the key constituents of microtubules, thus preventing the cell from making the spindles it needs to divide.

Before the emergence of vincristine and vinblastine, the diagnosis of Hodgkin's disease was virtually a death sentence. Now there is a 90% chance of survival with the treatment of vinca alkaloids and other chemotherapy. Madagascar periwinkle, the little ornamental plant, has been transformed into a lifesaving gift from Mother Nature.

Taxol

Paclitaxel, with the trade name Taxol, has had considerable success in treating ovarian and breast cancer since 1992. It was initially isolated from the Pacific yew tree as part of the National Cancer Institute-United States Department of Agriculture (NCI-USDA) plant-screening program. The NCI is one of the many divisions of the National Institutes of Health (NIH). Arthur Barclay, a Harvard-educated botanist, worked for the U.S. Department of Agriculture in the 1960s. In August 1962, Barclay, along with three graduate students, traveled to the Gifford Pinchot Forest in the state of Washington. The quartet found a little-known pacific yew tree, *Taxus brevifolia.* They collected samples of twigs, leaves, and fruits. Barclay labeled them as B-1046, because it was his 1,046th sample, and shipped them to the NCI.[15]

One of the NCI's contractors was the Wisconsin Alumni Research Foundation (WARF, most famous for their important anticoagulant, warfarin). WARF tested the B-1046 extracts and found them to be cytotoxic. The potential anticancer properties prompted Barclay to go back to the Northwest and collect an additional 30 pounds of the precious stem bark of *Taxus brevifolia.* In 1966, after being rejected by many other laboratories for fear of toxicity, the stem bark found its way to the hands of Monroe E. Wall, chief chemist of the Fractionation and Isolation Laboratory at the Research Triangle Institute in North Carolina. Wall initially put the Pacific yew tree project on the back burner because he and his colleague, Mansukh C. Wani, were focusing on camptothecin, which was very promising as a chemotherapeutic agent. Camptothecin was isolated from the Chinese *Camptotheca acuminate,* whose Chinese name is *Xi Su* (Joy Tree). Although camptothecin did not become a major cancer chemotherapy agent, it served as the prototype for irinotecan (Camptosar) and topotecan (Hycamtin), two very important cancer drugs, which were more soluble. Wani described the isolation of camptothecin as "the most exciting scientific event in my life."[16] He once commented on Wall's contribution to the Research Triangle Institute "the institute was nothing but four 'walls.' It was not until the fifth 'Wall' arrived that the chemistry programs, in the form of the Natural Products Laboratory, started moving."[16]

Camptotheca acuminate, the Chinese "Xi Su" (Joy Tree). Author's collection

To isolate the active principle from the Pacific yew, chemists in Wall's laboratory ground the stem bark into a fine powder and extracted the active ingredients with ethanol. Guided by a process called "bioactivity-directed fractionation," they were able to purify the crude extract so that the cytotoxic potency increased 1,000-fold. In 1967, they isolated the active principle as white crystals in a 0.014% yield from the dry bark. The molecule was later determined to have a molecular formula of $C_{47}H_{51}NO_4$ and a molecular weight of 839. Due to the complexity of the molecule, deciphering Taxol's structure was a long and frustrating endeavor. Wani described Wall as "a go-getter" and "a very dedicated scientist," whose "persistence was his greatest virtue, but patience was not."[16] Therefore, Wall directed Wani to work on it with lower priority, tinkering with it only once in a while. In the end, Wall's persistence and Wani's patience paid off. Importantly, Wani's expertise in recrystallization was crucial for obtaining a single crystal suitable for X-ray crystallography, a key technique for deciphering the structure of Taxol in combination with proton nuclear magnetic resonance (NMR) spectroscopy.

One of the perks of discovering a novel molecule is that you get to name it. Some scientists have certainly taken advantage of this and created many interesting names, such as Barbituric acid, Rebeccamycin, Amycin, Herbindole, Paulomycin, and Colabomycin. Wall, more conventionally, christened the molecule *Taxol,* with *tax* signifying the origin of the molecule from *Taxus brevifolia* and *ol* indicating that the molecule contained one or more alcohol functionalities. In 1971, Wall and coworkers published the structure of Taxol, which contained a unique 10-deacetylbacctin (10-DAB) skeleton and an amino-acid side chain. Because of the intricacy of the Taxol structure, they initially submitted the manuscript with the side chain attached at the wrong site, although they corrected it immediately prior to publication.

Because of its insolubility, scarce supply, and low potency, there was no overwhelming enthusiasm for Taxol until 1979. It was then that Susan Howitz,[17] an assistant professor at the Albert Einstein College of Medicine, discovered that Taxol had a completely novel mechanism of action, unlike those of any other drugs known at the time. Taxol exerts its action by stabilizing the microtubules, resulting in inhibition of mitosis and induction of apoptosis. Taxol stops malignant tumors from growing by interfering with the microtubules that are responsible for dividing the chromosomes during cell division.

This novel mechanism renewed interest in Taxol. The NCI took the torch from Wall, thanks to championing by the NCI's Mathew Suffness. In 1984, the NCI amassed enough positive data to commence the Phase I clinical trial of Taxol with about 30 patients. The drug was shown to be relatively safe. Despite tremendous difficulty in procuring enough Taxol—it took about 20,000 pounds of yew tree bark to isolate 1 kilogram of Taxol—the NCI moved forward with the Phase II trials in 1987. Taxol is notoriously insoluble in water—like brick dust—but the problem was overcome by the addition of ethanol and Cremophor EL, a surfactant, which later proved to be important in reversing multidrug resistance. Although tested in ovarian and renal cancers and melanoma, Taxol initially was found efficacious only for ovarian cancer. Eventually, its chemotherapeutic applications were expanded to breast cancer, non-small-cell lung cancer, and Kaposi's sarcoma.

Securing proof of concept from the Phase II trial was a triumph for the NCI, which had overseen the clinical development of Taxol from the beginning, more than 20 years previously. However, the NCI was not set up to take on the expensive and long Phase III trials that involved numerous disciplines such as oncology, pharmaceutical science, pharmacokinetics and drug metabolism, statistics, drug safety science, and more. Following a competitive selection process, Bristol-Myers Squibb (BMS), the only major U.S. pharmaceutical company to have made a bid, was awarded the molecule under the Cooperative Research and Development Agreement (CRADA) with the NCI in 1991. At the time, the commercial potential of Taxol had not fully manifested for the breast cancer indication. The NCI's choice of BMS over the French company Rhône-Poulenc made sense because Rhône-Poulenc already had a competing drug, Taxotere, discovered by Frenchman Pierre Potier, who invented Taxotere by a minor modification of Taxol (replacing the benzoyl group on Taxol with a *tert*-butoxylcarbonyl group).[18] Taxotere, more potent than Taxol, had annual sales of $1.54 billion in 2003.

The FDA approved Taxol for use in refractory ovarian cancer in December 1992, for breast cancer in 1994, and later for non-small-cell lung cancer and Kaposi's sarcoma. By 2000, Taxol was the best-selling cancer drug of all time, with annual sales of $1.6 billion.

The raw material for isolating Taxol became an extremely contentious issue in the late 1980s and early 1990s. Because it takes 100 years for the Pacific yew trees to become useful in terms of Taxol content, harvesting the yew trees for stem bark meant destruction of the forest. To make matters worse, the forest that harbors the yew trees is home to the endangered spotted owl. The battle raged for many years between the environmentalists, who wanted to save the trees, and cancer patients and oncologists, who were eager to get access to the drug. It ended abruptly in early 1993, when BMS started to use a semisynthetic route to make Taxol that did not use the Pacific yew trees at all. Instead, they extracted 10-deacetylbacctin (10-DAB) from *Taxus baccata,* the English yew, a common ornamental plant. BMS then used the side-chain installation process patented by Robert Holton at Florida State University to make Taxol. More than 3 tons of English yew needles, a renewable source, must be collected and processed in order to produce 1 kilogram of 10-DAB. The switch was worthwhile, because Taxol had cost as much as $600,000 per kilogram. Even now, it is still nearly $400,000 per kilogram. The worldwide market for Taxol is about 400 kilograms per year. Currently, Taxol is produced in large fermentation tanks using plant cells. Taxol is one of the best weapons in an oncologist's arsenal, and many even refer to times prior to 1994 as the "pre-Taxol" era. Interestingly, a major new use of Taxol is as a coating on stents to prevent restenosis, providing a larger market for its use than as just an anticancer drug.

Antimetabolites

An antimetabolite interferes with the formation or utilization of a normal cellular metabolite. It prevents cancer cells from metabolizing nutrients and other essential substances, thus blocking processes within the cell that lead to cell division. Most antimetabolites interfere with the enzymes involved in the synthesis of new DNA, and, as a result, many are derivatives of the building blocks of DNA itself, such as nucleotide-based inhibitors, or analogs of critical cofactors.

In 1944, Richard Lewesohn, a surgeon at Mount Sinai Hospital in New York, searched for cancer drugs using mice with transplanted tumors as animal models. He found that folic acid had promising antitumor activities. In 1947, Sidney Farber of Harvard Medical School took folic acid and two additional analogs of folic acid to Phase I trials for acute leukemia and several other malignancies. The results were completely unexpected: instead of arresting cancer cell growth, the folates accelerated the progress toward fatal termination. It occurred to Farber that if folic acid promoted cancer cell growth, then folic acid antagonists might be able to deter their proliferation. Coincidentally, American Cyanamid, a drug firm in New Jersey, had in hand 4-aminofolic acid, the first really potent folic acid antagonist. Farber used it to treat children with acute leukemia and saw their white cell count drop down to normal. In 1953, the FDA approved methotrexate, a potent folic acid antagonist that was less toxic than 4-aminofolic acid. A combination of methotrexate and prednisone, a steroid, doubled the survival rate of acute lymphoblastic leukemia (ALL). Methotrexate, a folic acid analog and an antimetabolite, has seen wide application in cancer treatment.

In 1952, Gertrude Elion and George Hitchings of Wellcome Research Laboratories introduced another antimetabolite, 6-mercaptopurine, which was an intermediate in their synthetic scheme that turned out to be better than the final drug they wanted to make.

Gertrude Elion, known as Trudy to her friends, was born on January 18, 1918. Her father and mother immigrated to America from Lithuania and Russia, respectively. Education was of paramount importance to the Elion family. Gertrude went to an all-female college, Hunter College, in New York City, which was free at the time for qualified students. She then enrolled at New York University, where she earned her master's degree in 1941, the same year that her fiancé, Leonard Cantor, died of subacute bacterial endocarditis, an inflammation of the heart lining. Sadly, the disease could have been cured by penicillin, which became available only 2 years after his death. Heartbroken, Elion never married, and she became determined to do research in drug discovery. But even with her master's degree in chemistry, Elion could not find employment in a research laboratory. She was once told "You are qualified. But we never had a woman in the laboratory before and we think you would be a distracting influence."[19] In 1989 Elion commented that "Maybe because I was young and 'cute' (after all, I was only twenty then), but I've learned over the years that when you put white lab coats on chemists, they look alike!"[20] For 7 years, she taught high school chemistry and physics, then worked in a food company as an analytical chemist, whose job description included checking the acidity of pickles, the color of mayonnaise, and mold on fruits. Her big break came during World War II, when laboratories were short of *man*power. In 1944, George Hitchings of the Wellcome Research Laboratories in Tuckahoe, New York, hired her. Together with Hitchings, Elion would invent six drugs and share the Nobel Prize with Hitchings and James Black in 1988, an honor rarely bestowed on industrial chemists. They received the award "for their discoveries of important principles for drug treatment."[21]

When Elion joined Hitchings's group in 1944, she was in the right place at the right time. George Hitchings earned his Ph.D. at Harvard University in 1933 working on DNA, with a focus on analytical methods used in physiological studies of purine, adenine, and guanine. Hitchings's group first prepared potential antimetabolites of purines and pyrimidines, which were all required for the synthesis of nucleic acids. Their realization of the importance of these building blocks was revolutionary, considering that the double-helix structure would not become known until 1953. They examined more than 100 pyrimidine analogs for their ability to inhibit the growth of *Lactobacillus casei.* In 1948, they discovered that 2,4-diaminopyrimidine, a folic acid antagonist, inhibited the growth of *Lactobacillus casei.*

In 1952, Elion synthesized 6-mercaptopurine (Purinethol), which was an intermediate in the synthetic route of 6-aminopurione. Elion decided to test the intermediate and found out that it had a better biological profile than the intended targets. Purinethol was one of the first anticancer drugs of any real importance. By interfering with the process of nucleic acid synthesis, it retarded the growth of sarcoma 180 in mice in a test that was becoming common. The favorable results of 6-mercaptopurine in mice led to a collaboration with the Sloan-Kettering Foundation in New York and the Chester Beaty Research Institute in England. It was found that 6-mercaptopurine worked better for childhood leukemia in combination with other cancer drugs, such as corticosteroids and folic acid antagonists, than when used alone. As a result, childhood leukemia became curable for a large population of patients.

Even without a Ph.D., Gertrude Elion invented six drugs, held 45 patents, and was the first woman to be inducted into the Inventors Hall of Fame. Elion commented, "I am happy to be the first and I doubt that I'll be the last!"[20] Her happiest moments came watching patients get well using her drugs and receiving letters from patients and their families. One letter read, "I have a little boy who was diagnosed two years ago with acute lymphocytic leukemia. Since that time, he regularly takes 6-mercaptopurine . . . with inexpressible gratitude for having contributed to the saving of one human life so very dear to me, and so many human lives, that I write to you. . . ."[20]

Another important antimetabolite is 5-fluorouracil (5-FU), a nucleotide analog, which was approved in 1962. It is still in use today despite the fact that its side effects, which include cardiovascular, central nervous system, and dermatological adverse reactions, can be so terrible that some oncologists dubbed it "5 Feet Under"!

Thalidomide

Thalidomide, one of the most notorious drugs in history, has recently witnessed its own reincarnation in cancer treatment, namely multiple myeloma, a hematological disease that afflicts about 50,000 Americans. Multiple myeloma is characterized by the neoplastic proliferation of plasma cells in the bone marrow.

In 1953, Wilhelm Kunz, a chemist at Chemie Grünenthal, a small German soap and toiletries company, attempted to synthesize antibacterial peptides. He isolated thalidomide as a nonpeptide by-product that pharmacologist Herbert Keller found to possess hypnotic effects. In 1956 it was extensively marketed in Germany and the rest of Europe as a sedative-hypnotic for pregnancy-related morning sickness. Tragically, many incidents of horrific phocomelia (the word's roots reached back to the Greek *phoke,* meaning "seal," and *melos,* meaning "limb") in newborn babies took place—thousands of babies were born without arms or legs, instead having tiny, useless, seal-like flippers. In November 1961, a German pediatrician and an Australian obstetrician independently linked phocomelia with thalidomide, which is a powerful teratogen, a chemical that causes birth defects. America was largely spared the tragedy thanks to Frances Kelsey, an FDA officer in charge of examining the drug, who withheld the approval of thalidomide due to concerns about its safety. Since the thalidomide debacle, testing for teratogenicity has become one of the cornerstones of the early drug-safety evaluation that pharmaceutical companies conduct, usually before conducting clinical trials in women with childbearing potential.

Forty years later, Celgene Corporation, a small drug firm in New Jersey, resurrected thalidomide despite its worldwide infamy. It was initially sold for the treatment of leprosy and wasting (severe weight loss) of AIDS patients. To deal with the horrific teratogenicity of thalidomide, Celgene devised and patented a controlling distribution system for dispensing the drug that requires, among other things, regular pregnancy tests for patients of childbearing age. The controlled distribution system, along with a similar restrictive distribution system for the antiacne agent Accutane, was one of the earliest risk management plans enacted.

In 1997, Bart Barlogie, an oncologist in Little Rock, Arkansas, tried thalidomide on an elderly man with multiple myeloma; the man went into a nearly complete remission. Encouraged by the result, Barlogie carried out a clinical trial with a larger population of multiple-myeloma patients. A few years later, Barlogie reported the results: about 30% of 169 patients who had relapsed after other treatments saw levels of myeloma protein (M-protein) decrease by 50% or more after taking thalidomide.

Since the late 1990s, many doctors have prescribed thalidomide "off-label" (once a drug is approved for use by the Food and Drug Administration [FDA], physicians can prescribe drugs on the market for indications other than the ones approved by the FDA) for multiple myeloma. Understandably, a battle was waged between multiple-myeloma patients clamoring for the drug and victims of the birth defect insisting on keeping the ban. Early in 2005, thalidomide was approved in Japan as an orphan drug for the treatment of multiple myeloma. The term *orphan drug* refers to a product that treats a rare disease that affects a small population. In the United States a drug is an orphan drug if it is used by fewer than 200,000 Americans.

Hormone Treatment

The four most common types of cancer are lung, breast (almost exclusively in women), colon/colorectal, and prostate (in men). In particular, breast cancer (second only to lung cancer in terms of fatality rate) strikes one in eight women.[22] There are about 200,000 annual incidents in the United States alone, and 25% of women with breast carcinoma will eventually die from it. The current arsenal for the treatment of breast cancer includes, as appropriate to tumor type and extent of spread, surgery (mastectomy or lumpectomy), radiation, chemotherapy, and hormone treatment.

In the late 1930s, Charles B. Huggins, a urologist at the University of Chicago, established the connection between prostate cancer and the male hormone testosterone. To test his theory, Huggins castrated a patient dying of prostate cancer and treated him with the female hormone estrogen. Estrogen, the primary female hormone, is secreted by the ovaries until menopause, when estrogen production ceases. The treatment halted the cancer, and the patient lived a productive and comfortable life for an additional 15 years. Huggins's experiments provided definitive proof that some cancers could be controlled by chemicals. In 1966, he shared the Nobel Prize in Medicine or Physiology with Peyton Rous (see page 5, this chapter) for his discoveries concerning hormonal treatment of prostatic cancer.

Tamoxifen, the most frequently prescribed anticancer drug in the world, was initially made as a contraceptive. Robert Robinson of Oxford University, the 1947 Nobel laureate in chemistry, synthesized a non-steroidal estrogen, diethyl stilbestrol dimethyl ether, which was listed as a poison in the United Kingdom in 1939 due to the fear of its being used as an abortion pill. Partially due to the influence of Robinson, who was a consultant to ICI (Imperial Chemicals, Inc., a British chemical firm), chemists in the company synthesized many derivatives of triphenylethylenes in the 1940s and 1950s. In the 1960s, the emergence of birth-control pills fueled the American sexual revolution. ICI joined the foray of discovery in contraceptives. Arthur Walpole, along with his endocrinologist colleague Michael J. K. Harper and chemistry colleague Dora M. Richardson, discovered ICI-147,741, the *trans* isomer of triphenylethylene, which would later become tamoxifen.

Tamoxifen is a partial estrogen receptor (ER) antagonist that also shows some partial agonist activity in selected organs (including bone, cardiovascular organs, and endometrium). It blocked the action of estrogen in some parts of the body while acting like estrogen in other parts of the body. More specifically, ICI-46474, a selective estrogen receptor modulator (SERM), was shown to be an effective contraceptive in rats. With great anticipation, ICI moved tamoxifen to clinical trials in humans. To everyone's surprise, tamoxifen induced ovulation in women, exactly the opposite of what it did to rats. Fortunately, Walpole had a specific interest in cancer therapy and included coverage of the control of hormone-dependent tumors in the patent of tamoxifen, which primarily focused on the management of the sexual cycle. Walpole's inclusion of cancer as an indication in his application proved to be exceedingly prescient. Despite management's concerns of small market prospects, Walpole succeeded in convincing ICI Pharmaceuticals to market tamoxifen in the United Kingdom as a breast cancer treatment (1973) and as an inducer of ovulation (1975). In 1985 Zeneca (ICI's pharmaceutical division) would engage in a legal wrangling with generic companies, which ended in Zeneca's favor.

V. Craig Jordan became interested in antiestrogens when he was a graduate student at Leeds University in England, where the examiner for his Ph.D. thesis was Arthur Walpole (then at ICI Pharmaceuticals). With Walpole's help, Jordan started the first systematic study of tamoxifen as an anticancer agent, first at the Worcester Foundation in Massachusetts (1972–1974) and then continuing pivotal animal studies at Leeds University during the remainder of the 1970s. This latter work was sponsored by ICI, which had no cancer research program of its own and was not inclined to do the studies. In 1974, Jordan demonstrated that long-term tamoxifen treatment was necessary to prevent and control breast cancer growth in rats. In a recent article in *Nature Review, Drug Discovery,* he recounted the discovery and development of tamoxifen as "a story of interpersonal relationships, rather than a planned effort by ICI Pharmaceutical Division—the manufacture of tamoxifen—to establish themselves as a major player in oncology."[22]

Clinical trials showed that tamoxifen was very well tolerated and exerted limited adverse effects. The side effects were hot flashes, risk of endometrial carcinoma, and blood clots with prolonged use. In 1978, the FDA approved tamoxifen for treatment of ER-positive metastatic breast cancer in the United States. Normally, estrogen-dependent breast cancer is treated with tamoxifen for 5 years, at which point drug resistance starts to develop. Tamoxifen has now become the most frequently prescribed anticancer drug in the world and is the gold standard for the treatment of all stages of breast cancer. Tamoxifen ultimately provided a treatment that has saved 400,000 women's lives. More remarkably, in 1998 the FDA also approved tamoxifen for use in women at high risk of developing breast cancer (those who have ER-positive tumors) as a risk-reduction medicine—patients taking tamoxifen are 45% less likely to get breast cancer. This was the first time an anticancer drug was being used specifically to prevent breast cancer in susceptible patients. Tamoxifen is now used as a first-line agent for male breast cancer as well. A newer SERM, raloxifene (Evista), marketed by Eli Lilly as an osteoporosis drug, renders a 58% reduction in breast cancer.

Today, the correlation between sex hormones and cancer has been well established. Breast cancer in particular is linked to estrogen abnormality. In 2004 hormone replacement therapy (HRT) for postmenopausal discomforts was linked to an increase in breast cancer. Estrogen, the key trigger for two-thirds of all breast cancers, can fuel the growth of breast cancer cells. This is one of the reasons that pregnancy should be avoided by breast cancer patients. Moreover, women with breast cancer fare far better after removal of their ovaries. Logically, if estrogen can cause breast cancer, it is likely that antiestrogens would be able to stave off breast cancer.

Although tamoxifen is still the first-line treatment for breast cancers, third-generation aromatase inhibitors have shown great promise in treating breast cancers. Aromatase is the enzyme that blocks the synthesis of estrogen in the body. It catalyzes the conversion of androgens into estrogens that contain a phenolic ring. Small-molecule aromatase inhibitors have shown great benefit to breast cancer patients. Aromatase inhibitors may be classified into two types. Type I aromatase inhibitors bind to the aromatase enzyme irreversibly and are called inactivators. In some cases they are dubbed as mechanism-based or "suicide" inhibitors when some of them are metabolized by the enzyme to reactive intermediates that bind covalently to the active site. Type I aromatase inhibitors are usually steroidal in structure, as represented by exemestane, formestane, and atamestane. Aminoglutethimide, also a first-generation aromatase inhibitor, was the first marketed aromatase inhibitor for the treatment of breast cancer by Ciba-Geigy in 1981. Type II aromatase inhibitors bind to the enzyme reversibly. Thankfully, the latest type II aromatase inhibitors, such as anastrozole and letrozole, possess exceptional specificity for aromatase P450 enzyme—so there are fewer selectivity-related toxicities with the drugs.

Protein Kinase Inhibitors, Monoclonal Antibodies

Protein kinase inhibitors, as targeted cancer drugs, are the brightest light to illuminate the darkness of cancer. In contrast to the carpet-bombing approach of old chemotherapy, they promise a kinder, gentler, and more effective method of cancer treatment.

Protein kinases are enzymes inside the cell that are capable of donating phosphate groups to target proteins. For instance, tyrosine kinases are involved in phosphorylation (adding a phosphate) to tyrosine. Protein kinases comprise a family of more than 518 members. They are responsible for signal transduction, turning on and off the switches for cancer cells to grow. Many protein kinases have been implicated in cancer. By blocking the functions of protein kinases, it may be possible to stop cancer growth. In the 1980s, Japanese scientists showed that staurosporine, an indole natural product, inhibited protein kinase C (PKC). In 1988, Alexander Levitzki and colleagues from Israel showed that selectivity could be achieved for epidermal growth factor receptor (EGFR). In a *Science* paper,[23] Levitzki and colleagues revealed that a series of simple dicyanobenzylidenes and carboxybenzylidenes could block EGF-dependent cell proliferation. They were among the first selective small-molecule EGFR kinase inhibitors.

Many scientists have made important contributions in elucidating the nature and functions of protein kinases. Sadly, at least one scientist's work was shown to be dubious. A quarter of century ago, protein kinase was associated with a scientific scandal known as the "kinase cascade," or the "signaling cascade."[24–30]

In 1980, Mark Brian Spector was a graduate student at Cornell University working in the laboratories of Efraim Racker, a winner of the National Medal of Science. The 24-year-old was exceptional in many ways. In his

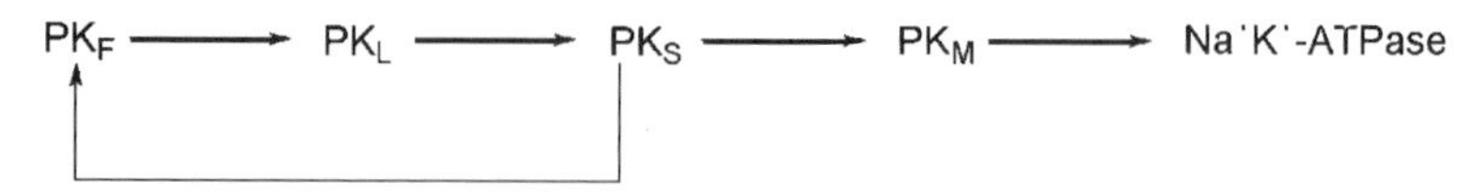

Spector's kinase cascade

spare time, he painted and wrote poetry. In the laboratory, he was ambitious, with an unmatched work ethic—it was not unusual for him to put in 18 hours a day. While investigating retroviruses, which make proteins that cause cancer in laboratory animals, Spector claimed that he had isolated four protein kinases (PK_F, PK_L, PK_S, and PK_M). Through a series of dazzlingly elegant experiments, Spector proposed the "kinase cascade" theory. Namely, these four protein kinases aligned like dominos so that, if any one of the protein kinases was phosphorylated, the next would be subsequently phosphorylated as well, in a cascading fashion. At the end, phosphorylation would take place for the sodium-potassium ATPase, an enzyme crucial to cell metabolism in which a defect would transform a healthy cell into a cancerous one.

The theory was straightforward, and it made perfect sense. It explained how a single viral protein could make a normal cell cancerous, connecting tyrosine phosphorylation with malignancy. In May 1980, Spector awed the audience with his protein kinase cascade theory in a meeting at the Cold Spring Harbor Laboratory in Long Island, New York. The publications by Racker and Spector in *Science*[24] and *Cell*[25] in 1981 galvanized cancer research. Many speculated that the brilliant graduate student would get his Ph.D. in a year and half (a process that normally takes 5) and possibly a phone call from Stockholm. Unfortunately, it was quickly proven too good to be true. Many reputable laboratories were eager to reproduce Spector's spectacular experiments, but all failed to do so. David Baltimore, a virologist who won the Nobel Prize in Physiology or Medicine in 1975 for his "discoveries concerning the interaction between tumor viruses and the genetic material of the cell,"[8] was not able to obtain samples from Spector to reproduce his results. In March 1981, Baltimore invited Spector to give a talk at the MIT Cancer Center. Instead of staying in the hotel where Baltimore offered to put him up, Spector insisted on sleeping in the laboratory after the seminar (so he could "help" with the experiments).

During collaboration with Spector, Volker Vogt, a young tumor virologist working on the floor above Spector's, discovered that Spector's sample contained radioactive iodine, ^{125}I, which should not have been there. At the NCI, Ed Scolnick, who would become Merck's science chief a few years later, revealed the same discrepancy. Typical of Scolnick's aggressiveness, he confronted Racker with his suspicion. Pretty soon, a pattern was established: these experiments worked only when Spector performed them or with his "help." A little further digging exposed that not only had Spector been admitted to Cornell without having earned a bachelor's or master's degree but also that some of the laboratory results he had obtained before entering Cornell also had the distinction of not being reproducible.

Racker, a psychiatrist turned biochemist, was 68 years old in 1981. He went to the laboratory himself to try to reproduce Spector's results. Curiously, Racker was able to repeat *some* of his experiments. It was evident that Spector had fabricated much of the data, but not *all* of them. Racker later lamented: "The tragedy is that faking this would require more hard work than doing the thing right."[29] Racker retracted four of their papers based on Spector's experiments and paid back a good portion of the grant money to the NCI. At the height of the controversy, the "kinase cascade" was among many cases of scientific fraud cited in a congressional hearing chaired by a young congressman from Tennessee, Al Gore. The greatest irony of the whole story was that Racker himself had uncovered a scientific fraud 20 years earlier. In the 1960s, Racker had not been able to reproduce a competitor's results despite numerous attempts. He revealed the major fraud in front of a big audience of biochemists with the perpetrator standing next to him at the podium.

Herceptin

Herceptin is a bioengineered human monoclonal antibody for breast cancer.[31] It was one of the first targeted cancer drugs. In 1962, Stanley Cohen at Vanderbilt University identified the epidermal growth factor (EGF), a growth factor that modulates the growth of epidermis. He discovered that EGF caused the eyes of newborn mice to open and their teeth to erupt several days sooner than normal. The EGF receptor inside the cell was later found to be a protein kinase. Cohen won the Nobel Prize in Physiology or Medicine for this work in 1986, along with his postdoctoral mentor, Rita Levi-Montalcini, who discovered one of the first growth factors, the nerve growth factor (NGF), in 1960. In

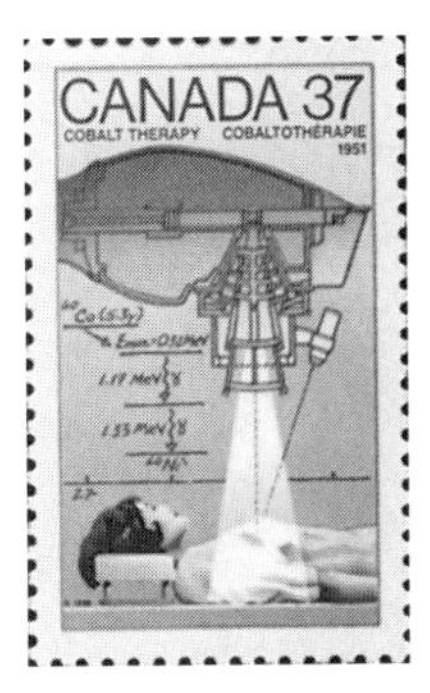

Radiation treatment of cancer © Canada Post Corporation

1979, Robert Weinberg's group at MIT discovered the neu oncogene from cancerous mouse cells in the brain.[7] On the West Coast, in 1984, Axel Ullrich of Genentech found a human gene: "HER-2," human epidermal growth factor receptor-2, which was soon discovered to be identical to Weinberg's neu oncogene. After identifying HER-2 as a cellular oncogene, Ullrich collaborated with Dennis J. Slamon, an oncologist at the University of California, Los Angeles, whose hobby was collecting human tumors. By a fantastic stroke of luck, Slamon established the linkage between HER-2/neu and certain breast cancers. Simply put, the more HER-2/neu a woman expresses, the more breast cancer growth she has. Therefore, by blocking the HER-2/neu enzyme with either an antibody or a small-molecule drug, one could stop the growth of cancer. An added bonus of the HER-2/neu enzyme is that the protein protrudes from the cell surface, making it an easy target for drugs to latch onto. Ironically, short on funding, Slamon could not afford to hire graduate students and postdoctoral fellows. An undergraduate student working in his laboratory, Wendy Levin, carried out the groundbreaking research.

The 1984 Nobel Prize for Medicine was awarded to Niels K. Jerne, Georges J. F. Köhler, and César Milstein "for theories concerning the specificity in development and control of the immune system and the discovery of the principle for production of monoclonal antibodies."[32] With the HER-2/neu protein in hand, using immunology techniques developed by Jerne, Köhler, and Milstein, Genentech produced a series of more than 100 monoclonal antibodies for the protein from mouse cells. The panel of more than 100 murine monoclonal antibodies was capable of inhibiting the HER-2+ cell line. The researchers named the most potent clone muMAb 4D5, indicating that it was a murine monoclonal antibody.

Chimera is a Greek word meaning something made from different parts. A chimeric antibody (chMAb), in this case, is part human and part mouse. Mouse protein, however, tends to cause the human immune system to respond with its own antibody to destroy the foreign one, not unlike the body's rejection response to organ transplant. In 1990 Genentech's Paul Carter and his colleagues achieved the task of "humanizing" the murine monoclonal antibody in a record 10-month period using a technique called *gene*

Immunology © Australia Post

splicing. They subcloned the hypervariable region of the antibody and derived a chimeric antibody, chMAb 4D5. Carter further humanized chMAb 4D5, using a technique called site-directed mutagenesis, by splicing a human gene into Chinese hamster ovary (CHO) cells, which Genentech brewed in giant batches in "bioreactors." The humanized monoclonal antibody was later named trastuzumab (trade name Herceptin), which was 95% human and 5% murine. It was a great achievement considering that the scientists had started with 100% murine and were able to retain the binding affinity with just 5% of the original components.

Phase I clinical trials in 1992 established Herceptin's safety profile. One side effect was cardiac toxicity in about 13% of patients, whereas 4% of the patients in the placebo group were struck with cardiac toxicity. (A *placebo,* meaning "I shall please" in Latin, was a mourner employed by people who had lost loved ones to cry and sing at funerals for a fee, thus protecting the bereaved from having to do so. In medicine, a placebo is an inactive drug, a sugar pill). The other side effect was mostly associated with infusion, which would be alleviated after the first infusion. The Phase II trial took place in 1993, followed by the Phase III trial in 1998. Genentech spent $200 million in developing Herceptin, which was approved by the FDA for treatment of certain breast cancers in 1998. A scant 6 years later, the annual sales of Herceptin reached $425 million. By keeping HER-2/neu overexpressed breast cancer at bay, Herceptin revolutionized oncology.

Revlon made significant financial contributions to Slamon's HER-2/neu research. The Revlon money ultimately helped to accelerate the Phase I trials. On May 5, 1998, the *Wall Street Journal* featured a cartoon of Dennis

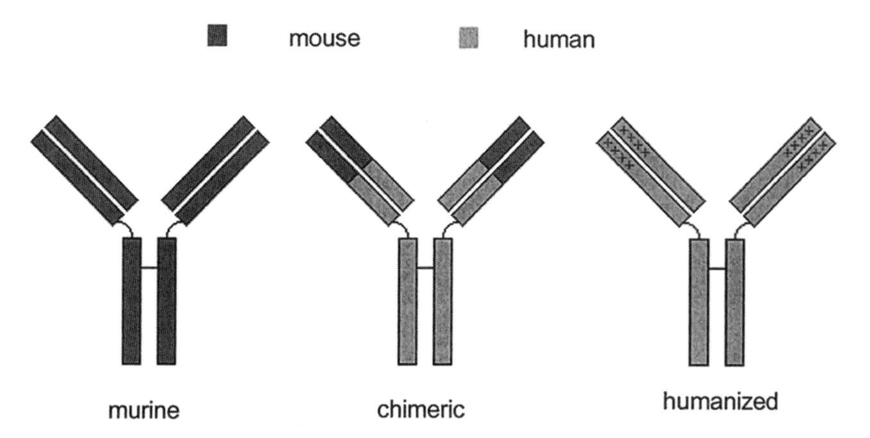

From murine to humanized antibodies. Diagram by author

Slamon with supermodel Cindy Crawford on one arm and actress Halle Berry on the other. Finally science met some Hollywood glamour.

Erbitux

Herceptin's success opened doors to another antibody of EGFR, Erbitux. Erbitux, approved in 2004 for treatment of colorectal cancer, is a human monoclonal antibody produced through bioengineering.

Synergy often allows collaborators to achieve far beyond what individuals could achieve alone. The discovery of C225, whose humanized version would later become Erbitux, is a good example.

In 1980, John Mendelsohn and Gordon Sato were both professors at the University of California at San Diego (UCSD). That fall, Sato approached Mendelsohn and proposed to combine his background in growth factors with Mendelsohn's background in immunology and the cell's growth cycle to figure out a way to block cancer cell growth. They zeroed in on EGFR, a sibling of HER-2/neu, because EGF had been implicated in one-third of all cancers by then. The Sato and Mendelsohn groups toiled in their laboratories for 2 years trying to clone a monoclonal antibody that would block EGFR. Even Sato's son, Denry, took part in the endeavor. The 225th monoclonal antibody, cloned from cells isolated from mouse's spleen, bound and inactivated EGFR. Reminiscent of Ehrlich's 606 (see chapter 2), they christened the antibody C225, in which C stood for chimera, indicating the antibody had part mouse and part human proteins. They later showed that the blockade of human EGFR by C225 inhibited proliferation of cultured cells and human xenografts in nude mice. Xenografts are human skin grafted on mouse skin.

In 1985, Mendelsohn moved to Memorial Sloan-Kettering in New York. During the next decade, Mendelsohn licensed C225 to an antibody biotech company in San Diego called Hybritech, which carried out a Phase I clinical trial and proved that C225 was safe enough to move on to the next phase. In 1992, Eli Lilly bought Hybritech and returned the rights of C225 to UCSD because it did not fit its business plan—oncology was not a major focus at Lilly. Subsequently, Mendelsohn spoke with many big pharmaceutical companies, but none were interested. Eager to see his "baby" move forward, Mendelsohn met with Samuel Waksal, the chief executive officer of a small Manhattan biotechnology company called ImClone, in April 1992.

Samuel Waksal had cofounded ImClone in 1985 with his brother Harlan. With his immunology background and his insight into biologics, Samuel Waksal grasped the value of C225 and agreed to develop it. During the clinical trials of C225, one dramatic event changed the fate of the antibody and ImClone. In March 1999, colorectal cancer patient Shannon Kellum was treated with a combination infusion of C225 and a chemotherapeutic, irinotecan (Camptosar). In December she saw her tumor shrink by 80%. The tumor had shrunk enough for doctors to surgically remove the rest of the cancer. This spectacular result catapulted ImClone's stock sky high overnight. In 2001, the clinical trials showed that tumor shrinkage was observed in 22.5% of refractory colorectal cancer patients treated with a combination of C225 and irinotecan. However, because the subsequent Phase II clinical trials were so poorly designed and executed, the FDA decided to refuse to review ImClone's application for marketing Erbitux. As a consequence, the FDA issued a refusal to file (RTF) letter to ImClone's biological license application (BLA) at the end of 2001 on the grounds of sloppy clinical trial design and data collection. ImClone and Bristol-Myers Squibb redesigned the clinical trials and, after an additional year, Erbitux was approved by the FDA for clinical use in February 2004.

With regard to the two original inventors of Erbitux, things have changed much as well. Mendelsohn is now the president of the M. D. Anderson Cancer Center in Houston. At one point, he sat on the boards of both Enron and ImClone, the famous pair of troubled companies. Despite being unlucky in choice of boards to sit on, his idea of tackling EGFR was partially responsible for the success of several targeted cancer drugs delineated here. Sato, having been born on a farm, is very much concerned with solving world hunger. He quit his lucrative teaching position and went to Africa to help the locals raise food, using his own money. The commercial success of Erbitux is welcome to him, for he intends to use the royalty that he is receiving to supplement his humanitarian endeavor.

Avastin

Because they are fast growing, tumors frequently need to develop their own blood supply to continue to grow. In 1971, Judah Folkman at Children's Hospital in Boston proposed a novel approach to staving off cancer using antiangiogenesis, a process that cuts off the blood supply to the tumor. He initiated efforts aimed at the isolation of a "tumor angiogenesis factor." In 1991, Folkman and his postdoctoral fellow, Michael O'Reilly, succeeded in isolating two proteins from mice: angiostatin and endostatin, both of which inhibited blood vessel growth. Folkman's antiangiogenesis insight was so revolutionary at the time that his grant application was turned down by the NIH. He persisted and, in time, saw angiostatin and endostatin produce dormancy in three types of tumors in mice without inducing the drug resistance frequently seen with traditional chemotherapy. In 1997, he published his results in the November 27 issue of the prestigious British journal *Nature.*[33]

In March of 1998, Gina Kolata, a *New York Times* science reporter, sat by James Watson at a dinner party. She asked the discoverer of the double-helix structure of DNA what was "hot" in biomedical research. Kolata later reported that Watson lauded Folkman's achievement by saying "Judah is going to cure cancer in two years."[34] On May 3, 1998, on the front page of the *New York Times,* Kolata published an article titled "A cautious awe greets drugs that eradicate tumors in mice." In addition to the quote from Watson, she also mentioned that James Pluda of the NCI stated that he was "electrified" by Folkman's lecture. Although the article was relatively balanced, the media quickly hyped the development as though a cure for cancer in humans were near. Phone calls flooded the Children's Hospital in Boston, the NCI, and the NIH, clamoring for the lifesaving drugs. Worse yet, some cancer patients refused to take their chemotherapy drugs, thinking that the nontoxic drugs would be available soon. EntreMed, the company that Folkman worked with to develop those two drugs, saw their stock soar from $12 to $82 overnight, before dropping back to $30. James Watson immediately issued a clarification statement, playing down the cancer-cure frenzy. He claimed that he was misquoted and that what he really told Kolata was "endostatin should be in clinical trials by the end of the year and that we should know about one year after that whether they were effective."[35] Regardless of the media hype, endostatin was proven not efficacious in humans, like many other experimental cancer drugs.

Avastin, however, proved to be the real thing. The approval of Avastin vindicated the antiangiogenesis approach to fighting cancer. Avastin, a humanized monoclonal antibody against vascular endothelial cell growth factor (VEGF) receptor, was discovered by Genentech's Napoleone Ferrara in

1989. It is the first marketed drug that works through the antiangiogenesis mechanism.

Napoleone Ferrara joined Genentech in 1988. His official assignment was to work on cardiovascular diseases. In 1989, Ferrara isolated a pituitary gland protein in the course of his research on cardiovascular diseases. He named the protein vascular endothelial cell growth factor (VEGF). Ferrara recounted, "At that time, no one thought this would be therapeutic, but Genentech has this great policy that allows people to pursue their own interests."[36,37] As a consequence, Ferrara was able to follow what he really believed in. In 1993, he and colleagues developed a mouse antibody that blocked VEGF receptors and shrank tumors in mice. It was a pleasant surprise, because the common wisdom was that one needed to block many pathways to inhibit angiogenesis. Like airplanes, tumors have backup systems for backup systems.

Armed with the spectacular animal results, Ferrara convinced management that VEGF blockade was a viable approach to treating cancer. They then set out to humanize the mouse antibody, a more challenging feat. They finally succeeded in making bevacizumab (Avastin), a humanized variant of the anti-VEGF antibody that was 96% human and 4% murine, using a technique called "site-directed mutagenesis" of a human antibody framework.

In clinical trials, Avastin did indeed work as a regulator of angiogenesis, effectively killing cancer cells by choking off their blood supply. It increased survival by 4.9 months for 20% of colon cancer patients tested when used in conjunction with chemotherapy. Thanks to Avastin's astounding specificity, the side effects are minor, notably modest hypertension and a few serious cases of lung bleeding. Mechanistically, the resulting hypertension is not surprising, as VEGF induces nitric oxide, which is involved in blood pressure regulation (see chapter 3). In February 2004, Avastin was approved for colorectal cancer. Avastin had annual sales of $224.2 million that year. On March 15, 2005, Genentech announced that, in a large clinical trial involving 878 patients, Avastin helped certain lung-cancer patients live longer. The Genentech stock shares soared 25%, to $55. Now hopes are that Avastin might be effective against other solid tumors, such as lung and breast cancer.

Building on the success of Herceptin, Avastin, and Rituxan, licensed from IDEC, Genentech has risen to become a leader in cancer biological drugs.

Protein Kinase Inhibitors, Small Molecules

The success of the aforementioned monoclonal antibodies for cancer has opened the door to small-molecule protein kinase inhibitors such as Iressa, Tarceva, and Gleevec as oral drugs. Monoclonal antibodies left much to be desired, as they are very expensive to make, manufactured by growing an enormous number of genetically identical ovary cells from hamsters in huge vats. In addition, because they are large proteins that do not survive in the gut, they can be administered only by infusion. Oral drugs would, on the other hand, be cheaper to make and easier to administer.

Iressa

Like Herceptin, AstraZeneca's Iressa is also an EGFR inhibitor at the intracellular domain level. Unlike Herceptin, which has to be administered by infusion, Iressa is an oral small-molecule EGFR tyrosine kinase inhibitor (TKI). On the cellular level, antibodies such as Herceptin cannot penetrate the cell membrane. They simply bind to the receptor protein on the exterior of the cell and disable the function of the kinase. In contrast, small-molecule inhibitors such as Iressa enter the cell membrane and bind to a specific motif of the receptor protein in the interior of the cell. Iressa works by hitting the signal transduction pathway of EGFR and by blocking binding of ATP at the ATP binding site, causing disruption of molecular signals that in cancer patients turn normal cells into tumors. As such, it is not expected to damage healthy cells or cause side effects as harsh as those associated with standard chemotherapy treatments.

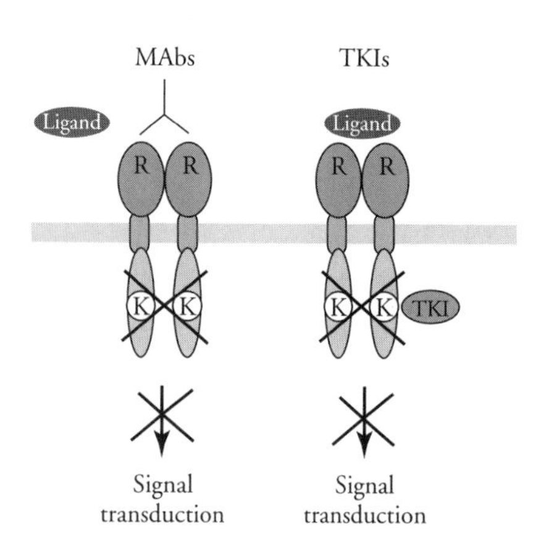

Monoclonal antibodies (Mabs) vs. Tyrosine Kinase Inhibitors (TKIs). Diagram by Vivien H. Li

AstraZeneca began the project with an older EGFR-TKI, which was pretty potent but was metabolized quickly in the human body. The scientists figured out that the major metabolic pathway was oxidation. Blocking the two oxidation sites and further optimization of the molecule with a basic 6-alkyl side chain produced Iressa, which had improved in vivo activities, as well as physical properties. Its favorable pharmacokinetic attributes allowed a once-daily oral dosing. In addition, it is at least one hundredfold more selective against other tyrosine kinases such as *erb*B-2, KDR, c-flt, or serine/threonine kinases such as PKC, MEK-1, and ERK-2.

In clinical trials, about 10% of patients with non-small-cell lung cancer treated with Iressa saw their tumors shrink dramatically. The response of women, Asians, nonsmokers, and adenocarcinoma patients was best. The quality of life is significantly enhanced in comparison to old-fashioned chemotherapy. In an accelerated review, the FDA approved Iressa for treatment of non-small-cell lung cancer in 2003. Unfortunately, at the end of 2004, a study involving nearly 1,700 patients showed that patients taking Iressa did not see survival rates increase compared with those on standard chemotherapy. A possible explanation is that Iressa targets only a specific molecule, namely EGFR, that spurs cancer cells but that seems to work only in patients whose tumors have a certain gene mutation that is more common among nonsmokers, women, and Japanese patients. So if you are a female Japanese nonsmoker with advanced lung cancer, Iressa might work wonders for you. In an unusual move, on June 17, 2005, the FDA announced that it would allow only a few thousand lung cancer patients who believe Iressa is helping them to have the drug. Any new patients who wished to try Iressa would have to do so in a strictly controlled study after September 15, 2005.

Tarceva

Tarceva, another oral small-molecule EGFR inhibitor, was approved in 2004 for the treatment of non-small-cell lung cancer. Its mechanism of action is the same as that of Iressa. It blocks the tumor cell growth by inhibiting the tyrosine kinase activity of the EGFR, thereby blocking the EGFR signaling pathway *inside* the cell.

Tarceva's journey from bench to market is as legendary as it is fascinating. In the early 1990s, Oncogene Sciences, Inc. (now OSI Pharmaceuticals), was founded by scientists at the Cold Spring Harbor Laboratories in Long Island, New York. Its main business was to screen compounds for other big pharmaceutical companies. In 1995, OSI entered an alliance with Pfizer to look for cancer drugs that work by blocking the EGFR signaling pathway. Medicinal chemists at Pfizer's Groton laboratory worked with OSI to secure novel chemical matter that they could patent. But EGFR research was so hot that AstraZeneca and Parke-Davis had already patented the first two compounds that OSI came up with. The third one, CP-358774, which would later become OSI-774 and Tarceva, was patentable. Pfizer and OSI moved it ahead to clinical trials. In 1999, Pfizer bought Parke-Davis, its partner in marketing Lipitor. Because Parke-Davis already had an EGFR inhibitor in development, Pfizer had to give away the rights of Tarceva to OSI *at no cost* in order to fulfill the Federal Trade Commission's (FTC) antitrust requirement. Overnight, OSI went from owning 6% of the rights of Tarceva to 100%. Nonetheless, OSI was a small biotechnology company and needed a partner with financial prowess, as well as experience, to conduct the long and expensive clinical trials. Almost every big pharmaceutical company was a suitor to OSI at the time due to Tarceva's promising data. In the end, Genentech won the bid by agreeing to share Tarceva's development with OSI rather than simply taking it over, like other big companies.

In clinical trials, Tarceva prolonged the lives of patients with advanced non-small-cell lung cancer, 6.7 months versus 4.7 months on standard chemotherapy—a result not seen with Iressa. Tarceva was launched in 2003 and sent OSI stocks sky high.

Gleevec

Leukemia, widely known as blood cancer, is actually a class of cancer that includes blood, bone marrow, and liver cancers. In Chinese, leukemia literally means "the white blood disease," indicating the abnormally high white blood cell counts observed in leukemia patients. Among many subtypes of leukemia, a rare but particularly ferocious form is chronic myeloid leukemia (CML), in which *myeloid* indicates "marrow-related." CML is one of the four main types of leukemia, and it strikes about 5,000–7,000 patients a year in the United States. It mainly affects adults between the ages of 50 and 60 and is characterized by abnormally high white blood cell counts. Before the emergence of Gleevec, the standard treatments were interferon, a biologic that had to be injected; Ara-C, an intravenous chemotherapy agent; and hydroxyurea, an oral chemotherapy agent.

As early as 1973, reports linking CML with an oncogene called the Bcr-Abl protein started to emerge. Abl is an enzyme called Abelson tyrosine kinase, named after its discoverer, Herbert Abelson (see page 5). In

1986–1987, David Baltimore and his postdoctoral fellow, Owen Witte, published two articles in *Science,* identifying Bcr-Abl as a tyrosine kinase. Like EGFR, Bcr-Abl is an enzyme that carries out signal transduction through the transfer of phosphate groups to specific amino acids (tyrosine in this case) on a protein. As a consequence, the cells that receive the signal begin dividing uncontrollably, thus triggering cell proliferation and then leukemia. The hallmark of chronic myeloid leukemia is the expression of Bcr-Abl, an activated form of the tyrosine kinase Abl. In effect, Baltimore and Witte identified the Bcr-Abl gene as the cause of CML. It then made sense that blocking the Bcr-Abl enzyme and stopping the faulty signal transduction would stave off the production of white blood cells. CML was one of very few cancers that had been directly linked to a single oncogene (i.e., Bcr-Abl). Most cancers are associated with at least two, and often more, oncogenes. In reality, many hurdles still existed on the road to an oral drug for combating CML. There are more than 518 subtypes of protein kinases, all of which bear an uncanny resemblance to each other. To create a molecule that selectively targets Bcr-Abl required not only ingenuity but also fantastic luck. An unselective inhibitor would have undesired toxicity. In addition, because most protein kinase inhibitors resemble the adenosine molecule on adenosine triphoshate (ATP), they are mostly flat molecules and are notoriously insoluble. Insoluble compounds cannot penetrate the cell membranes. The Bcr-Abl protein is located on the interior of cells, whereas the EGFR protein is extracellular, extruding out of the surface of cell membrane, making it more accessible. To bind to the Bcr-Abl protein, the drug has to penetrate the cell membrane, or it will be completely useless.

Alex Matter of Ciba-Geigy in Basel, Switzerland, embarked on an ambitious oncology program that targeted kinases in the late 1980s. He found a willing ally in America: Brian Druker of Oregon Health and Science University (OHSU), who was the staunchest advocate of targeting the Bcr-Abl kinase for CML. Yet, no matter how good the biology, a chemist is always needed to make and bottle the drug before it can be given to the patient. This was where Jürg Zimmermann came into the picture.

Zimmermann, born in 1957, earned his Ph.D. under Dieter Seebach at the Eidgenössische Technische Hochschule (ETH, Federal Institute of Technology) in Zürich, Switzerland. He carried out his postdoctoral training in Australia and Canada before joining Ciba-Geigy as a medicinal chemist in 1990. As luck would have it, his first assignment was to work on the protein kinase inhibitor project in oncology with Matter. In order to stop the action of Bcr-Abl kinase, a molecule that replaced ATP was needed to trick the enzyme into thinking the phosphorylation had been accomplished already. Zimmermann was studious and productive, making 10 compounds a week at the bench. He used a small-molecule protein kinase C (PKC) inhibitor with a phenylamino pyrimidine core structure as his starting point. Many old drugs on the market had the same core structure. Because they resembled the adenosine's flat structure, they had a good chance of occupying the same "pocket" on the enzyme. His team used an iterative trial-and-error approach, maximizing selectivity without sacrificing potency and solubility. They added an amide and a methyl to a phenyl ring to increase the potency and selectivity. In order to increase water solubility and bioavailability, they installed a piperazine ring, which was later shown by X-ray crystallography to provide additional binding to the Bcr-Abl. In the end, Zimmermann synthesized STI571 (STI here stands for "signal transduction inhibitor") in August 1992, a mere 2 years after he started his first job. Years later, Zimmermann commented, "The crazy thing was that even though we had computer-assisted modeling data, we eventually fell back on pen and paper."[38] He also confessed "Astonishingly—we can hardly admit today, all the experiments were done on the side and were received with a certain indifference."[38] The reason was that CML was such a rare disease that a drug to treat it was not so attractive from a commercial perspective.

After STI571 was made, protein binding and later cellular assays by the biologists revealed that not only was it a potent blocker for the phosphorylation of the Bcr-Abl kinase but it was also selective against most of the major kinases that they screened. In 1993 at Ciba-Geigy, Brian Druker also tested STI571 in protein, cellular, and animal models. He observed that STI571 killed CML cells without harming the healthy ones. STI571 also showed sufficient efficacy in animal models.

In 1996, just when Novartis (Ciba-Geigy merged with Sandoz in March 1996 to become Novartis) was getting ready for the Phase I clinical trial, a crushing piece of data came in. Severe liver toxicity was observed in dogs, as well as in rats, which was a big blow to the project. However, Druker argued that the disparity between dogs and humans is huge and that testing in a species closer to humans would be more relevant. Indeed, monkey toxicity studies revealed that STI571 had no significant liver toxicity at reasonable doses. Although the researchers had not intended to gauge the efficacy in Phase I clinical trials, they showed spectacular results. An added bonus was that the half-life of the drug was found to be 5 times longer than in animal models, allowing a once-daily regimen. STI571 arrested CML cells and left healthy cells alone.

Novartis, encouraged by the unprecedented efficacy observed in Phase I, took an unorthodox approach and moved the production with a fast-track status. The 12-step synthesis of STI571 in the manufacturing process was accomplished in an astonishingly short time—they prepared 1.6 tons of STI571 in 12 months, as opposed to the normal 24-month period. The successful Phase II in 1999 showed that the drug worked on all three phases of CML: the chronic, accelerated, and blast phases. Novartis was able to submit a new drug application (NDA) in March 2001 armed with just the Phase II results. For their part, the FDA approved it on May 10, 2001, in a record time of 2½ months. STI571 went through clinical trials faster and received approval faster from the FDA than any other cancer therapy in history. Novartis sold STI571 under the trade name Gleevec, which brought in $1.6 billion in sales in 2004. It was one of the best sellers for Novartis, second only to Diovan (an angiotensin II receptor antagonist for the treatment of hypertension).

With Gleevec on the market, numerous studies have taken place to further understand the drug. One study indicated that Gleevec actually blocks a panel of eight protein kinases, including Bcr-Abl, platelet-derived growth factor receptor (PDGFR), c-Kit, and five additional kinases. The fact that Gleevec also inhibits another tyrosine kinase receptor, the c-Kit receptor, that is associated with gastrointestinal stromal tumors (GIST), opened the door to its use in treating GIST.

The medicinal chemist who discovered Gleevec, Jürg Zimmermann, is still working at Novartis in Basel. He moved to the core technology division of Novartis to become the head of combinatorial chemistry in 1998. He figured that the chance of his discovering another drug was next to nil, considering how many lucky factors came into play with Gleevec. Discovering *one* drug is surely a better achievement than most medicinal chemists have to show at the end of their careers.

Deservedly, Brian Druker has been showered with accolades for his instrumental role in identifying and shepherding Gleevec from laboratory to patients. He remains modest and generous in sharing credit, stating "I am grateful for the recognition by my colleagues of my research contribution and the opportunity to help so many patients . . . Even though my lab was critical to the development of Gleevec, you should realize that a large part of the development of Gleevec occurred at Novartis." When asked about research expenditure for Gleevec, Druker honestly noted that Novartis was responsible for the bulk of the development cost. This includes costs of formulation and production as well as the cost of clinical trial.[39]

Because of the redundancy of cell functions, it is probably not surprising that a group of kinases, as opposed to just one single kinase, have to be blocked in order to achieve desired effects. If only one kinase were blocked, the body most likely would find another route to achieve cell proliferation. The protein kinase inhibitors are not obliterating cancer cells despite their phenomenal benefits in shrinking tumors. It turns out that tumors seem to have more pathways than there are field mice in a meadow. Gleevec, the golden child of targeted cancer drugs, is tarnished by showing drug resistance shortly after its emergence. About 15–20% of patients with CML developed resistance within 3 years. The Bcr-Abl enzyme changes its shape after it wrestles with Gleevec. As a result, Gleevec keeps bumping into it but is unable to bind, as it did in the original enzyme.

Gleevec revolutionized the treatment of CML as an oral drug with relatively few and more tolerable side effects in comparison with Ara-C, hydroxyurea, and interferon. The remarkable clinical effectiveness of Gleevec validated the promise that targeted cancer therapy with a kinase inhibitor is possible. It has become a new paradigm in biomedical research.

I have devoted this chapter to only a few classes of cancer drugs. Two additional important new drugs deserve mentioning. One is rituximab (trade name Rituxan), by IDEC and Genentech, which was approved by the FDA for the treatment of B-cell non-Hodgkin's lymphoma in 1997. It was the first monoclonal antibody approved for therapeutic use in cancer. Mechanistically, Rituxan is a chimeric IgG1 monoclonal antibody that specifically recognizes the CD20 surface marker present on more than 80% of non-Hodgkin's lymphomas. The other one is bortezomib (trade name Velcade), by Millennium Pharmaceuticals, approved by the FDA in 2003. It is a small-molecule proteasome inhibitor for treating multiple myeloma. The most striking feature of Velcade is that it contains a boronic acid functionality; boron is rarely seen in drugs.

Fight cancer © The United Nations Postal Administration

During the past decade, we have seen the emergence of targeted cancer drugs, which zero in on tumors and leave healthy cells relatively alone. Targeted cancer drugs are changing the face of cancer treatment. Because of their greater selectivity and fewer side effects, they also improve the quality of life for cancer patients, although it is unlikely that these new cancer drugs will eliminate chemotherapy for the time being. Like most other cancer drugs, they seem to work better in combination. Each targeted cancer drug is effective for only a small portion of the targeted classes of cancer. But the completion of the human genome project has unleashed a tremendous possibility—that a personalized genetic approach will topple the remaining cancer strongholds one by one. And it is not impossible that cancer may become a manageable disease, like cardiovascular diseases and diabetes. If a cancer goes into remission for 5 years, it is considered cured. Therefore, oncology is one of the most exciting areas of drug discovery. Many more targeted cancer drugs are surely to follow.

One of the challenges for targeted cancer drugs is to solve the problem of drug resistance. Tumors seem to have many pathways, and they can switch course if they encounter a drug that targets their preferred pathway. Indeed, because cancer cells have many redundancies built into their growth, several pathways must be blocked to keep the tumors in check. Sutent, a promising cancer drug developed by Pfizer, with the initial indication for Gleevec-resistant GIST, and with other indications, including metastatic renal cell carcinoma, was approved by the FDA in 2006.

During the past decade, stem-cell research has shed light on cancer formation and cancer treatment. Stem cells, immature cells that can replicate, or renew themselves, play a crucial role in all aspects of biology. They are able to differentiate, or mature, into all the cells that an organism or particular organ system need. There are two types of stem cells. Embryonic stem cells can give rise to all the tissue types in the adult organism. Adult stem cells residing in a number of adult tissues are important to tissue self-renewal and repair. As far as cancer is concerned, adult stem cells, not the politically controversial embryonic stem cells, are involved. In 2000, Irving L. Weissman, a professor of cancer biology at Stanford University, discovered leukemia stem cells. In 2003, Michael F. Clarke and Max S. Wicha at the University of Michigan discovered cancer stem cells in breast cancer. Now, scientists are trying to figure out what makes cancer stem cells different from other cells in the tumor. Then it will be possible to find ways to kill cancer stem cells while leaving the normal cells alone. Even now, stem cells from umbilical cords have been used in more than 3,000 young patients to treat their leukemia. In China, more than 200,000 people have donated their stem cells, which have saved about 200 patients who have had haematopoietic stem cell transplant operations. As the stem cell research progresses, more and more cancer patients will be helped.

In the same vain, stem cell research has begun to reveal the great potential that stem cells possess in the treatment of cardiovascular and neurodegenerative diseases and type I diabetes. But stem cell research is just in its infancy; scientists in the field will gain better and deeper understanding in the future, which in turn may eventually translate into the development of lifesaving medicines.

CHAPTER 2

Drugs to Kill Germs

The task of chemistry is not to make gold or silver, but to prepare medicines.

—Paracelsus

Lister and Carbolic Acid

Surgical standards before antiseptics starkly contrasted to the surgical art today. Conditions were especially atrocious for amputations and for compound fractures in which the bones penetrated the skin and were exposed to the air. Patients who did not die from the surgery often died of postsurgical infections and subsequent blood poisoning.

James Young Simpson, a Scottish surgeon and obstetrician who was the first to use chloroform as an anesthetic (see chapter 7), once said of surgical operations: "A man laid on the operating table in one of our surgical hospitals is exposed to more chance of death than the English soldier on the battlefield of Waterloo."[1] The mortality rate in hospitals after surgeries was 40–60%. During the American Civil War, the surgical fatalities were just as horrific as those from combat. A commonly used antiseptic in the battlefield was exceedingly corrosive nitric acid (HNO_3—ouch!).

However, in 1867, Joseph Lister's use of carbolic acid, whose chemical name is phenol, as an antiseptic changed the prospect of surgery. In Greek, *septic* means "rotten." Antiseptics, in turn, are substances used to treat a

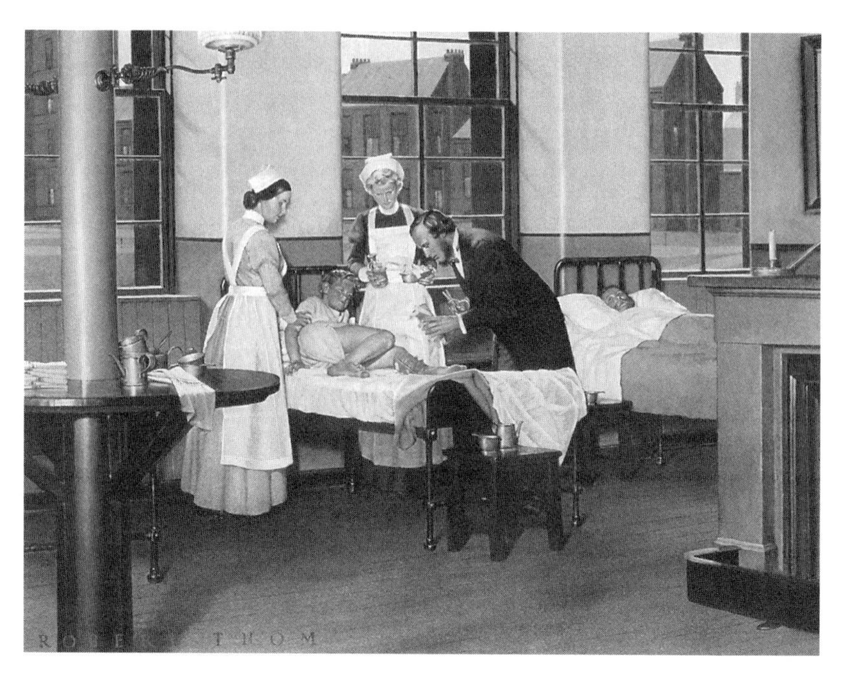

Joseph Lister performing surgery © Warner-Lambert

person to prevent the occurrence of infection. They are also known as germicides.

Joseph Lister (1827–1912) was born to a Quaker family in southern England. His father, Joseph Jackson Lister, was a wine merchant and a well-known microscopist. In his youth, Joseph Lister practiced surgery under the tutelage of James Syme in Edinburgh and married Agnes, his mentor's daughter. He had to give up his religion because Quakers at that time did not allow marriages outside the faith. That turned out to be a worthwhile sacrifice, because his marriage brought him lifelong joy.

Lister became a surgeon at Glasgow Royal Infirmary in 1860. He was acutely conscious of the appalling conditions in the infirmaries and determined to do something about postsurgical infections. In 1865, Lister was introduced to Louis Pasteur's exploits with germs by Thomas Anderson, chair of the chemistry department at Glasgow. Afterward, he personally repeated all the experiments that Pasteur published. However, simple and direct applications of the Pasteurization process would not be ideal during surgery—after all, boiling patients would not be acceptable.

Meanwhile, Lister read in a newspaper that the city of Carlisle used carbolic acid to deodorize sewage. Fields treated with carbolic acid did not

Louis Pasteur © Warner-Lambert

have parasites that caused diseases in cows. In addition, carbolic acid was also proven to be effective in controlling typhoid. All of those observations together led Lister to believe that carbolic acid might be effective in killing germs. Carbolic acid was an ingredient isolated from coal tar, a waste from coal gas production. Since its discovery in 1785, coal gas was generated by heating coal in the absence of air in a sealed system. The resulting methane was used for illuminating streetlights, cooking, and other domestic purposes. In addition to methane, the coal-burning process also left sulfur dioxide and copious amounts of oily tar, which was a nuisance to get rid of. After distillation, the carbolic oil was then further purified by washing with a slaked lime—calcium hydroxide, $Ca(OH)_2$—solution to give carbolic acid. Lister's invention included two parts: spraying carbolic acid in the air in the operating room and covering the wound with carbolic acid–soaked dressings. Spraying carbolic acid in the air is no longer done because the benefit does not justify the harm—carbolic acid is caustic to skin and body tissues, especially lungs. Lister himself admitted that carbolic acid was "a necessary evil incurred to attain greater good."[2] Initially, Lister's antisepsis was received in both England and the United States with fierce resistance from established surgeons. They viewed spraying and dressing of carbolic acid as cumbersome and harmful to their hands and lungs. It took many years and several improvements for the benefits of antisepsis to gain wide acceptance. Lister himself gave up using carbolic acid and used boracic acid (i.e., boric acid, H_3BO_3) as a better and milder antiseptic. Once antiseptics became widely used, the surgical mortality rate dropped dramatically. Lister lived to witness his discovery bear fruition and revolutionize surgical science.

Lister was one of the first to apply germ theory to surgery, and he is regarded as the father of antiseptics. Antisepsis was one of the two most significant discoveries in medicine during the nineteenth century. The other was anesthesia. Today, asepsis (meaning "free of germs") has largely replaced antisepsis in the operating rooms.

Lambert Pharmaceutical Company (one of the predecessors of Warner-Lambert, which was later acquired by Pfizer) developed a mouthwash in the 1880s as a general oral antiseptic for relief of bad breath. However, the product did not take off until it was branded as Listerine (after Lister) in the 1920s. Lambert Pharmaceutical Company grew 40 times in 7 years thanks to the popularity of Listerine. Some of Lister's family members are still receiving royalties from the profits of Listerine. A similar oral antiseptic was marketed with the brand name Pasteurine. Somehow it never caught on.

As an antiseptic, carbolic acid works by solubilizing the phospholipids in cell membranes, thus disrupting the cell membranes. Other phenols, such as thymol and quinolol, are effective antiseptics, too. H. D. Dakin, a British-American biochemist, developed a new and effective antiseptic, Chloramine-T, also known as Chlorazen or Dakin's solution, which works by slow release of chlorine in water and is frequently used by surgeons in dressing wounds.

In addition, phenol is also a crucial ingredient for making aspirin, Bakelite, and trinitrophenol, an explosive. During World War I, phenol was the star of a sensational scandal historically dubbed "the great phenol plot." In 1915, shortly after the war began, the main supplier of phenol, Britain, imposed an embargo on Germany that severely limited Bayer's aspirin production. Meanwhile, also suffering from phenol shortage, Thomas A. Edison (1847–1931) decided to make it himself to support his own enterprise of phonograph records. Because Edison's plants produced 6 tons of phenol per day, he was willing to sell half of the output. A German spy named Hugo Schweitzer engineered a plot to purchase Edison's surplus phenol for Bayer's aspirin manufacturing. When he was accused of stealing American chemicals for Germany, Schweitzer falsely claimed that he intended to use it as a disinfectant. The *New York Times* calculated that 3 tons of phenol a day would translate to 240,000 pounds of disinfectant, enough to supply every single American with a new 2-ounce bottle each week.

Dr. Ehrlich's Magic Bullet

Although Joseph Lister's antiseptic, carbolic acid, saved numerous lives during surgical operations, it worked only on superficial surfaces. If the germs penetrated the wound, as frequently happened during wartime, the application of antiseptics on the exterior would not annihilate the germs completely; the result was often lethal infections. Therefore, a drug that functioned systemically would be preferred for killing bacteria.

From the end of fifteenth century onward, Europe was ravaged by syphilis, an epidemic venereal disease that affected the human race for centuries. A conservative estimate of the syphilis infection rate in Europe during this period was 10%. The Europeans also brought syphilis to the New World. Late-stage syphilis would affect the central nervous system and cause brain damage, which resulted in general paralysis and insanity (GPI). Ivan the Terrible (1530–1584) was proven to be syphilitic after the Soviet Union exhumed his tomb and found typical lesions in his bones. Ivan ruled Russia wisely and humanely from 1551 to 1560. The madness, paranoia, and cruelty that were the consequences of his cerebral syphilitic effects resulted in the slaughter of thousands of people, including his own son and heir, whom he stabbed with a steel-pointed staff. Another famous syphilis victim was Winston Churchill's father, Lord Randolph Churchill, who was sent to an insane asylum after his odd behaviors manifested apparently due to late-stage syphilis. In the sixteenth century, Paracelsus (1493–1541) treated syphilis with mercury (it was actually mercury sulfide, HgS) with some success, although it only lessened the symptoms and had numerous toxic side effects. The major side effects of mercury treatment included constant salivation, severe indigestion, tooth decay and tooth loss, loss of weight, weakness, emotional disturbance, and even death. There was a famous saying during that era: "A night of Venus and a lifetime

Paul Ehrlich and Sachachio Hata © Warner-Lambert

of mercury."[3] Paracelsus was the first known physician to use a specific chemical to treat patients.

In 1910, Ehrlich's 606 emerged as the first effective drug against syphilis.

Paul Ehrlich (1854–1915) was appointed a professor at Charité Hospital in Berlin in 1887. He worked in his own personal laboratory, but being Jewish, he was not paid for his work. In collaboration with his friends, immunologist Emil von Behring (1854–1917) and Shibasaburo Kitasato, he developed a horse serum antitoxin to quell diphtheria. The vaccine saved thousands of children's lives during the 1891 diphtheria outbreak. Behring won the first Nobel Prize for Physiology or Medicine in 1901; Ehrlich won in 1908.

Although most people would have been content to rest on their laurels, Ehrlich only became more resolved, and his most important contributions came after he received the Nobel Prize in 1908. In 1910, working with his Japanese associate Sachachio Hata, Ehrlich experimented with numerous chemicals to find a "magic bullet" to fight syphilis. Most of those chemicals were derivatives of atoxyl, a highly toxic arsenic compound. They experimented with 605 chemicals without success, and they finally triumphed with the 606th compound, arsphenamine. Dubbed Ehrlich's 606, or Ehrlich-Hata 606, arsphenamine killed syphilis microbes. After the drug was tested safely on dogs, two of his assistants volunteered as human guinea pigs, and Ehrlich's 606 was found to be relatively safe in humans.

Arsphenamine was patented under Ehrlich's name to Hoechst in 1911, which marketed it using the trade name of Salvarsan, an event that heralded the beginning of the modern antibacterial era. Initially, arsphenamine was injected once, using a 900-milligram (mg) dose. Because it was produced as a hydrochloric (HCl) salt, an alkaline solution had to be added in order to dissolve it. Later, Ehrlich found out that it was more efficacious to inject 600-mg doses consecutively, which stopped a syphilis infection without severe toxic effects. Soon Salvarsan as a cure for syphilis became well known. Physicians and desperate patients clamoring to procure the lifesaving "magic bullet" besieged the Hoechst manufacturing site near Frankfurt. The scene would be replayed over and over again throughout the years, with patients seeking experimental drugs for tuberculosis, cancer, and AIDS. Although Salvarsan was less toxic than previous treatments, it still had quite serious side effects. In 1912, it was replaced by a better derivative, Neosalvarsan, the 904th compound that Ehrlich tested. Salvarsan and Neosalvarsan had a tremendous impact in fighting syphilis, wiping out half of the syphilis infections in Europe in a mere 5 years (syphilis was not completely eradicated until the introduction of penicillin in the 1940s). Ehrlich's friend Emil von Behring was initially doubtful of the concept of chemotherapy but was transformed from a skeptic into the staunchest supporter after the beneficial results of Salvarsan and Neosalvarsan became apparent. On a personal level, two intellectual giants, von Behring and Ehrlich, rekindled their friendship after having been estranged by differing opinions on chemotherapy.

In addition to his discovery of serum antitoxin for diphtheria, Salvarsan, and Neosalvarsan, Ehrlich's perspective on scientific discoveries also had a long-lasting impact. His philosophy of scientific discoveries was summarized as 4 G's:[4]

Geld (money)
Geschick (brain)
Geduld (patience)
Glück (luck)

Salvarsan was one of the first antibacterials. In addition to being one of the founders of immunology, Ehrlich is also regarded as the father of chemotherapy, which he defined as "The use of drugs to injure an invading organism without injury to the host."[4] The history-altering discovery of Ehrlich's 606 was vividly depicted in the 1940 movie *Dr. Ehrlich's Magic Bullet.*

Domagk and Sulfa Drugs

Josef Klarer (1898–1953), a chemist at the Bayer Company of I. G. Farben (I. G. Farbenindustrie Aktiengesellschaft) in Germany, synthesized a brilliant orange-red dye, 2′,4′-diaminoazobenzene-4-sulfonamide, in 1932.[5,6] By the end of the year, Gerhard Domagk (1895–1964), then the head of the bacteriology laboratory of I. G. Farben, was experimenting with the different dyes available to him in search of drugs. By injecting dyes in mice infected with *Streptococcus pyogenes* bacterium, Domagk discovered that 2′,4′-diaminoazobenzene-4-sulfonamide, later branded as Prontosil, was effective in killing the bacterium without toxic effects. Another Bayer chemist, Fritz Mietzsch (1896–1958), prepared the salt of Prontosil, enabling a liquid formation that was more amenable to injection.

In November 1932, Domagk's 6-year-old daughter, Hildegarde, became ill with an infection from the prick of an embroidery needle contaminated with streptococcal bacteria. The infection quickly spread to her lymph nodes, and blood poisoning became severe. Other doctors recommended amputation of her arm, but even that would not afford a good chance of survival. Near death and unresponsive to other treatments, Hildegarde was injected by Domagk himself with a large dose of Prontosil. She made a miraculous recovery.

Domagk published his discovery in February 1935. To historians, the timing has always been controversial. The patent for Prontosil as a dye expired around the mid-1930s, just when its antibiotic properties were discovered. It was highly possible that Domagk was trying to find a compound that was patentable. One could argue that many lives could have been saved during the years that the discovery was sitting at I. G. Farben. We may never know who made the critical decision to withhold the publication within the company. Without patent protection, everybody would be able to make Prontosil, which was exactly what all the big drug firms around the world did.[7] Prontosil quickly became widely prescribed for streptococcal infections.

In late 1935, a French husband-and-wife team, Jacques Tréfouël and Therèse Tréfouël, discovered that Prontosil was not active in vitro. The real active ingredient for its antibacterial activity was sulfanilamide (mistakenly being called sulfonamide even today), which was generated from in vivo metabolism of Prontosil. Prontosil is the prodrug of sulfanilamide, which had been produced in tons by the dye industry for decades without anyone looking into its antibacterial properties. Sulfanilamide is vastly superior to Prontosil: not only is it much cheaper to manufacture, but it is also a white powder, so it does not change the color of patients' skin. Patients treated with Prontosil always had telltale red skin, similar to cooked lobsters.

Before World War II erupted, scientists around the world nominated Domagk to the Nobel Committee in Sweden. He was awarded the Nobel Prize for Physiology or Medicine in 1939. In those years, Hitler forbade Germans to accept the Nobel Prize in retaliation for the Nobel Committee's having bestowed the Nobel Peace Prize on Carl von Ossietzky (a Jewish pacifist writer and journalist jailed by the Nazis) in 1935. In October, Domagk wrote a letter to the Karolinska Institute in Stockholm to acknowledge the honor. The Gestapo threw Domagk in jail until he signed a letter prepared by the Ministry of Education declining the prize; only then was he allowed to go back to his laboratory. From then on, Domagk suffered a debilitating depression from which he never completely recovered.

In 1947 Domagk did receive his Nobel Prize, but not the money (which was diverted back to the Nobel Foundation). Domagk received many letters from patients and doctors expressing their gratitude for his discovery of Prontosil. In contrast, the chemist who first synthesized it, Josef Klarer, received none.

Gerhard Domagk
© Guyana Post Office Corporation

Sulfanilamide was associated with one of the most notorious medical blunders in American history, which became the impetus for United States Food, Drug, and Cosmetic Act of 1938. In 1937, the magic sulfa drug was very much in demand for the treatment of streptococcal infections in the United States. Because no patent protection existed, big drug firms, including Squibb, Winthrop, Eli Lilly, and Parke-Davis, all rapidly began mass production of sulfanilamide. Due to sulfanilamide's insolubility in ethanol, there was no syrup or liquid form that was easier for children to take. It was instead sold as a powder, as capsules, or as tablets. The S. E. Massengill Company in Bristol, Tennessee, decided to pursue a syrup form. The president, Samuel E. Massengill, assigned the task to his chief chemist, Harold C. Watkins, who discovered that sulfanilamide dissolved in diethylene glycol (DEG) with ease. He added some pink raspberry flavor and water and branded his concoction "Elixir Sulfanilamide." With no check for toxicity and no clinical trials, Massengill's 200 eager salesmen rushed to sell it to doctors and pharmacies all across the country. But things soon went horribly wrong. Children who took Elixir Sulfanilamide to treat a minor sore throat became terribly ill, experiencing excruciating pain. In the end, 107 patients died of horrible kidney failure. Their kidneys were often enlarged to twice the size of normal kidneys. Several years later, the 108th victim of Elixir Sulfanilamide emerged: the unemployed former head chemist of S. E. Massengill Company, Watkins, shot himself while cleaning his handgun.

Although sulfanilamide was thoroughly tested for safety in animals and humans and was proven to be reasonably safe, DEG is highly toxic. DEG, similar to antifreeze (ethylene glycol), is a deadly poison. Mouse studies carried out at the University of Chicago confirmed that DEG caused kidney failure. After the FDA learned about the tragedies, it confiscated all existing Elixir Sulfanilamide in the biggest operation in the FDA's history. It even jailed a Massengill salesman until he revealed his client list. In all, the FDA probably saved more than 4,000 lives. Unfortunately, due to lack of proper legislation, Massengill and his company were prosecuted only for "mislabeling," because anything called an elixir had to contain ethanol.

In 1938, steered by Franklin (1882–1945) and Eleanor (1884–1962) Roosevelt, Congress passed the Food, Drug, and Cosmetic Act, a legislation that had been sitting in Capitol Hill for over a decade. It called for tougher regulations for the sales of food, drugs, and cosmetics. As a consequence, the FDA has to approve a drug before it goes on sale, and it is illegal to sell certain drugs without prescription. The act greatly increased the accountability of drug firms regarding the safety of drugs. Largely thanks to the 1938 act, the United States was spared the thalidomide horror (see chapter 1), which produced many deformed babies in Europe prior to 1962.

Spurred by the success of Prontosil, scientists around the globe explored derivatives of sulfanilamide. In the end, more than 5,000 analogs were prepared in the decade immediately following Domagk's initial publication. Domagk's laboratory alone screened more than 2,000 sulfanilamide analogs. Unfortunately, the rate of return was not so high. Fewer than 20 sulfa drugs, including sulfapyridine and sulfadiazine, were proven to be clinically useful. In 1936, sulfanilamide was known to have saved the life of Franklin Roosevelt, Jr., the son of the President of the United States. And sulfapyridine, in turn, was used to help Winston Churchill (1874–1965) recover from pneumonia shortly after the Teheran summit. During World War II, wounded soldiers were treated with sulfanilamide powder topically before being given sulfadiazine orally. Sulfa drugs, a German invention, fortuitously aided the Allies.

The mechanism of action (MOA) of sulfanilamides is through folate antagonism. Because the structure of sulfanilamide is similar to that of *para*-aminobenzoic acid, an essential ingredient for cell synthesis, it interrupts bacteria growth. Sulfa drugs were successful in treating gangrene, leprosy, and strep infections. Later they were found to be effective in treating childbed fever and meningitis. However, they have a narrow spectrum of activities and are effective against only a limited number of diseases. Furthermore, many bacteria soon developed resistance to sulfa drugs. (They are used today with limited capacity in some corners of the world.) An antibacterial with a broad spectrum and high efficiency in killing bacteria with minimal toxic side effects was badly needed. Sulfa drugs would soon be outshined and mostly replaced by the "wonder drug"—penicillin.

Fleming, Florey, Chain, and the "Wonder Drug" Penicillin

It is hard to dispute that penicillin is the greatest discovery in medical history. Alexander Fleming (1881–1955) actually first discovered penicillin in 1928 in England, 4 years before Domagk's Prontosil. However, more than 15 years elapsed until Howard Florey and Ernst Chain isolated enough penicillin and demonstrated its curative effects in both mice and humans. Penicillin, fittingly dubbed a "wonder drug," quickly replaced Ehrlich's 606 and Domagk's sulfa drugs as the most widely used antibiotic. It works for treating Gram-positive bacterial infections, including strep and staph infections, pneumonia, gangrene, and meningitis, as well as gonorrhea (now, however, resistant) and syphilis. Best of all, it has a low toxicity—although some people are allergic to penicillin, sometimes with lethal consequences.

The penicillins are actually a group of more than a dozen or so closely related analogs, which are the secondary metabolites of microorganisms of the genus *penicillium.* In addition to possessing a thiazole moiety and an amino-acid side chain, they universally have the labile β-lactam, and thus they are prone to ring opening by nucleophiles (see Appendix A, page 250, for the structure of penecillin). That is when and how its effects manifest.

The discovery of penicillin was one of the most fascinating episodes in the history of medicine. By 1928, 47-year-old Scottish bacteriologist Alexander Fleming had lived an unremarkable life.[8] In 1922, he discovered a lytic agent, lysozyme, which he isolated from mucus of his running nose. Lysozymes are the bacteriolytic agents that act on the carbohydrate moiety of certain bacteria, lysing them, that is, causing disintegration. Lysozyme was fundamentally important but of no therapeutic value because it attacked only harmless bacteria. In the summer of 1928, Fleming took a 2-week vacation. Unlike a diligent researcher, he did not rinse his petri dishes in Lysol to clean up. Instead, he left them on his desk, and he forgot to close the windows of his laboratory.

In an extraordinary stroke of good luck, when he came back, he noticed that the colonies of *stapholycoccus aureus* culture had been dissolved (lysed, to use the biological term) around some molds. The thick greenish molds evidently inhibited the growth of the bacterium. After consulting his colleagues, Fleming found out that the mold was *Penicillium notatum.* A professor downstairs was doing experiments with the mold, and it had been inadvertently blown into Fleming's *stapholycoccus aureus* culture dish. Years later, almost nobody else could reproduce Fleming's exact results in the laboratory; it turned out that London had been cold for 9 days then become very hot for 6 days. Barring any of the coincidences, penicillin would not have been produced in Fleming's petri dish.

By another splendid stroke of good fortune, because of Fleming's experience, curiosity, and acute perception, he did not scrub the "ruined" petri dish into buckets of antiseptics. Instead, he proceeded to culture that mold and isolated an extract that he named *penicillin,* which proved to be extremely potent against bacterial growth. He even used the extract to treat the eye infection of one of his colleagues. Fleming published a paper in 1929 in the *British Journal of Experimental Pathology* and left the subject to rest.

One could only speculate as to why penicillin did not catch the attention of the medical community. One of Fleming's attributes might have delayed the public's recognition of penicillin. At 5 feet 5 inches, Alec Fleming, nicknamed "little Flem," was not a man of great stature (although today he is remembered as a giant in man's conquest of infectious diseases). His speech was quiet and halting, and he did not exude much conviction. Although he himself was convinced that penicillin was a major scientific breakthrough, his seminars and lectures on the topic were neither persuasive nor compelling and left audiences unimpressed. As a consequence, his scientific publications on penicillin just sat in the library, collecting dust, for nearly a decade.

Another reason might better explain the delay. The philosophy of Fleming's boss, Almroth Wright (nicknamed "Almost Right"), and his aversion toward chemotherapy deterred the possible realization of penicillin's therapeutic potential as an antibacterial. For Almroth Wright, vaccination and antitoxin were the proper direction for medical research. Ironically, Wright's philosophy was reinforced by the toxicity of Ehrlich's 606, Salvarsan. He believed that Salvarsan killed both bacterial and healthy cells with equal ferocity.

However, things became completely different when penicillin was developed and became a wonder drug. Fleming became the idol of the masses and the spokesperson on the discovery of penicillin. In contrast, Howard Florey, who led the Oxford team that made enough penicillin to treat human infections, shunned the media and public attention.

After Fleming's 1929 paper, 6 years would elapse before Howard Walter Florey (1898–1968) took notice.[9,10] In 1935, Florey, a 37-year-old Australian, was appointed as the prestigious head of the Sir William Dunn School of Pathology at Oxford University to replace the renowned Georges Dreyer. One of Florey's research interests was lysozyme, Fleming's first major discovery. Florey did not have much luck with lysozyme, which had little therapeutic value despite its importance in fundamental science. However, Florey's fortune was about to change when he hired Ernst Chain, a biochemist, to join his research group.

Howard Florey © Australia Post

Ernst Boris Chain (1907–1979) fled Germany after the rise of the Nazi party in 1933. He was exuberant, excitable, arrogant, brusque, and yet very insecure as a Jewish immigrant living in England. With a mustache and hair that stood up, he bore an uncanny resemblance to the young Albert Einstein. Like Einstein, Chain was helped tremendously by his musical talent. He was a welcome guest because his piano performance was invariably the highlight of any gathering. As a matter of fact, he once seriously contemplated becoming a professional piano player. In 1937, Florey hired Chain for his department, providing the first chance for stability for Chain after years of turmoil in his life. Despite Florey's reserved personality, he took an immediate liking to Chain's artistic talents, boisterous manner, and, most of all, his enthusiasm toward his research.

Harold D. Raistrick, a biochemist at the London School of Tropical Medicine and Hygiene, briefly explored the properties of penicillin in 1932. In 1938, with Chain's prodding, Florey resurrected Fleming's penicillin project. Florey mentioned that in Raistrick's experiments penicillin had appeared to be unusually unstable. Overhearing the comments, Chain characteristically quipped that Raistrick might not be such a good chemist. He added that it must be possible to produce it in stable form. A stroke of luck took place at the very beginning of their work. While Chain was reading Fleming's article in the *British Journal of Experimental Pathology,* he inquired of a laboratory technician whether she knew of the existence of a mold of *Penicillin notatum.* She immediately replied that there was a sample, supplied by none other than Alexander Fleming to Georges Dreyer, Florey's predecessor. Chain prompted Florey to agree to redirect their attention to look into the penicillin's antibacterial effects. Both Florey and Chain admitted that when they started their research on penicillin, they were more interested in satisfying their academic curiosity and in securing long-term funding of research than in saving lives.[11] With regard to their respective contributions to the penicillin resurrection, perhaps Chain described it best: "My part of this project was the isolation and study of the chemical and biochemical properties of this substance, Florey's the study of their biological properties."[11]

Like his predecessors, Chain found that penicillin was stable in water only in the form of a salt in near-neutral conditions with pH ranging between 5 and 8. It decomposed readily in an aqueous solution of higher acidity or basicity. Chain worked with microbiologist Norman Heatley and figured out a back-extraction process. Chain's contribution also included the application of a freeze-dry process in purification, that is, lyophilization. Purification of penicillin was also aided by column chromatography.

Fortunately, the team chose Swiss albino mice rather than guinea pigs to test penicillin's toxicology. Guinea pigs do not tolerate penicillin, for unknown reasons. In another stroke of luck, the first penicillin sample used to test toxicity was shown to be nontoxic, although the sample contained only 1% penicillin. Imagine what the outcome would have been if they had chosen to test penicillin in guinea pigs or if any of the impurities had been toxic.

The results of mouse studies were astonishing: mice infected with *streptococci* died within a few days, whereas infected mice treated with penicillin lived for 6 weeks and beyond. They tested penicillin in humans and observed similar results. Because such a minute amount of penicillin was available, they had to extract penicillin secreted from patients' urine samples and reuse it.

In the early 1943s, Fleming visited Florey's penicillin team at Sir William Dunn School of Pathology. Although he was taciturn during the majority of his visit, his comment to Florey—"I am here to see what you are doing with my penicillin"[12,13]—sowed the seeds for future rivalry between Fleming and the Oxford team.

Alexander Fleming, Howard Florey, and Ernst Chain © Warner-Lambert

In July 1941, pummeled by the Luftwaffe bombardment during the Battle of Britain, England was not safe enough to afford the investigation and manufacture of penicillin. Florey turned to America for help. Under a $5,000 grant from the Rockefeller Foundation, Florey and Heatley boarded an airplane carrying the *Penicillium* mold and some penicillin samples in a briefcase. Not having been invited, Chain was not happy. More hurtful still was the fact that he was not informed of the trip until Florey and Heatley were on the way to the airport.

Florey and Heatley constantly worried that change of temperature would destroy their precious mold and their penicillin samples would lose activity. Also afraid that the Nazis would confiscate the precious mold if they were captured, Florey and Heatley actually rubbed the *Penicillium* mold into their coats. Fortunately, they all survived the trip. Arriving in America, Florey and Heatley brought the mold to the Northern Research Laboratory (NRL) of the Department of Agriculture in Peoria, Illinois. There, Heatley stayed and helped to optimize the penicillin production process. An American microbiologist, A. J. Moyer, suggested the use of an effective yet inexpensive nutrient supply—corn steep liquor, an abundant by-product of the wet-corn milling industry. Later, it was discovered that it was only the presence of phenylacetic acid in corn steep liquor that enabled the increase of penicillin production. Because phenylacetic acid could be easily prepared chemically, the once inflated price for corn steep liquor plummeted to nearly nothing again.

In an effort to find a better yielding strain of mold and encouraged by the government, citizens near Peoria brought all kinds of molds to NRL to test their penicillin-producing capacity. A lady named Mary Hunt ("Moldy Mary") brought a rotten cantaloupe, which turned out to contain a strain called *Penicillium chrysogenum*. The strain doubled the penicillin production output over the original strain. At the end, the optimized process increased the penicillin output by twelvefold.

In order to supply the Allied troops with enough penicillin, a mass production process with great efficiency was desperately needed. After all, Florey's initial fermentation process, developed at Oxford, had a yield of a mere 0.0001%. The American War Production Board (WPB) and private pharmaceutical companies took on the challenge. "The Big Three," Merck, Squibb, and Pfizer, were the first ones to participate. Later, many pharmaceutical companies made significant contributions to the production of penicillin. Although Hoffmann-La Roche and Ciba in the United

States were eager to participate in the penicillin enterprise, they were denied the opportunity because "they were of foreign ownership and under foreign management." WPB feared that the technique would be inadvertently leaked to enemies.

Pfizer, in particular—which was then more a chemical company producing citric acid than a bona fide pharmaceutical company—made its name through the penicillin endeavor.[14] Taking advantage of the fermentation expertise accrued from its citric acid production, Pfizer invested $3 million to build 14 ten-thousand-gallon fermentation tanks. They utilized a "deep-tank fermentation" process involving sterilized air continually pumped through the tank. In the end, Pfizer produced 90% of the penicillin that went ashore with the Allied forces on the beaches of Normandy in June 1944. Sadly, without the name recognition of a drug maker and without a sales force, Pfizer sold penicillin in bulk to the more established and prestigious drug firms, Lilly, Parke-Davis, and Upjohn, which then distributed the product under their own labels. The humiliating experience served as a rude awakening. When Pfizer discovered its first drug, oxytetracycline (Terramycin; see page 68), in 1950, the company hired their own sales force and sold it themselves. Pfizer recently bought the original *penicillium* mold plate for $50,000 and donated it to the Smithsonian Institution, where it still resides.

During World War II, France was under the yoke of Nazi occupation. The Tréfouëls (who had discovered that Prontosil was a prodrug) produced penicillin by surface culture using Fleming's strain number 4222 of *Penicillin notatum.* The Rhone-Poulenc Company adapted the process but did not produce any significant amount of penicillin. In early 1944, German occupiers requested the Fleming strain from Institut Pasteur and were given a false one. At the end of the war, Hitler awarded his personal physician, Theodor Morrell, the Iron Cross for "his" discovery of penicillin. Historians considered the episode a complete farce. As a matter of fact, the German medical community regarded Morrell so poorly that Hitler's accolade to him might have even impeded further possible progress in Germany's quest for penicillin.

Penicillin profoundly revolutionized the art of medicine. Fleming, Florey, and Chain shared the Nobel Prize for Physiology or Medicine on December 11, 1945.

Regrettably, the relationship between Florey and Chain deteriorated. Being an immigrant from Germany, Chain always felt ill at ease in England.

The discord between Florey and Chain actually went back to the beginning of their penicillin work, when Chain spent £60 to upgrade their cold room. Sixty pounds was a considerable sum then, and Florey never let Chain forget his extravagant spending. More significantly, with his industrial background, Chain was a proponent for patenting penicillin, whereas Florey, as a physician abiding by Hippocrates' oath, would not dream of profiting from a medical discovery. Their American counterparts had no such qualms. A. J. Moyer quickly patented his process, as did Merck, Abbott, and Pfizer. Seeing Britain pay Americans for their own discovery of penicillin embittered Chain to no end. Another issue was Florey's silence with regard to the credit for the penicillin discovery. Chain perceived that Fleming had stolen the spotlight and that the credit was solely given to Fleming. The lack of recognition deprived the Oxford team, and thus Chain himself, of the prestige. This deterred their ability to acquire research funding, which would have provided equipment and autonomy on the scale that Chain wanted.

It became increasingly difficult for two Nobel laureates and intellectual giants to work in the same department. Things became considerably worse in 1949, when Chain decided to leave Oxford and take a position in Italy. In his departing letter to Florey, Chain wrote:

> My dear professor, I am very sorry that our personal relationship has deteriorated so much during the last years; I think the reason for it is mainly the general imperfection of human nature. I have always regretted this development and I hope that as time goes on the unpleasant episodes—which after all, were not so frightfully important when looked at from a broad viewpoint—may gradually sink into oblivion and we shall remember only the exciting and unique events of the time of our collaboration which a curious fate has destined us to experience together.[11]

In another letter to Florey, Chain reiterated:

> I shall remember with great pleasure, and always with gratitude, the first years of our association in which the foundation for the subsequent work was laid, and shall try to forget the bitter experience of the later years which I am sure will shrink into insignificance as time goes on.[11]

Florey's reply letter identified the culprit in their rivalry: "I only trust that the difficulties of running laboratories will not disillusion you too soon, but I am sure it is best that you should run your own show."[11]

As the penicillin saga continued, Dorothy Crowfoot Hodgkin (1910–1994) at Oxford solved the structure of penicillin using X-ray crystallography at the end of 1945. The elucidation of the structure of penicillin paved the way for chemists to develop its analogs and other β-lactam antibiotics. The medical success of penicillin spurred a flurry of research activities around the globe before 1945 in synthesizing it. In the United States, R. B. Woodward at Harvard worked on it extensively, whereas Sir Robert Robinson and John Cornforth at Oxford in the United Kingdom did a tremendous amount of synthetic work in collaboration with Florey and Chain. A named reaction in organic chemistry, the "Cornforth rearrangement," was actually a direct result of Cornforth's research effort toward penicillin synthesis.[15]

Despite intensive collaboration (and competition) among more than 1,000 English and American organic chemists in 39 major laboratories, the chemical synthesis of penicillin was not accomplished during World War II. After the war, almost all of them abandoned their efforts, deeming it impossible. Only one group kept working on it: John C. Sheehan (1915–1992) at MIT completed the first total synthesis of penicillin V in 1957, after 9 years of hard work and frustration.[16,17] The key step was the cyclization of the corresponding penicilloic acid to β-lactam using DCC (dicyclohexyl-carbodiimide), a peptide-coupling reagent that Sheehan himself popularized.

Waksman, Schatz, and Streptomycin

"The Lord hath created medicine out of the earth; and he that is wise will not abhor them" (Ecclesiastes 38:4). It was indeed the case for streptomycin, which was discovered from soil samples, like many other lifesaving antibiotics.

The wonder drug penicillin works well for most Gram-positive bacteria infections, but it does not work for infection by tubercle bacillus. Tuberculosis (TB) is a slowly growing bacterial infection of the lung, which often disseminates to other parts of the body. A telltale sign of having contracted tuberculosis is coughing up bright red blood—oxygen-rich blood from an artery bleeding into the lung. It is often also accompanied by chest pain and fever. Tuberculosis is such a slow killer that it was also known by the terms *consumption* and *wasting*. When the English poet John Keats discovered that he had tuberculosis, he exclaimed: "The drop of blood is my death warrant. I must die."[18] Indeed, tuberculosis germs that had been hiding had already severely damaged his lungs by then. In those days, tuberculosis meant a death sentence to many because there was no cure, earning it the name "the great white plague." It is a disease as old as humankind and can be traced back to the Stone Age. Tubercular lesions have been found in Egyptian mummies dating back to 3700 B.C. In ancient times tuberculosis was rife and was especially dangerous for women during childbirth. The scourge of the great white plague was so horrendous that it claimed one out of seven lives—2 billion lives over the past two centuries. The international symbol for tuberculosis is "‡," symbolizing graves that it helped to create. Keats (1795–1821), Polish composer Frédéric Chopin (1810–1849), American first lady Eleanor Roosevelt (1884–1962), and British movie star Vivien Leigh (1913–1967) were all inflicted with and died of tuberculosis. So many artists were inflicted with tuberculosis at the time that it was considered romantic and attractive to be tubercular. In fact, it was most likely that tuberculosis bacteria spread in the air in urban settings more readily than in rural areas.

Before the emergence of effective antibiotics, the best that physicians could do for tuberculosis patients was to recommend better hygiene, diet, rest, and exposure to fresh air and sunshine. To cater to affluent patients, many tuberculosis sanatoriums or pavilions were established on wooded mountaintops or in deserts to provide fresh air. The method had its own scientific basis, because some tuberculosis patients are infected by inhaling the bacterium in the air. In addition, brutal medical interventions included collapsing lungs and inducing pneumothorax.

Robert Koch (1843–1910), a German country doctor, was profoundly inspired by the work of Louis Pasteur. His involvement with bacteriology began after his wife pointed out that microbes proliferated on discarded meat. Koch, a genius of scientific experimentation, was the first to use dyes to stain bacteria. Later, he discovered the bacterium that causes tuberculosis, tubercle bacillus, using a stain that he invented with a dye.

In 1910, the year Koch passed away, 22-year-old Selman Abraham Waksman (1888–1973) immigrated to the United States from a small Jewish town in the Ukraine. He studied agriculture at Rutgers College and earned his

Ph.D. in biochemistry from the University of California, Berkeley. In 1918, he obtained a position at Rutgers as an assistant professor. His academic career initially concentrated on soil microbiology. It was not until 1939, when he was already 51 years old, that he started to focus his attention on discovering antibiotics in soil. After he heard about what was being done with penicillin at Oxford, he was quoted as saying, "Drop everything! See what these Englishmen have discovered a mold can do. Let's focus on getting antibiotics from soil."[19] At first they isolated a small-molecule antibiotic, actinomycin, and then streptothricin. Although both killed Gram-negative bacteria, they were so toxic that they also killed test animals. The third time was the charm, however. Waksman's student Albert Schatz isolated streptomycin, an amino-sugar antibiotic, on October 19, 1943.

Albert Schatz, only 23 years old in 1943, came back to school in June that year after having served in Florida as an army bacteriologist. Born to an impoverished Russian Jewish immigrant family (like Waksman), and having gone through the depression, Schatz was fiercely motivated. He lived in the laboratory in the basement and screened hundreds, if not thousands, of actinomyces. Waksman's office was on the third floor, and he assigned Schatz's laboratory to be as far away from him as possible for fear of contagion with tuberculosis. Indeed, the strains that Schatz worked on were highly contagious, and Schatz tested positive for tuberculosis when he graduated.

In October 1943, he isolated two colonies of actinomyces called *Actinomyces griseus,* later renamed *Streptomyces griseus.* One was cultured from an agar plate given to him by a laboratory mate, Doris Jones, who obtained the strain from the swab of a healthy chicken's throat. *Streptomyces griseus* produced an antibiotic that was very effective in killing Gram-negative bacteria, which were not touched by penicillin. Waksman christened the antibiotic *streptomycin.* With assistance from Merck for large-scale production and from the Mayo Clinic for animal testing and clinical trials, streptomycin was proven to be both safe and effective in curing tuberculosis. Astonishingly, only 3 years elapsed from its discovery to the first successful treatment of a human patient.

Merck initially experienced a recurring difficulty with the purification of streptomycin at the beginning of its development. After fermentation, streptomycin was purified by adsorption on charcoal, then by elution to remove the material from the carbon. However, the product was always contaminated with histamine, and the drug produced histamine-like allergic reactions in patients—elevated blood pressure, pain, and allergic rashes—until a chemist solved that problem overnight.

John C. Sheehan, who later distinguished himself as the first chemist to synthesize penicillin G in 1957 at MIT, was a group leader at Merck in the 1940s. Streptomycin, being an amino sugar, is extremely soluble in water but virtually insoluble in certain organic solvents that are immiscible in water. Mimicking an old German process for purifying cane or beet sugar, Sheehan mixed the impure sample in a separatory funnel charged with water and phenol. The brown color immediately passed into the phenol layer. After separating the phenol layer, the clear water solution was freeze-dried, which furnished the purest streptomycin they had yet seen.

Streptomycin was the first drug to be effective against Gram-negative bacteria. It was particularly interesting at the time because of its activity against human *tubercle bacillus,* which made it the first specific agent effective in treating tuberculosis.

The attribution of credit for the streptomycin discovery is one of the most contentious issues in medical history. The betrayal and acrimony generated during the fight far exceeded that involving the discovery of insulin (see chapter 6), if that was even possible. Immediately after the discovery, Waksman listed Albert Schatz, Elizabeth (Betty) Bugie, and himself, in that sequence, as authors of their paper in *Proceedings for Experimental Biology and Medicine* at the end of 1944.[20] At that time, Elizabeth Bugie was a graduate student whom Waksman assigned to confirm Schatz's results independently. When the patent application was initiated, Waksman convinced Bugie to give up the status of a coinventor because she was leaving the scientific field. As a consequence, Schatz and Waksman became the inventors of streptomycin. The U.S. Patent Office granted U.S. 2,449,866, titled "Streptomycin and Process of Preparation," to Schatz and Waksman on September 21, 1948. It has become one of the top 10 patents that changed the world. Rutgers licensed streptomycin to Merck and later to all eligible pharmaceutical firms. Schatz and Waksman signed away the royalties to Rutgers for $1 each; part of the royalties financed the building of a microbiology institute that still bears Waksman's name, the Waksman Institute of Microbiology.

Streptomycin was touted as the second "miracle drug," after penicillin. Waksman became a hero in medicine. Physicians all over the globe considered him one of the greatest benefactors of mankind. After 1944, Waksman was the person who received the media, giving speeches and lectures.

Everyone perceived that Waksman was the only person who was responsible for the streptomycin discovery. Schatz felt slighted, as he perceived that Waksman was trying to take the full credit for their discovery. He was hurt that he learned of the tremendous success of streptomycin in curing tuberculosis at the Mayo Clinic from newspapers rather than from Waksman. Schatz hurriedly finished his Ph.D. in 1946 and went on to work at the New York State Department of Health in Albany. Understandably, he was later outraged to learn that somehow Waksman had procured $350,000 in royalties for himself personally before the remainder was transferred to Rutgers. Schatz filed a lawsuit in a New Jersey Court. Waksman chose to settle out of court to avoid the humiliation of being interrogated. In addition to 3% of the royalties and a sum of $125,000 payable to Schatz, Waksman issued a statement admitting: "Albert Schatz was entitled to credit legally and scientifically as co-discoverer of streptomycin." Waksman retained 10% of the royalties, and 7% was distributed to everybody involved in the streptomycin discovery, including the dishwasher! In his autobiography, Waksman wrote: "As I look back upon the year 1950, I consider it the darkest one in my life."[21]

George Merck, then the president of Merck and Company, signed away the exclusive rights to develop streptomycin. This exceptional gesture by Merck enhanced his company's reputation, not only for its science but also for its humanitarian concern. As a result of competition between many drug firms, the price of streptomycin became very affordable.

Elizabeth Bugie married Francis Joseph Gregory, a codiscoverer of actinomycin D. She received a minor 0.3% of the royalties from streptomycin. Her daughter is now a microbiologist as well. Bugie lived probably the happiest life among all parties involved in the streptomycin saga.

Not being able to secure a single academic job in the United States after 50 applications, Schatz taught in Chile for a few years before returning to the United States. Evidently, the fallout from the lawsuit had made such a bad impression that all the universities blackballed him. He was understandably consumed by a bitter sense of injustice. Thankfully, history slowly started to recognize what really happened with regard to the streptomycin discovery. In an effort to do justice, Rutgers University awarded Schatz the "Rutgers's medal," the highest honor bestowed by the university, in 1994.

Waksman alone was awarded the Nobel Prize in 1952 in Physiology or Medicine. He was as passionate a writer as he was a scientist. In addition to publishing more than 400 research papers, he also authored or coauthored 28 books—quite an achievement for an immigrant who barely spoke English when he first set foot on American soil. In addition, Waksman also coined the term *antibiotics,* meaning "against life"—drugs that inhibit the growth or even destroy bacteria and other microorganisms.[21–23]

The epitaph on his gravestone says: "Out of the earth shall come thy salvation."

The success of streptomycin incited tremendous research in this field. Using the Rutgers technique, more than 30 antibiotics were isolated from soil samples in the ensuing 20 years. Many important drugs were identified in uncanny places. In 1948, Giuseppe Brotzu in Sardinia, Italy, isolated cephalosporin from a fungus strain of *Cephalosporium acremonium* in the sea near a sewage outfall from the city of Cagliari. Jean Borel of Sandoz in Basel, Switzerland, isolated cyclosporine A from a soil sample brought from Wisconsin in 1970. Also in 1970, two chemists at Eli Lilly isolated erythromycin from a soil sample from the Philippines. Vancomycin, effective against Gram-positive microbes, was isolated by scientists at the Lilly Laboratories in the mid-1950s. Vancomycin quickly became a very important lifesaving antibiotic, serving as the last line of defense against bacterial infections for almost half a century.

Another drug, *para*-aminosalicyclic acid (PAS, synthesized by Karl Rosdahl), for treating tuberculosis was conceived and discovered by Swedish scientist Jorgen Lehmann in 1943. In 1952 Lehmann would be passed over for the Nobel Prize for discovering a cure for tuberculosis; it was awarded to Waksman alone.

In 1951, Herbert Hyman Fox, of the Hoffmann-La Roche Laboratories, and Harry L. Yale, of the Squibb Institute for Medical Research in the United States and Bayer in Germany, almost simultaneously developed and introduced the tuberculosis drug isoniazid. For Yale, isoniazid was not intended as a drug but rather as an intermediate for another drug. Had he selected any of the other possible routes, he would not have prepared isoniazid. Fortunately, all intermediates in any synthesis were also evaluated, and isoniazid was shown to be 15 times more efficacious than streptomycin. The difference is that, whereas streptomycin is administered by intravenous injection, isoniazid is taken orally, an additional advantage. More important, isoniazid worked for most tuberculosis patients, even the ones who were not responsive to both streptomycin and PAS. It was credited with reducing the incidence of tuberculosis in the United States from

188 deaths per 100,000 annually in 1904 (the leading cause of death) to 4 deaths per 100,000 in 1953.

Because the synthesis of isoniazid had already been carried out and published by a couple of Prague chemists in 1912, no one could take out a patent on it. As a result, isoniazid was cheaper to make than streptomycin and PAS. Interestingly, isoniazid not only has been used in combination with streptomycin, but the two molecules have also been chemically (covalently) fused together to make a hybrid drug called streptonicozid.

With the discovery of streptomycin, PAS, and isoniazid, for the first time humans were given effective weapons to fight the great white plague. Waksman's autobiography, *My Life with Microbes*, was optimistically and prematurely subtitled *The Conquest of Tuberculosis.*[21,22] Unfortunately, tuberculosis is still rife in developing countries; it ranks as the second biggest killer after AIDS and infects one-third of the world's population. More than 2 million people still die of tuberculosis each year. Moreover, tuberculosis also acts in lethal synergy with the HIV virus, doubly infecting many patients, weakening their immune systems, and hastening death.[24] With the advent of HIV/AIDS, the incidence of multiple-drug-resistant tuberculosis is increasing, too. Mankind has a long way to go in stamping out tuberculosis.

Duggar, Conover, and Tetracycline

The discoverers of tetracycline antibiotics had their own share of luck. In the 1940s, American Cyanamid's Lederle Laboratories at Pearl River, New York, started to screen all kinds of soil samples to look for antibiotics that would possess a better safety profile than streptomycin for treating tuberculosis. In 1945, 73-year-old botanist Benjamin M. Duggar was a consultant for Lederle and led their screening efforts in the hunt for antibiotics. Coincidentally, a sample from the University of Missouri, where Duggar had taught botany 40 years earlier, yielded an antibiotic that was later named chlortetracycline. Lederle has sold chlortetracycline under the brand name of Aureomycin since 1948. Thanks to great oral bioavailability, Aureomycin won a good share of the antibiotics market. Nowadays, Benjamin Duggar is considered the pioneer of tetracycline antibiotics.

The discovery of the ensuing tetracycline antibiotics could be described as nothing but serendipitous. In the late 1940s Charles Pfizer and Company became deeply worried about competitors' resurgence in the antibiotics arena, especially when the price of penicillin plummeted.[14] Like many pharmaceutical companies at that time, Pfizer plunged into research on newer antibiotics. Every imaginable means of soil sample collecting was resorted to. Everyone's help was enlisted: travelers, missionaries, explorers, airline pilots, students, housewives, and Pfizer sales agents were encouraged to pick up a teaspoon of earth, seal it in a packet, and mail it back to the company for a small reward. Soil samples rushed in from the most unlikely places: the jungle of Brazil, the tops of mountains, the bottoms of mine shafts, cemeteries, deserts, and even the ocean. Balloons were sent up to collect soil that was airborne. More than 100,000 soil samples were screened.

In 1949, a yellow powder with strong antibiotic properties was isolated from a soil sample and labeled as PA-76 (PA stands for Pfizer Antibiotic). This sample provided a broad-spectrum antibiotic, oxytetracycline, that proved to be both safe and effective against a range of bacteria that caused more than 100 infectious diseases. The soil organism was named *Streptomyces rimosus,* and the compound was generically known as oxytetracycline. Backtracking revealed that the soil sample was collected at the Terre Haute factory in Indiana owned by Pfizer. Pfizer sold oxytetracycline under the brand name Terramycin; *terra* means "earth" in Latin, and *Terre Haute* simply means "high land."

Terramycin was the first drug discovered by Pfizer while the company was still smarting from its painful experience with penicillin. In 1950, the company hired their own sales force and sold Terramycin themselves. They also initiated an aggressive advertising campaign in medical journals, which was controversial because the advertising expense was twice as much as the expenditure for discovery and development of Terramycin ($4 million for discovery and development).

In addition, Pfizer formed a team to elucidate the chemical structure of oxytetracycline. They enlisted the help of R. B. Woodward at Harvard. In 1952, Pfizer and Woodward jointly published their elegant work on the structure of oxytetracycline in the *Journal of the American Chemical Society.*[25] Meanwhile, a member of the team, Lloyd Conover, shocked his colleagues by preparing another powerful antibiotic chemically from chlortetracycline. Under carefully controlled catalytic hydrogenation conditions, using hydrogen gas and palladium on charcoal, Conover converted Lederle's chlortetracycline to tetracycline by stripping the chlorine atom and replacing it with a hydrogen atom.

This was truly revolutionary. Before that, it was generally believed by all practitioners that "natural" antibiotics produced by microbial metabolism were the only ones that possessed desirable biological properties. Conover demonstrated that chemical manipulations could afford active antibiotics, as well. Tetracyclines became the most prescribed broad-spectrum antibiotics in the United States within 3 years. Conover's discovery created a brand-new field of medical research—semisynthetic antibiotics. It also sparked a wide-scale search for superior structurally modified antibiotics, which has provided most of the important antibiotic discoveries since then. Lloyd Conover has been inducted to the American Hall of Fame for Inventors. Only 98 people have earned that honor, including Thomas Edison and the Wright Brothers.

Quinolones, Zyvox, and Beyond

George Y. Lesher (1926–1990) was born in Norman, Illinois.[26] He studied chemistry at the University of Illinois and received a master's degree in science from Dartmouth College in 1952. Lesher joined Sterling Winthrop Research Institute at Rensselaer, New York, where he stayed for the remaining 38 years of his illustrious career. The precursor of Sterling Winthrop Research Institute was the American subsidiary of the renowned German pharmaceutical giant Bayer. The major thrust of research at Sterling Winthrop was the ubiquitous aspirin. For the first 4 years of his tenure, Lesher took advantage of the vicinity of the Rensselaer Polytechnic Institute (RPI) and earned his Ph.D. in organic chemistry.

While pursuing better antimalarial drugs, chemists at Sterling-Winthrop synthesized many quinine-like compounds during and after World War II. In 1946 they isolated a by-product, nalidixic acid, during their attempts to synthesize chloroquine. During routine screening nalidixic acid was found to be an antibacterial agent. However, nalidixic acid did not become popular until 1962, when Lesher introduced it into clinical practice for kidney infections. It was also used to treat urinary tract infections because it was excreted via urine in high concentration. Shortly after, the quinolone antibacterials derived from nalidixic acid flourished, rendering thousands of 4-quinolones, as represented by pipemidic acid.

Lesher discovered this novel class of therapeutic agents. Even now most pharmaceutical firms are still reaping the benefits of Lesher's discovery.

More important, numerous human beings are benefiting from quinolone antimicrobial agents. Sadly, Lesher died at age 64 as a result of a tragic canoe accident near Albany, New York.

Nalidixic acid and pipemidic acid are considered the first-generation quinolone antibacterials for their moderate activity toward susceptible bacteria and poor absorption by the body. They possess oral activity only against Gram-negative bacteria and suffer as a class for their inability to affect Gram-positive strains. Furthermore, the bioavailabilties are too low (partially due to high protein binding) to treat systemic infections such as pneumonia and skin disease.

In the early 1980s, fluorinated quinolone (fluoroquinolone) antibacterials were discovered to possess longer half-lives (staying in the system longer) and better oral efficacy than the first-generation quinolones. These so-called second-generation quinolone antibacterials are exemplified by norfloxacin (the first fluoroquinolone discovered in 1980) and ciprofloxacin (Bayer sells it under the trade name of Cipro).[27]

The second-generation quinolone antibacterials display a broader spectrum of antibacterial activity, increased potency, decreased potential for resistance, and less toxicity. They have become the first line of attack for the clinical treatment of a variety of infectious diseases in contemporary medicine. Since 1980, more than 10,000 fluoroquinolones have been synthesized and their antibacterial activities explored.

The discovery of Cipro is a good example of the role that persistence plays in discoveries. The scientist who initially worked on the series was pulled out of the project because management lost faith in his approaches. But he persisted in his pursuit, kept working on it as a "submarine" project without his superior's endorsement, and finally succeeded in making Cipro.

Cipro ascended to stardom in the wake of bioterrorism threats after September 11th, 2001. In October 2001, two postal workers in a Washington, DC, distribution center died of anthrax infections caused by anthrax powder found in two letters that were broken by a distribution machine. It took more than 2 years and $134 million to decontaminate the distribution center building. Anthrax is a bacterium that, when inhaled, travels to the lungs and begins to disseminate and produce toxins, which can be lethal if left untreated. In 1876, Robert Koch in Germany identified the microbe that causes anthrax, a disease indigenous to sheep and cattle that also can be spread to humans. Louis Pasteur developed a vaccine (from dead or weakened sheep) to ward off anthrax that saved millions of sheep across

France. Ciprofloxacin (Cipro) is approved by the FDA for the treatment of anthrax. Indeed, a 60-day regimen is effective in treating the inhaled form of anthrax after an individual has been exposed. However, it is not the only antibiotic that can treat an anthrax infection. Tetracyclines such as deoxycycline and β-lactams such as penicillin work as well, although the older antibiotics are more prone to drug resistance because they have been in use for a longer period of time. Widespread use of Cipro has been discouraged to avoid development of drug resistance.

Like all fluoroquinolone antibacterials, ciprofloxacin causes articular damage in juvenile animals. Consequently, it is not recommended for children or pregnant women. Nonetheless, more data have emerged for pediatric applications thanks to high antibacterial effectiveness and convenience in oral administration.

It is worth mentioning that the third generation of quinolone antibacterials are still being actively investigated due to the rapid development of resistance by bacteria to existing antibacterial drugs. Examples of the third generation of quinolone antibacterials include fleroxacin and tosufloxacin. They are endowed with sufficiently long half-lives to enable a once-daily regimen, along with enhanced activity toward a variety of bacteria.

Another important category of antibacterial agents is the oxazolidinones, as exemplified by linezolid (Zyvox).[28] The genesis of Zyvox began in 1978, when a Dupont patent described some novel oxazolidinones for controlling fungal and bacterial plant pathogens. Although made for agricultural purposes, the pharmaceutical arm of Dupont picked up one of these compounds during routine antibacterial screening. This compound and two subsequently optimized drug candidates, DuP 721 and DuP 105, did not materialize as marketed drugs due to unacceptable toxicities. In early October of 1987, the 27th Interscience Conference on Antimicrobial Agents and Chemotherapy (ICAAC) took place in New York. Dupont scientists at the conference presented their findings on DuP 721 and DuP 105. Steven J. Brickner, a medicinal chemist working at the Upjohn Company in Kalamazoo, Michigan, was intrigued by many attributes of this class of antibacterial. He immediately began an exploratory oxazolidinone project on returning from the meeting. Upjohn encouraged its employees to spend up to 10% of their time exploring "blue sky" projects, not official projects but of special interest to the scientist. Low and behold, the 10% "free time" paid off. Brickner and his lab began work on the oxazolidinones, and soon they had identified several new series of potent oxazolidinones. The personal interest soon became a full-blown project. But news came through the grapevine that Dupont had dropped their entire oxazolidinone project due to toxicity in preclinical species. Richard Piper, a pathologist at Upjohn, played a key role in demonstrating that Brickner's lead compound at that time was not toxic when chronically administered to rats. Racemic Dup 721, tested in parallel, was not well tolerated and even resulted in lethal toxicity. This was a critical demonstration that allowed Brickner's project to proceed. He believed, correctly, that the team would be able to minimize or even eliminate toxicity via extensive structure toxicity relationship (STR) studies. Working with two other groups, led by Michael R. Barbachyn and Douglas K. Hutchinson, they quickly made headway. In the spring of 1993, eperezolid and linezolid (Zyvox) were prepared (the first samples of both compounds were prepared by Brickner personally), and clinical trails for both compounds commenced in 1995. Although the trials were successfully completed and both compounds found safe in Phase I, linezolid was more advantageous than eperezolid in terms of its pharmacokinetics. Although eperezolid would require a three-times-daily regimen, linezolid needed only twice-daily dosing. Linezolid was therefore moved forward and won FDA approval on April 18th, 2000.

Zyvox is the first marketed member of a novel class of oxazolidinone antibacterial agents—the first of a new class of antibiotics in about 40 years. Its mechanism of action is inhibition of the initial phase of bacterial protein synthesis. It is also an MAO-B inhibitor but without significant blood pressure liability (see chapter 5), thus bringing an interesting closure to the origin of MAO inhibitors for depression, which came about from the improvement in mood in patients with tuberculosis who were treated with the original MAOI agents. Due to this unique mechanism of action, there has been no reported cross-resistance between oxazolidinones and other protein-synthesis inhibitors. Zyvox has inhibitory activity against a broad range of Gram-positive bacteria, including vancomycin-resistant enterococci (VRE), penicillin-resistant *Streptococcus pneumoniae* and methicillin-resistant *Staphylococcus aureus* (MRSA). This is especially important at this time as MRSA, a serious and often fatal infection, is becoming increasingly resistant to all other available therapy, including Vancomycin.

Conclusions

In this chapter, I have chronicled only several categories of historically significant antibiotics. Now, dozens of classes and hundreds of antibiotics are at our disposal. Significantly, antibiotics are probably the only class of drugs that frequently cure diseases, making them one of the most powerful weapons in the medical armory. Unfortunately, the abundance of antibiotics has its own perils. As bacteria start to develop resistance, more and more antibiotics become obsolete, and there is a constant need to replenish the arsenal against germs. Take tuberculosis as an example. The great white plague was kept under control in developed countries for some time with such drugs as streptomycin, PAS, isoniazid, and other antibiotics. However, not only is tuberculosis still rife in poor countries, but it has also resurfaced in developed countries. In fact, a tuberculosis epidemic took place in New York City in 1989. A global epidemic of multi-drug-resistant tuberculosis has become a time bomb. We have a long way to go toward total eradication of tuberculosis.

CHAPTER 3

Cardiovascular Drugs

From Nitroglycerin to Lipitor

All we know is definitely less than all that still remains unknown.

—William Harvey, 1625

Cardiovascular diseases are the leading cause of death worldwide and are projected to remain in the lead through 2025. Heart-related diseases include angina, arrhythmia, atrial fibrillation, congestive heart failure, hypertension, atherosclerosis, myocardial infarction (heart attack), and sudden cardiac death. More than 300,000 Americans suffer sudden heart attacks each year. In addition, one of the more important recently identified drug-induced cardiac events, which has occasionally resulted in drugs being withdrawn, is drug-induced torsade des pointes. This is a rare, fatal arrhythmia that has been associated with some drugs that prolong the QT interval of the electrocardiogram (ECG).

Hypertension is America's number one chronic disease. Fifty million Americans, one in six, suffer from high blood pressure. Similarly, high blood pressure affects about one-sixth of the world's population (1 billion people) worldwide—mostly in the developed world. If uncontrolled, it can lead to heart attack, heart failure, stroke, and other potentially fatal events.

Great strides have been made during the past 50 years in conquering cardiovascular diseases. Cardiopulmonary resuscitation (CPR) was developed by a group of researchers at the Johns Hopkins University in 1961. The 1960s also saw the emergence of beta-blockers. Calcium channel blockers, angiotensin-converting enzyme (ACE) inhibitors, and statins appeared in the 1980s and the 1990s. Angiotensin II receptor blockers (ARBs) also emerged in the 1990s.

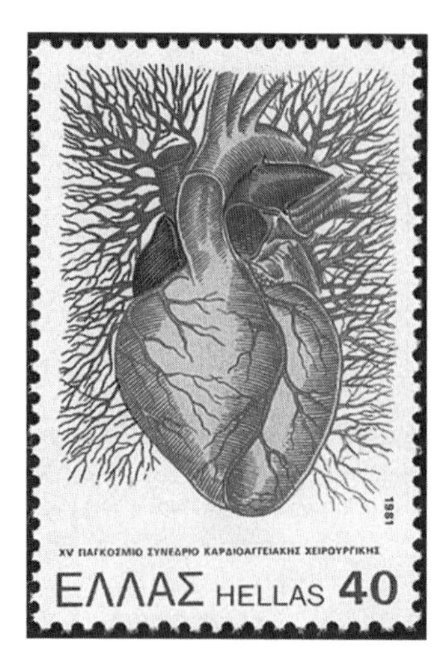

The heart © Greek Post

Harvey and Blood Circulation

The heart, about the size of a person's fist, beats about 2.8 billion times in a lifetime, pumping blood and oxygen through the body. Although its function was shrouded in mystery for centuries, mankind has come a long way in understanding how the heart works anatomically and physiologically, although we haven't made much progress in understanding its "emotional" nature.

Greek philosopher and anatomist Aristotle (384–322 B.C.) was the founder of biology.[1] He was very interested in human and animal anatomy, especially the cardiovascular systems in higher animals. In his books he described, for the first time, the human blood system with an emphasis on the deeper-lying vessels. He incorrectly believed that the heart was the organ in which emotions were generated, whereas the function of the brain was to cool the blood. More than 500 years later, the German-born Roman physician Galen (130–200 A.D.) made two revolutionary discoveries about the cardiovascular system. First, Galen discovered that the heart is a mass of muscle that serves as a pump of blood. Second, he found that arteries do not carry air, but blood. Galen, at times serving as the physician for the Roman army, gladiators, and Roman emperors, had gained first-hand experience in human anatomy by observing torn human bodies. Galen erroneously theorized that blood was made in the liver (which is actually true fetally) from food and that the vital spirit came from the lung. In addition, the futile, if not fatal, practice of bloodletting to treat disease was largely based on Galen's theory. Nonetheless, he was so highly revered as a scholar that his concepts were not challenged for the ensuing 1,500 years!

It was then left for the Englishman William Harvey (1578–1657) to discover the circulation of blood.

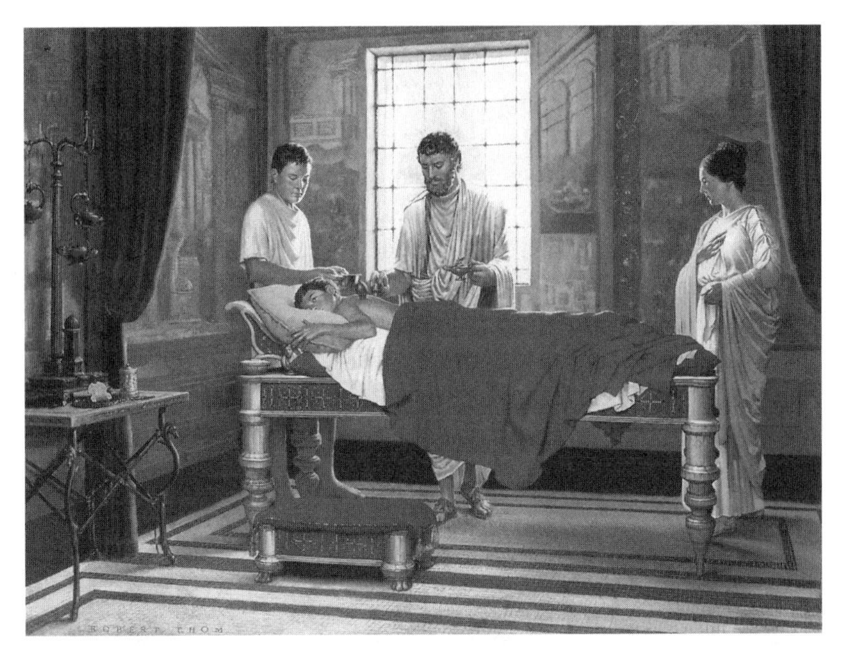

Galen, influence for forty-five generations © Warner-Lambert

Harvey was a short, slight, and dark-complexioned physician with flashing eyes and a wealth of nervous energy. Harvey was interested only in the quest for scientific knowledge; he had little interest in what was going on around him. In 1635, King Charles I was engaged in a battle at Edgehill. Harvey, the royal physician, sat under a hedge calmly reading a book. In his elder years, his only regret was that one of his manuscripts on anatomy had been lost in a fire; not a word about his "dear little wife" who had passed away some years before.[2]

Like Claude Bernard, the most well-known vivisectionist, Harvey spent most of his life dissecting living animals. Snakes, rats, geese, dogs, snails, turtles, fish, and shrimps all became his scientific subjects. King Charles I encouraged him by granting access to his royal herd, so that Harvey was able to investigate the anatomy of deer. He remarked about his fascination with anatomy: "It is true the examination of the bodies of animals has always been my delight, and I have thought that we might hence obtain an insight into higher mysteries of nature, but there perceive a kind of image or reflex of the omnipotent Creator himself."[2]

Harvey devised the concept of blood circulation when he was very young. In notes taken in his early career, he had already stated: "The

William Harvey experimenting circulation of blood © Warner-Lambert

movement of the blood occurs constantly in a circular manner and is the result of the beating of the heart."[2] Fearing the world might not be ready for such a revolutionary idea, he prudently accumulated evidence over several decades. In addition, he gradually spread his discovery at different lecture engagements. Finally, as a coup de grâce, he published his most celebrated monograph, *Exercitatio Anatomica de Motu Cordis et Sanguinis in Animalibus* (*Anatomical Exercise on the Motion of the Heart and Blood of Animals*), in 1628. Harvey tactfully avoided antagonizing Galen's theory. Instead, he slowly but surely introduced his revolutionary discovery on circulation of blood: blood flows in a continuous circle from the arteries to the veins and back again. The circulation of blood is considered one the greatest discoveries made in physiology.

Sobrero, Nobel, and Nitroglycerin

Angina pectoris is the feeling of tightness, heaviness, or pain in the chest, caused by lack of oxygen in the muscular wall of the heart. Before the discovery of nitroglycerin (glyceryl trinitrate), standard treatments for angina pain brought on by coronary artery disease were brandy, opium, ether, and chloroform. These anesthetics temporarily produced partial stupefaction to relieve the pain. Worse yet, at one time, surgical removal of part of the thyroid was used to relieve angina.

For over a century, nitroglycerin has been a lifesaver for many patients who suffer from angina. It is also an explosive that is famous for making DuPont a major chemical company and for making Alfred Nobel wealthy enough to begin his Nobel Prizes with money earned therefrom. Even today, nitroglycerin is the most commonly used treatment for angina pectoris.

Nitroglycerin was first synthesized in 1847 by Italian chemist Ascanio Sobrero (1812–1888).[3] In his youth, Sobrero apprenticed in the laboratories of Theophile-Jules Pelouze (1808–1867) in Paris and Justus von Liebig (1803–1873) in Giessen, Germany. He later returned to Torino, Italy, to establish himself as an independent scientist. Sobrero initially investigated the nitration of cotton, in which polysaccharide is the main ingredient. Later, after 1 year's experimentation, he successfully synthesized nitroglycerin from nitration of glycerol using a cold mixture of nitric acid and sulfuric acid. The reaction was extremely exothermic. If the container was not cooled during the reaction, detonation of nitroglycerin was likely to ensue. In fact, Sobrero's face was badly scarred from one of the explosions during his experiments with nitroglycerin.

In those days, chemists routinely tasted new chemicals as they were prepared. Many of them documented compound taste as part of their scientific records. This practice may explain why so many prominent chemists suffered poor health in that era. Liebig even declared, "A chemist with good health must not be a good chemist."[4] In his publication, Sobrero described the taste of nitroglycerin as sweet, pungent, and aromatic but warned that ". . . great precaution should be used, for a very minute quantity put upon the tongue produces a violent headache for several hours."[5]

Four years after Sobrero's discovery, Alfred Nobel (1833–1896)[6] learned about nitroglycerin's explosive properties from his Russian teacher, N. N. Zinin. Back in Stockholm, he and his father Immanual began experimentation to look for controlled detonation of nitroglycerin. After numerous failures and accidents, they succeeded by using a porous silica gel to absorb the unstable nitroglycerin, creating dynamite. The patented detonator made the Nobel family extremely rich.

Nobel did not forget about Ascanio Sobrero, to whom he owed his fortune. He hired Sobrero as a consultant in his Swedish-Italian firm in

Avigliana. In 1879, he erected a bust of Sobrero outside that plant. After Sobrero's death, Nobel further demonstrated his gratitude for the Italian by awarding his widow a lifetime pension.

Despite extraordinary wealth, Alfred Nobel suffered poor health most of his life, especially in his elder years. He never married, and loneliness was his lifelong companion. He constantly complained of indigestion and headaches, which may have been symptoms of his chronic depression. In 1890, Nobel complained of intense pain from angina pectoris, and guess what his physician prescribed for him? Nitroglycerin. Today we know that nitroglycerin is a "prodrug": inactive by itself, it becomes active in the body (in vivo) by releasing nitric oxide. Nitric oxide then travels to the ischemic tissue of the cardiac arteries and dilates the smooth muscle.

In 1896, just 7 weeks before his death, Nobel wrote in a letter to a friend: "Isn't it the irony of fate that I have been prescribed nitroglycerin to be taken internally! They call it Trinitrin, so as not to scare the chemist and the public."[5]

The genesis of nitroglycerin as a powerful vasodilator was also prompted by the observations made in explosive production plants. Munitions workers often experienced facial flush and severe headache when they returned to work after being away from the factory over the weekend. They dubbed the headaches "Monday disease." Because those munitions workers were exposed to copious amounts of nitroglycerin, their circulatory systems were well vasodilated. Following the weekend, when the bodies became adjusted to a more normal state, sudden exposure to nitroglycerin created the "Monday headache." Soon, savvy workers would bring nitroglycerin with them outside the plant. Some found that wearing their work clothes during the weekend would alleviate their "Monday disease."

The Chinese pioneered anginal pain treatment using potassium nitrate many centuries ago. Coincidentally, it worked via the nitric oxide mechanism as well. They recorded the use of potassium nitrate as a cure for anginal pain in a medical manuscript published in 800 A.D. (very likely one of the earliest printed books) titled *Diamond Sutra*. The medical manuscript recommended that patients place *nitre* (potassium nitrate) under the tongue for anginal pain and promised "instant relief." Although potassium nitrate itself is inactive biologically, it is likely that nitrate reductase is present in some bacteria, as there are numerous bacteria under the tongue. Nitrate reductase is an enzyme that converts nitrate into nitrite, which, in turn, releases nitric oxide under acidic conditions. Nitric oxide then induces vasodilation.

Despite more than 130 years of successful clinical use of nitroglycerin for the treatment of angina, its mechanism of action was not known until the 1980s. The mode of action was discovered by pharmacologists Ferid Murad, Robert Furchgott, and Louis Ignarro.

In the 1970s, Murad was at the University of Texas Medical School at Houston, studying the action of vasodilator molecules such as nitroglycerin and their effects on guanylated cyclase activity. In the 1980s, Furchgott, at the Health Science Center at Brooklyn of the State University of New York, recognized that soluble guanylated cyclase could be activated by free radicals such as nitric oxide (NO).[3,5] Meanwhile, Ignarro, at the University of California School of Medicine at Los Angeles, confirmed that endothelial-derived relaxing factor (EDRF) was indeed nitric oxide. In 1998, the Nobel Prize in Physiology or Medicine was awarded to Murad, Furchgott, and Ignarro for their discoveries concerning nitric oxide as a signaling molecule in the cardiovascular system. A fourth pharmacologist involved in these discoveries, Salvador Moncada of the Wellcome Research Laboratories in Beckenham, England, was passed over, although he had suggested that EDRF was identical to nitric oxide 6 months before Ignarro did. The unfortunate outcome was likely the result of at least two factors: (1) Only three laureates per year may be named for each category of the Nobel Prize; and (2) Moncada did not confirm his suggestion with experimental evidence.

In addition to being administered sublingually, nitroglycerin transdermal patches and pastes are also frequently applied topically by patients to relieve anginal pain. These patients are prone to have sudden heart attacks as well. The emergency doctor must be careful to remove the nitroglycerin patches so that explosion will not take place when resuscitation is carried out using electric shocks. Topical application of nitroglycerin often results in penile erection in men. Nitroglycerin has been used topically for quite some time for this effect. It is likely that nitric oxide again provides smooth muscle vasodilation.

In addition to nitroglycerin, many organic nitrates are used to treat angina pectoris. These organic nitrates are typically prepared by nitration of polyols (molecules with many alcohol functionalities). Examples include isosorbide dinitrate (ISDN), pentaerythritol tetranitrate, and erythrityl tetranitrate.

Amyl nitrite has also played a significant role in history for the treatment of angina pectoris. Amyl nitrite is a clear yellowish liquid with a strong chemical smell. It has been used to relieve heart pain in people who suffer from coronary artery disease. Amyl nitrite is used as an inhaler or as cloth-covered glass capsules. Amyl nitrite is also a prodrug that releases nitric oxide, which then exerts a vasodilating effect. Unfortunately, amyl nitrite has also been abused. Because the glass vials break with an audible "pop," amyl nitrite earned the street names of "poppers" and "snappers," among others. People use it for recreational purposes and believe the highs of the general anesthetic effects can enhance sexual experience by prolonging and intensifying orgasm. Although amyl nitrite is metabolized rapidly in the body, it is not a good idea to abuse it, because it often results in rapid blood pressure drop. For the treatment of angina, nitroglycerin is still the most common drug because it is conveniently administered and has fewer side effects than amyl nitrite.

Digitalis purpurea L., foxglove. Photo by author

Withering and Digitalis

Digitalis was discovered more than 200 years ago and is still in use today. Digitalis cardiac glycosides, popularized by British doctor William Withering in the 1780s, are still widely used to treat heart failure. It is undoubtedly one of the most valuable cardiac drugs ever discovered.[7–12]

More than 400 cardiac glycosides have been found in nature thus far. Digitalis, whose main ingredients include digtoxin and digoxin, is isolated from the leaves of the foxglove plant. Those digitalis cardiac glycosides work by augmenting the contractile force of heart muscle to improve the efficiency of the heart.

Foxglove plants, *Digitalis purpurea*, produce beautiful bell-shaped flowers that bear numerous nicknames: fairy's glove, witch's glove, and virgin's glove.[9] In the Middle Ages, herbalists used foxglove for the treatment of dropsy (now known as edema)—a fluid accumulation in body tissues caused by heart, kidney, or liver failure. In 1785 William Withering published an account of the action of dried foxglove leaves on the weakened heart. His book propelled digitalis into prominence in the medical community.

William Withering (1741–1799)[10,11] was born in Wellington, Shropshire, England. His father was a well-known apothecary, and quite a few of his family members were either surgeons or physicians. Carrying on the family tradition, Withering studied medicine at Edinburgh University and earned his M.D. in 1766. He practiced medicine mostly in Birmingham, where the doctor treated both the rich and the poor. Because of his dedication to his patients, regardless of their level of income, he would never make more than £2,000 per year, whereas many of his peers had annual incomes exceeding £5,000.

Withering's personality was characterized by extraordinary accuracy and caution. His botanical treatise *A Botanical Arrangement of All the Vegetables Naturally Growing in Great Britain with Description of the Genera and Species According to the System of the Celebrated Linnaeus: with an Easy Introduction to the Study of Botany* was renowned for its precision and remained a standard in the field for the next hundred years. Mankind is indebted as well to Miss Helena Cooke, Withering's girlfriend, who introduced him to botany, which in turn led to his investigation of digitalis. Because she was not able to collect her own flowers due to her heart condition, her doctor volunteered to do it for her. In order to impress her with his knowledge, Withering began to study the "lovable science" of botany.[10] He must have done a great job because she decided to marry him a few years later.

The digitalis story begins with failure. Withering had been treating a dropsy patient for quite some time without success. In the end, he gave up and told the patient that she would not live long. A few months later, the patient came back alive and well. She informed Withering that she had, in desperation, sought an herbal tea from a Gypsy lady. Withering tracked down the Gypsy lady and bought the secret formula for her herbal tea for 3 guineas ($15). Through careful investigation, Withering correctly concluded that among 20 or so ingredients in the herbal tea, purple foxglove was the active ingredient. Withering spent the next decade exploring the curative effects of digitalis. He treated more than 150 patients with various concoctions of digitalis and carefully chronicled the responses. His clinical evaluations were so meticulous and accurate that nobody even came close to his astonishing precision until the emergence of modern clinical trials. In 1785, Withering published *An Account of the Foxglove and Some of Its Medical Uses: with Practical Remarks on Dropsy and Other Diseases,* in which he summarized his extensive clinical trials and described the symptoms of digitalis toxicity. He stressed the importance of choosing dose and determined the correct dose of digitalis. His work paved the way for wide use of digitalis as a cardiac drug. Digitalis, as well as many cardiac glycosides, is toxic when taken in large doses. If the dose is too low, then it is ineffective. Digitalis is effective only when administered at a near-toxic dose, so finding the correct dosage is very important.

Withering wrote many poems concerning nature. He was to write about digitalis:[10,11]

The Foxglove's leaves, with caution given,
Another proof of favouring Heav'n
Will happily display;
The rapid pulse it can abate;
The hectic flush can moderate
And, blest by Him whose will is fate,
May give a lengthened day.

Withering died on October 6th, 1799, after having suffered for 20 years with bronchiectasis, a disorder in which the bronchial tubes become enlarged and distended, forming pockets where infection may gather. Foxgloves were appropriately carved on his headstone.

Digitalis was originally prepared from the powdered leaves of foxglove. Today, digitalis isolated from foxglove leaves is further crystallized to afford cardiac glycosides such as digoxin and digtoxin. Digoxin and digtoxin are pure crystalline substances and are therefore more easily quantified. Digitalis has a narrow therapeutic margin; thus stable tablets of uniform potency are very important. Doctors concerned with toxicity favor digoxin over digtoxin because it is eliminated more rapidly. It is believed that a rapidly eliminated drug represents a reduced hazard.

Vogl and Diuretics

Diuretics are drugs that cause high urine flow. Digitalis, like caffeine, is a diuretic. Tea, coffee, and alcohol are popular household diuretics. Currently in the United States, diuretics are the most prescribed drugs for heart conditions. The National Institutes of Health (NIH) has a branch called the Joint National Committee on Prevention, Detection, Evaluation, and Treatment of High Blood Pressure. Part of the committee's job is to evaluate hypertension drugs and make recommendations to American consumers. In 1993 and again in 1997, the committee recommended that diuretics and beta-blockers should be the first line of treatment for hypertension. These two classes of drugs are older and less expensive and have proven efficacy and safety.

Serendipity played an important role in the discovery of the first mercurial diuretic, merbaphen (Novasurol). Alfred Vogl credited "a series of fortunate errors and coincidences" for the discovery that completely revolutionized the treatment of congestive heart failure.[13]

On October 7, 1919, Alfred Vogl was a third-year medical student in the First Medical University Clinic (the Wenckebach Clinic) in Vienna. A patient, Johanna, was admitted for congenital syphilis, and Vogl was the clinical clerk in charge of her case. While his *Materia Medica* (knowledge of drugs) was still immature, Vogl wrote a prescription for a 10% solution of mercury salicylate in water for hypodermic injection. To his embarrassment, the pharmacy informed him that the compound was insoluble in water. At that time, a retired army surgeon was present and happened to have received a new mercurial antisyphilitic, Novasurol. Vogl gratefully took the army surgeon's suggestion and used Novasurol instead for the injection.

To Vogl's astonishment, Johanna's urine output reached 1,200 mL in 24 hours, as compared to her prior average daily urine output of 200 to 500 mL. Similar results were observed for patients with both syphilis and congestive heart failure. By removing fluid, the pressure on the heart was removed. Mercurial diuretics revolutionized the treatment of severe edema from congestive heart failure and were the primary treatment for this disease until the late 1950s and the emergence of thiazide diuretics. Alfred Vogl later immigrated to the United States and taught at the New York University College of Medicine.

The discovery of chlorothiazide had its share of serendipity as well. Sir James Black, winner of the 1988 Nobel Prize in Physiology or Medicine, once said, "The most fruitful basis for the discovery of a new drug is to start from an old drug."[14] Chlorothiazide was one of many such examples. Because mercurials were notoriously toxic, the sulfanilamides (the sulfa drugs; see chapter 2) quickly replaced them when sulfanilamides were found to have diuretic properties. Diuretics such as furosemide and chlorothiazide resulted from chemical modifications of sulfanilamides.

In 1957, Merck chemist Frederick C. Novello was synthesizing diuretic agents and wanted to make some analogs of an older sulfa drug, dichlorophenamide.[15] Surprisingly, the reaction gave the ring formation product rather than the linear derivatization product. The bicyclic ring formed was a benzothiadiazine derivative. Although disappointed by not getting what he had intended, Novello submitted the compound to the screening program anyway. It proved to be a potent diuretic without elevation of bicarbonate excretion, an undesired side effect. Chlorothiazide was groundbreaking. It was the first nonmercurial orally active diuretic drug whose activity was not dependent on carbonic anhydrase inhibition, such as acetazolamide.

Currently the most frequently prescribed diuretic is hydrochlorothiazide, discovered by Ciba scientists led by George deStevens.[16] Hydrochlorothiazide is a "me-too" drug, a copycat drug that resembles the prototype, based on Merck's chlorothiazide. In 1957, deStevens became aware of the research of Frederick Novello on the synthesis of disulfonamides in general, chlorothiazide in particular. DeStevens initially changed a six-membered ring on chlorothiazide into five-membered saccharin derivatives, which were inactive. But changing a double bond on chlorothiazide into a single bond gave him a hydrochlorothiazide that was tenfold more potent than the prototype. Hydrochlorothiazide was introduced into medical practice in 1959, and within a short time it became the drug of choice in the treatment of mild hypertension.

Hydrochlorothiazide dominated the market until the emergence of angiotensin-converting enzyme (ACE) inhibitors in the 1990s. Diuretics are still the gold standard for the treatment of hypertension. In addition, diuretics are widely used in certain developing countries as a popular cheap substitute for Viagra—an empty bladder seems to help to achieve penile erection.

Snake Venom and ACE Inhibitors

ACE inhibitors are drugs that inhibit the function of the angiotensin-converting enzyme (ACE).[17–19] They are widely used in treating hyperten-

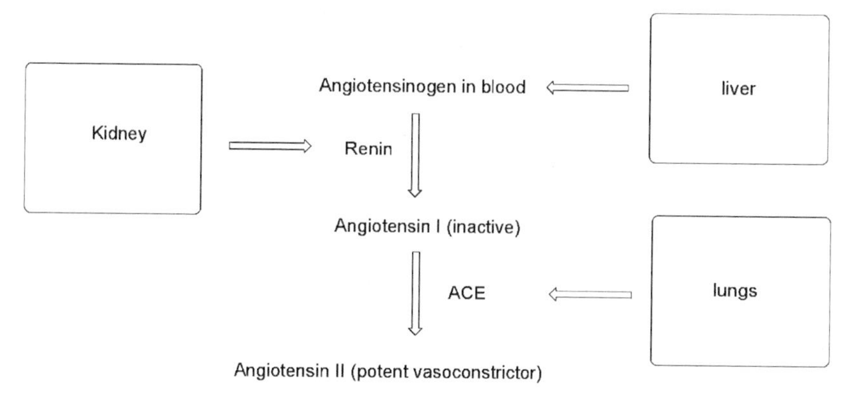

The renin-angiotensin system (RAS). Diagram by author

sion, congestive heart failure, and heart attacks. ACE inhibitors owe their genesis to snake venom—the first ACE inhibitor, teprotide, was isolated in 1971 from a poisonous venom extract of the Brazilian pit viper *Bothrops jararaca.*

As early as 1898, Finnish physiologist Robert Tigerstedt and his student Per Gunnar Bergman discovered that crude extracts of the kidney contained a long-acting presser substance, which they named *renin.* Over the next 100 years, the complex renin-angiotensin system (RAS) and its ramifications for hypertension became well elucidated. In this system, an enzyme that is involved in converting angiotensin I to angiotensin II is called angiotensin-converting enzyme, or ACE. Angiotensin I is inactive as far as modulating blood pressure is concerned, whereas angiotensin II is a potent vasoconstrictor. As a matter of fact, angiotensin II turned out to be the most potent blood-pressure-raising substance in the human body. As a consequence, inhibition of angiotensin-converting enzyme would provide vasodilation and lower the blood pressure.

In 1965, John Vane (1982 Nobel laureate in medicine) was a professor at Oxford University.[17,18] A Brazilian postdoctoral student of his, Sergio Ferreira, brought with him a dried extract of the venom of the poisonous Brazilian pit viper, *Bothrops jararaca,* which was the fruit of his Ph.D. thesis. The venom poison is known to reduce the blood pressure of its victims. In 1967, Vane persuaded his colleague Mick Bakhle to test Ferreira's snake venom extract on an in vitro preparation of the ACE and found it to be a potent inhibitor. Injection of the snake venom into humans would surely result in fatal consequences; therefore the venom itself was not a viable drug. Vane was a consultant for the pharmaceutical company Squibb Institute, and he suggested that they study the snake venom extract. His idea was received with only lukewarm enthusiasm from Squibb's marketing staff. The active principle of the snake venom extract consisted of peptides, which are not orally bioavailable because the acidic environment of the stomach induces a breakdown into constituent amino acids. Obviously, there would be a smaller market for a hypertension drug that had to be injected.

Fortunately, two bench scientists, biochemist David Cushman and organic chemist Miguel A. Ondetti, remained enthusiastic. In time they isolated a nonapeptide (a peptide with nine amino acids) that had a longer duration of action. They christened it *teprotide.* Squibb synthesized one kilogram of teprotide at the cost of some $50,000, a lofty sum at that time. Injection of teprotide into volunteers reduced blood pressure and confirmed that it was an ACE inhibitor in humans. With brilliant insight, Cushman and Ondetti chopped the teprotide molecule into some bare minimal fragments and obtained drugs with better oral activities. The breakthrough came when they replaced the carboxylate group with a thiol (–SH) and achieved a 2,000-fold increase in potency in ACE inhibition. The drug became the first *oral* ACE inhibitor, captopril. Squibb has sold it under the brand name Capoten since 1978. Captopril was the first commercially available ACE inhibitor and a market success; it contributed tremendously to the management of hypertension. Squibb arrived at captopril from only 60 compounds logically synthesized and tested. Ironically, Squibb had also set up a random screen for ACE inhibitors and tested more than 2,000 compounds from the Squibb library. None were active.

Captopril has a rapid onset of action, reaching maximum activity in 15–30 minutes, but the plasma half-life is only 2 hours. Therefore, captopril must be taken more than once a day. Other shortcomings of captopril are rashes and diminution or loss of taste perception, which may be attributed to the presence of the thiol functional group. Cough and angioedema are also side effects of ACE inhibitors.

The two discovers of captopril, Cushman and Ondetti, made significant contributions to medical science, not only by having discovered captopril but also by pioneering a revolutionary approach to drug discovery in the process. They exploited a three-dimensional protein structure to design an oral active drug. Their work ushered in a new technology called structure-based drug design (SBDD), which is now used throughout the pharmaceutical industry. In 1999, Cushman and Ondetti were honored with the Albert Lasker Award for Clinical Medical Research.

To make a better ACE inhibitor by improving captopril, a group of Merck scientists led by Arthur A. Patchett started to replace the thiol group with a carboxylate.[19] The loss in potency of the carboxylate was compensated for by modification of the molecule elsewhere. They arrived at a very potent molecule, enalaprilat, which suffered poor oral bioavailability. They simply converted the acid into its corresponding ethyl ester, creating enalapril, a prodrug of enalaprilat, with excellent oral bioavailability. Enalapril is a prodrug because, when it enters the gut, it becomes hydrolyzed to the active drug, enalaprilat. Although enalapril is a "me-too" drug of captopril, it is better absorbed by the stomach. One advantage of a prodrug is the delay in onset of action, which can be beneficial for a drug that treats blood pressure. The longer duration of action allows a once-daily dosage. It is also devoid of the side effects associated with the thiol group, including bone marrow growth suppression (due to a decrease in circulating white blood cells), skin rash, and loss of taste. In 1981, Merck successfully completed the clinical trials, gained approval, and sold enalapril using the brand name Vasotec, which became their first billion-dollar drug in 1988. Another popular ACE inhibitor is Pfizer's Accupril (quinapril hydrochloride).

Although early ideas about hypertension suggested that ACE inhibitors would be useful only in certain circumstances, this proved to be an oversimplification. ACE inhibitors have become an important class of drugs for controlling the commonly encountered form of hypertension. Moreover, ACE inhibitors have been proven beneficial to patients suffering from heart attacks, congestive heart failure, and possibly even atherosclerosis. A pharmacoeconomics study published in July 2005 calculated that the U.S. government would save money by giving away ACE inhibitors to seniors because the cost of the drugs is far less than the cost for treating the later-stage cardiovascular diseases.

Because angiotensin II is a potent vasoconstrictor, blocking its action would result in vasodilation. Dupont (a company that got its start with nitroglycerin) exploited the angiotensin II receptor in the early 1980s. But Dupont was a chemical company and did not have expertise in clinical trials and marketing. They formed a 50-50 joint venture with Merck, establishing Dupont-Merck Pharmaceuticals. Angiotensin II receptor antagonist losartan (Cozaar) was the fruit of Dupont-Merck. Cozaar, launched in 1995, quickly established itself as one of the most important drugs for the treatment of high blood pressure. Cozaar, along with other drugs of this class, proved to be superior to ACE inhibitors because it did not cause the irritating cough that occurs in a small percentage of patients who take ACE inhibitors. Novartis's top seller, valsartan (Diovan), is an angiotensin II receptor antagonist that generated $2.4 billion in sales in 2003. Other well-known angiotensin II receptor antagonists are Sanofi-Synthélabo's irbesartan (Avapro), Astra-Zeneca's candesartan (Atacand), Sankyo's olmesartan medoxomil (Benicar), Sovay's eprosartan (Teveten), and Boehringer Ingelheim's telmisartan (Micardis).

Black and Beta-Blockers

The introduction of beta-blockers is widely considered one of the most revolutionary, conceptually, in our quest to conquer human ailments.[14, 20–31] The approach of rational drug design also revolutionized how drugs are discovered.

Adrenaline (epinephrine) and noradrenaline (norepinephrine) are secreted by the adrenal gland. These hormones bind to their corresponding receptors and elicit biological responses. Stimulation of the sympathetic nervous system due to *fright* leads to preparation of the system for *fight* or *flight.* Increase of adrenaline results in dilation of bronchi, dilation of pupils, constriction of peripheral (outside the central nervous system) blood vessels, and so forth. The cardiotoxic effects of catecholamines were determined in late 1940s and early 1950s. As a consequence, an excess of adrenaline can cause heart attack and hypertension.

Adrenaline and noradrenaline stimulate adrenergic receptors. As early as 1948, Raymond P. Ahlquist at the Medical College of Georgia speculated that there were two types of adrenergic receptors (adrenoceptors for short), which he termed alpha-adrenoceptor and beta-adrenoceptor—later further subdivided as beta-1 and beta-2. The beta-adrenoceptors belong to a family of G-protein-coupled receptors (GPCRs). G-proteins, in turn, are guanine nucleotide-binding regulatory proteins.

Because Ahlquist's theory was so revolutionary at the time, he found himself having difficulty publishing his carefully reasoned and thoroughly researched paper. He later commented: "The original paper was rejected by the *Journal of Pharmacology and Experimental Therapeutics,* was a loser

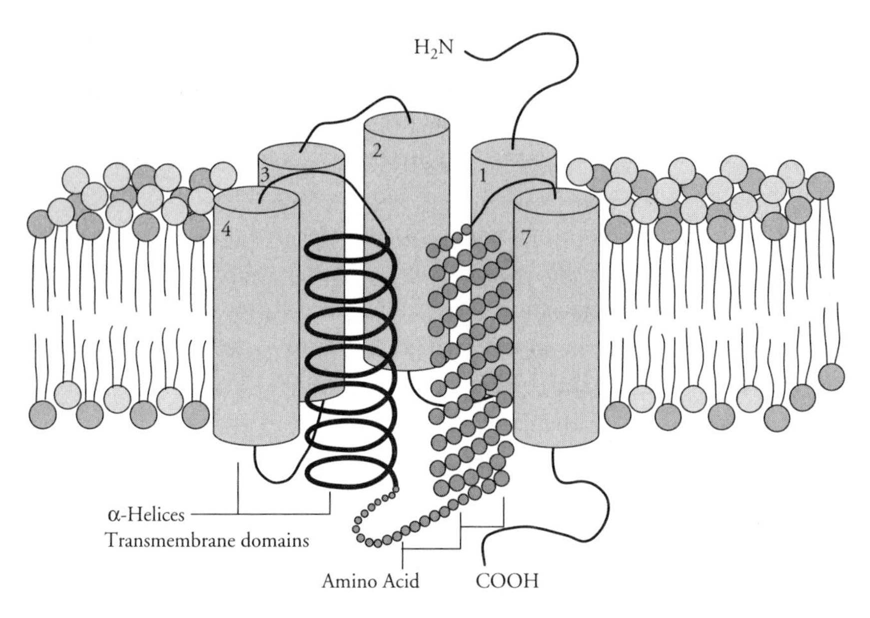

G-Protein-coupled receptor with 7 trans-membrane domains. Diagram by Vivien H. Li

in the Abel Award competition and finally was published in the *American Journal of Physiology* due to my personal friendship with a great physiologist, W. F. Hamilton."[21]

Ahlquist's hypothesis was largely ignored for the first 10 years after its publication. In 1958, C. E. Powell and I. H. Slater of the Eli Lilly Company were searching for a long-acting and specific bronchodilator to compete with isoprenaline. They prepared DCI (the dichloro analog of isoprenaline) and demonstrated that it inhibited the relaxation of bronchial smooth muscle elicited by isoprenaline and also the cardiac actions of isoprenaline. As an added bonus, DCI was found to be the first beta-adrenoreceptor blocking reagent, also known as beta-blocker. The finding that DCI selectively blocked beta-receptors has proven to be a significant advance in human pharmacotherapy.

James W. Black was born on June 14, 1924, in Scotland. In 1958, Imperial Chemical Industries (ICI) hired him from the Physiology Department at Glasgow University as a senior pharmacologist, although he had had no previous training in pharmacology. Black started to look for an antianginal agent that would reduce sympathetic stimulation of the heart and thereby decrease myocardial oxygen demand. To do this he applied his knowledge of how chemical messengers such as adrenaline and noradrenaline bind to the two major types of cell receptors in the organs of the body: alpha-adrenoceptors and beta-adrenoceptors. He spent a decade attempting to understand how cells talk to each other chemically. Black and colleagues investigated how these messengers and receptors might be manipulated to produce a desired medical result. He looked into the pharmacological properties of DCI, which not only possessed all the characteristics of a beta-blocker but also had a marked stimulant effect on the heart, an intrinsic sympathomimetic action (ISA). This led, in time, to the development of what came to be known as beta-blockers—substances that beat adrenaline to its target and, by acting as false messengers, bring about the intended pharmacological effect.

Beta-blockers interfere with the body's release of adrenaline. They are false messengers that infiltrate the messenger/receptor mechanism in human cells that trigger the release of adrenaline, thus calming the hearts of people who suffer from high blood pressure or tachycardia. In other words, beta-blockers act primarily by blocking the stimulation of the beta-receptors—the nerve endings that affect heart rate and the force of contraction. Their actions cause a decrease in the amount of blood pumped by the heart and thus lead to the lowering of blood pressure.

In 1962, Black and his colleagues succeeded in making a beta-blocker that was devoid of the stimulant effect on the heart: pronethalol. Unfortunately, pronethalol was withdrawn from further development when it was found to cause thymic tumors in mice. ICI eventually produced the drug propranolol (trade name Inderal), which possessed a better efficacy and safety profile. Not only was propranolol more potent than pronethalol, but it was also devoid of the carcinogenic properties in mice. Propranolol is now widely used in the management of angina, hypertension, arrhythmia, and migraine headaches. Two additional beta-blockers, atenolol (trade name Tenormin) and practolol (trade name Dalzic), were later discovered and marketed by ICI. Beta-blockers were not just a new class of drugs; they represented a revolutionary approach to pharmaceutical research. Black changed the process of drug discovery from one of hunting to one of engineering—employing rational drug design to discover novel compounds that nature had not thought of. In 1963, Black moved to Smith-Kline French, a small and little-known drug outfit. From 1963 to 1972, he developed the drug cimetidine (Tagamet), which decreases the secretion of acid in the stomach, thereby promoting the healing of peptic ulcers. Tagamet was the first blockbuster drug in pharmaceutical history—it generated more than $1 billion per year in sales and transformed SmithKline French into one of the biggest pharmaceutical companies in the world.[22]

Never a content soul, in 1972 Black took the prestigious position of Professor of Pharmacology at University Hospital, King's College. He was awarded the Nobel Prize in Physiology or Medicine in 1988, along with Gertrude (Trudy) Elion and George Hitchings, both from Glaxo-Wellcome Pharmaceuticals, for their discoveries of important principles for drug treatment.

Beta-blockers have made a major impact on the treatment of cardiovascular diseases since their discovery. Almost every single major pharmaceutical company was involved with beta-blockers. About 2,000 patents have been filed, and approximately 20 major drugs are marketed as beta-blockers. ICI alone tallied three major beta-blockers: propranolol, atenolol, and practolol. In addition, there are Hässle's alprenolol, Astra's metoprolol, Sandoz's pindolol, Ciba's oxprenolol, May and Baker's acebutolol, Synthélabo-Searle's betaxolol, and many more.

In addition to their primary uses in treating hypertension and cardiac impairments, beta-blockers have also been used in anesthesia to control a racing or irregular heart. Some beta-blockers even possess interesting central nervous system effects. Propranolol, for instance, causes the brain to enhance the memory of emotionally charged events, which would otherwise have been suppressed. Propranolol has been reputed to erase painful memories as well—a magic potion indeed. Somehow, propranolol also enables the person taking it to look back at stressful past events in a calm manner.

Beta-blockers are frequently used by performing artists and athletes to reduce anxiety. Because anxiety is associated with increased activity of the sympathetic nervous system and increased levels of catecholamines, beta-blockers have been proven to be effective in lowering anxiety levels. It may be, in part, that the sensation of the racing heart causes the person to feel or at least to be aware of his or her anxiety and that blocking one symptom ameliorates the other. Whereas performing artists may take beta-blockers to reduce stage fright, it is illegal for athletes to take them during competitions. In 1986, the National Collegiate Athletic Association (NCAA) listed beta-blockers as prohibited performance-enhancing drugs. The use of beta-blockers is prohibited in marksmanship, archery, ski jumping, freestyle skiing, sailing, synchronized swimming, diving, and pentathlon in the Olympics. The International Olympic Committee (IOC) considers the use of adrenergic antagonists, including alpha- and beta-blockers, a serious offense and rarely accepts any excuses.

In their quest for beta-blockers, medicinal chemists prepared more than 100,000 compounds. The pharmaceutical industry would not be what it is today without beta-blockers, the knowledge and profit gained catapulted the industry to a new height.

Fleckenstein and Calcium Channel Blockers

Calcium is one of the most abundant elements in the human body. Our bones are mostly made of calcium carbonate—which may be related to the fact that calcium is one of the most abundant terrestrial metals on earth. Calcium plays a key role in our physiology and pathology. More than a century ago, calcium was discovered to be necessary in maintaining the contractile activity of the heart. Intracellular injection of calcium evokes the contraction of a muscle. It is required for both sperm motility and the fertilization response. David J. Triggle, a professor at the State University of New York at Buffalo, once said "in a very real sense we are conceived in a moment of calcium-related enthusiasm, we die from an excess of calcium and we are laid to rest under a tombstone of calcium carbonate."[32] Indeed, calcium is a pathological contributor to the process of cell death.

Unlike many new classes of drugs, most calcium channel blockers were discovered before the concept of calcium channel antagonism was discovered. Therefore, the knowledge of channel classification or structure contributed nothing to this drug discovery process. Calcium channel blockers, also known as calcium channel antagonists or calcium entry blockers, are widely used in the treatment of high blood pressure, angina, and rapid heartbeat (tachycardia), including arterial fibrillation.

Prior to the appearance of the calcium channel blockers, nitroglycerin and beta-blockers were the major drugs for treating angina pectoris. Unlike beta-blockers, whose structures are very similar, the four earliest known calcium channel blockers (CCBs)—verapamil, nifedipine, diltiazem, and perhexiline—were not structurally related. How they worked was not known until the 1960s, when Albrecht Fleckenstein discovered that they all had the same mechanism of action. When a calcium channel blocker enters the opening of a calcium channel, the drug gets stuck, like a fat man caught halfway through a porthole, thus preventing calcium ions from getting through the channel. Calcium channel blockers slow or block the movement of the calcium ion into muscle cells on the walls of blood vessels, thus reducing contraction of blood vessels and lowering blood pressure.

In November 1963, Albrecht Fleckenstein was a professor at the Physiological Institute at the University of Freiburg in Germany.[33] Two German pharmaceutical companies, Knoll and Hoechst, asked him to look at two newly synthesized coronary vasodilators that had unexplained cardiodepressant side effects. Those two drugs were prenylamine and verapamil, which were discovered the old-fashioned way—the pharmacologist simply measured the biological responses, such as vascular smooth muscle dilation and tension, solicited by drugs they tested. Fleckenstein and his colleagues observed that both prenylamine and verapamil exerted a negative inotropic (decreased contractility) effect on the heart, in addition to the expected coronary dilator response. Purely by chance, they discovered that calcium counteracted this negative inotropism. With remarkable and somewhat enviable intuition, they went on to conclude that this negative inotropism was due to an ability of these drugs to block excitation-induced calcium influx. In 1966 Fleckenstein then coined the term *calcium antagonists,* because both drugs mimicked the cardiac effects of simple calcium withdrawal. Unfortunately, the results of their initial papers, mostly written in German, failed to cross national boundaries.

At the end of October 1967, a symposium was held on the island of Capri in the Mediterranean Sea to discuss the mechanism of action of prenylamine and verapamil. Fleckenstein shocked the audience with the calcium concept of prenylamine action. He demonstrated that prenylamine was able to block calcium-dependent excitation-contraction coupling in heart muscle. Also present at the symposium was Winifred G. Nayler from Melbourne, Australia. She independently developed a similar concept in Capri, which postulated that prenylamine inhibited the calcium permeation across cellular membrane. Almost no one among the participants of the symposium seemed to believe their calcium stories. But Fleckenstein and Nayler became staunch allies in the fight for calcium. Nayler dedicated her book *Calcium Antagonists* to Fleckenstein 20 years after that symposium, when the calcium channel antagonism gained ubiquitous acceptance worldwide.[34,35]

In 1968, Ferdinand Dengel, the chief chemist of the Knoll Company, gave Fleckenstein the 600th compound that he had synthesized in his career.

The compound was initially given the name D 600, and it turned out to be much stronger than verapamil on both myocardium and smooth muscle. One year later, Kroneberg, the leading pharmacologist of Bayer Company, handed Fleckenstein Bay-a-1040 and Bay-a-7168. Both compounds were strong coronary vasodilators that exerted significant negative inotropic effects on the myocardium. Fleckenstein also found out that the mechanism of action of those two drugs appeared to be similar to that of verapamil and D 600. Later, Bay-a-1040 and Bay-a-7168, both having the 1,4-dihydropyridine-core structure, were given the generic names nifedipine and niludipine, respectively. Curiously, Kroneberg and Bayer kept the chemical formulas secret for more than 3 years. Nifedipine and niludipine, in turn, heralded one of the most important classes of calcium antagonists: 1,4-dihydropyridines.

A naturally occurring calcium antagonist, Tanshinone IIA, is isolated from the roots of the plant *Salvia miltiorrihiza Bunge*, which is widely distributed in most of China. The Chinese have used its roots for centuries for medicinal purposes, particularly for the treatment of "chest pain brought on by exertion." Tanshinone IIA was found to be a calcium channel blocker. It is amazing how "trial and error" worked so marvelously for the Chinese herbalists in the discovery of these remedies for ailments.

Bayer's nifedipine (Adalat) was the prototype 1,4-dihydropyridine calcium antagonist. It is a short-acting calcium channel blocker taken several times a day. In contrast, amlodipine (Norvasc) by Pfizer is another 1,4-dihydropyridine calcium antagonist with a high bioavailability and a long half-life in plasma; thus it can be taken once daily. The long duration of action makes it ideal for long-term treatment of hypertension. Moreover, it is safe and gradually reduces blood pressure. This is a good attribute because reducing blood pressure too quickly can cause fainting spells, which happens with some of the 1,4-dihydropyridine calcium antagonists. All of those features have made amlodipine the most prescribed antihypertensive agent in the world, with annual sales of $4.3 billion in 2003. Worldwide sales of calcium channel blockers in 2003 totaled $6 billion that year.

Endo and Statins

Cholesterol is essential to our lives. It is most abundant in our brains and is a crucial ingredient for the synthesis of sex hormones. Cholesterol in the human body comes from two sources. One is the intestinal absorption of dietary cholesterol. The majority is generated endogenously (inside the body, primarily in the liver and intestine) to meet the body's need if the diet is lacking sufficient cholesterol.

In the 1950s and 1960s, many studies, especially the famous Framingham Heart Study, demonstrated that high cholesterol is a major risk factor for the development of coronary artery diseases. Since 1948, researchers from Boston University have been carrying out systemic epidemiology studies of Framingham's residents, including routine health monitoring and testing of blood samples. Following the Framingham Heart Study, medical research focused on how to reduce cholesterol levels.

A genetic disease called familial hypercholesterolaemia (high cholesterol) provided the ideal means for scientists to study the impact of cholesterol on coronary artery diseases. People with this condition are born with exceedingly high levels of cholesterol; thus they experience significant cardiac diseases and are more prone to falling prey to cardiovascular ailments. It was found that they lack the LDL (low-density lipoprotein) receptor responsible for removing cholesterol from the blood. In order to decipher how cholesterol worked, a viable animal model that bore resemblance to familial hypercholesterolaemia was needed. After all, one could not just use people with this genetic defect as human guinea pigs! Serendipitously, a Japanese veterinarian noticed that a male rabbit in his colony had 10 times the normal concentration of cholesterol in its blood. By appropriate breeding, he produced a unique strain of rabbits with high cholesterol levels. These rabbits promptly developed coronary artery diseases and served as an ideal animal model for studying human familial hypercholesterolaemia. Taking advantage of these rabbits, scientists gained a greater understanding of the linkage between cholesterol and the corresponding receptors that prevented their tissues from taking up low-density lipoproteins. They also found that this receptor was primarily located in the liver and adrenal glands.

In 1961 pharmaceutical firm Richardson-Merrell received approval from the FDA to market their new anticholesterol drug, triparanol. But triparanol caused cataracts and other severe side effects in monkeys. Soon these side effects began to manifest in humans, too. Ultimately, Richardson-Merrell paid $200 million to 500 civic litigants and an $80,000 fine. Two scientists and one executive pleaded no contest to charges of making fraudulent statements to the FDA.

An encouraging development in the treatment of hypercholesterolaemia has been the introduction of statins, which are potent competitive inhibitors of the enzyme 3-hydroxy-3-methyl-glutaryl-CoA (HMG-CoA) reductase, the rate-determining enzyme in cholesterol biosynthesis. The process by which the body makes cholesterol is a long cascade. The cataracts caused by Richardson-Merrell's triparanol generated many concerns over the safety of inhibiting cholesterol biosynthesis, but further work led to the conclusion that triparanol inhibited one of the last steps in the biosynthetic cascade and that cataracts were formed by deposition of nonmetabolizable steroids in the cornea. The hypothesis then arose that inhibition early in the pathway might lead to a safe method of inhibiting cholesterol biosynthesis.

An early step, also the slowest and thus a rate-limiting step, involves the reduction of HMG-CoA to mevalonate, which is then transformed into cholesterol after several steps. The crucial reduction process is accomplished by an enzyme, fittingly called HMG-CoA reductase, which, in turn, is the rate-controlling enzyme in the biosynthesis pathway for cholesterol. Therefore, it is reasonable to believe that if one could block the function of HMG-CoA reductase, the chain reaction for the cholesterol production would be suppressed.

In 1971 Japanese researcher Akira Endo was pursuing a belief. Because many microorganisms require cholesterol for growth, he believed that he could find microbial products that inhibit HMG-CoA reductase. As a consequence, those microbial products might reduce levels of plasma cholesterol in humans.[36]

Inspired by Alexander Fleming's success with penicillin, Endo and colleagues at Sankyo Laboratories tested more than 6,000 microbial strains over a 2-year period for their ability to block lipid synthesis. One strain that produced an active compound was *Penicillium citrium,* which belongs to the same genus of fungus that produced penicillin, *Penicillium notatum.* Little did Endo know then that he had discovered the first of what would be known as the statin class of drugs, the biggest moneymaker to date for the pharmaceutical industry—nearly $20 billion a year in 2004. Although there are numerous subspecies of the *Penicillium* genus, it seems profound that penicillins and statins, two of the most important classes of modern drugs, are produced from the same genus.

The active compound that Endo isolated was mevastatin, the first statin. Endo's group extracted a whopping 600 liters of the culture filtrate and ended up with only 23 mg of mevastatin in crystal form.

Almost simultaneously, A. G. Brown and his colleagues at Beecham Pharmaceutical Laboratories also isolated mevastatin (which was initially named compactin), also from a *Penicillium* fungus, *Penicillium brevicompactum*. Brown and coworkers initially discovered compactin as an antifungal agent. Its structure was elucidated by X-ray crystallography. Unfortunately, tests in rats at Sankyo showed that mevastatin had no efficacy in lowering plasma cholesterol levels even at very high doses. It turns out that rodents are poor models of lower plasma cholesterol because statins induce a massive amount of HMG-CoA reductase enzyme in rat livers. Higher vertebrate species, such as dogs, rabbits, and monkeys, are better for this purpose in reflecting human response.

At the beginning of 1976, Endo started to test mevastatin on laying hens whose eggs were known to have high cholesterol levels. Treatment with mevastatin decreased the cholesterol level in eggs by 50%. Encouraged by this observation, Endo's group subsequently tested mevastatin on dogs and monkeys and obtained satisfactory results. Endo published his discovery in 1976, but mevastatin never made it to the market, possibly because tumors were observed in the dogs treated with mevastatin. After further investigation, mevastatin was modified by microorganism into pravastatin (Pravachol), an HMG-CoA inhibitor with a ring-opened dihydroxycarboxylic acid side chain, first identified as a metabolite of mevastatin in dog urine. In 1989, Sankyo comarketed Pravachol with Bristol-Myers Squibb.[37]

Between 1972 and 1974, Michael S. Brown and Joseph L. Goldstein at the University of Texas Southwestern Medical School in Dallas investigated how the liver processed cholesterol. They identified the key biochemical steps that involved HMG-CoA in the regulation of cholesterol through LDL-receptors, for which they received the Nobel Prize in 1985. Later, they also demonstrated that statins could dramatically reduce the level of LDL, the "bad cholesterol."

Roy Vagelos was the head of research and development at Merck at the time.[38] Merck hired him when he was a professor and chairman of the Department of Biochemistry at Washington University in St. Louis in 1975. Vagelos brought to Merck a strong academic culture in which science and scientists dominated the company. He also brought with him several academic friends from Washington University, including Alfred W. Alberts, who would become the "product champion" of statins. Alberts had a unique career path. He did his graduate work in cell biology at the

University of Maryland. In 1959, he got a job at the National Institutes of Health (NIH), working for Vagelos in lipid research, and became an ABD ("all but the dissertation," meaning that he completed all requirements to get his Ph.D. but did not write a dissertation). In 1965 Vagelos was offered a job as the chair of biochemistry at Washington University, succeeding Nobel laureate Carl Cori. Alberts showed great loyalty and followed Vagelos to St. Louis. There, Alberts helped Vagelos run his research group while Vagelos busied himself building a world-class department. Because of his value to the students, the department, and biochemistry in general, Alberts was unanimously recommended for an associate professorship with tenure. Michael Brown, a consultant for Merck at the time, convinced scientists there in 1976 to begin a screening program to look for other statins from Merck's natural product efforts. A breakthrough came in November 1978. Working with his laboratory assistant Julie Chen, Al Alberts identified lovastatin as a potent inhibitor of HMG-CoA reductase from the culture broth of *Aspergillus terreus,* the 18th microorganism among thousands of soil samples tested. Nothing with a better profile was ever discovered from the screen. Three months later, the structure of lovastatin was elucidated. It differs from mevastatin by the addition of one methyl group. Merck quickly moved lovastatin into clinical trials in April 1980.

In 1982, the clinical trials were suddenly halted; Merck scientists heard rumors that Sankyo Laboratories' mevastatin development was discontinued because tumors were popping up in dogs. It was not until 2 years later that development of lovastatin resumed in the United States for patients with severe familial hypercholesterolaemia. Lovastatin was approved for marketing with the brand name Mevacor in 1987. The initial tablet color for Mevacor was yellow, resembling the color of butter, perhaps not a good image for cholesterol reduction. It was later changed to blue.

Meanwhile, Merck synthesized simvastatin (brand name Zocor), an analog of lovastatin with an extra methyl group on the side chain. Simvastatin is 2.5 times more potent than lovastatin. Zocor quickly became a blockbuster drug, with peak sales of $5 billion in 2003. In 2004, Merck and Schering-Plough came up with a drug called Vytorin, containing Zocor and Zetia (ezetimibe, a newer cholesterol-absorption drug).

There are two types of statins. Lovastatin (Mevacor) is a fermentation-derived statin, whereas atorvastatin (Lipitor) is a synthetic statin.

Bruce D. Roth began working for Parke-Davis Research of Warner-Lambert in 1982 as a medicinal chemist. He became the project cochairperson for the statin project in 1984, with biologist Roger Newton as the chair. In order to replace the decalin core structure of lovastatin, Roth recalled a methodology for pyrrole synthesis that he developed during his postdoctoral training at the University of Rochester. Applying that method and incorporating some molecular attributes (especially the side chain) of lovastatin, Roth and his colleagues synthesized atorvastatin in 1985. The drug showed cholesterol-lowering efficacy in animal models equivalent to that of lovastatin. Unfortunately, atorvastatin would have been the fifth statin on the market. Despite this, Parke-Davis decided to move forward and pushed the drug ahead into clinical trails. In the human trials, atorvastatin was extremely efficacious, better than any statin known at the time. The drug became Lipitor, launched in 1997 by Warner-Lambert, with Pfizer as its comarketing partner. The initial projection of annual sales was a meager $300 million, as it was a "Johnny come lately." But Lipitor's spectacular efficacy vaulted it into a billion-dollar drug within a year. The success planted the seeds of Pfizer's acquisition of Warner-Lambert in 2000. In 2005, Lipitor was the largest-selling drug ever, with worldwide sales of $12.9 billion, the first drug to sell over $10 billion worth in one year. In 2004, Pfizer combined amlodipine (Norvasc) and atorvastatin calcium (Lipitor) under the name Caduet, the first medicine to treat two different conditions in one pill. It lowers both blood pressure and cholesterol at the same time, allowing physicians to help patients reduce their risk of developing cardiovascular disease.

Around 2003–2004, head-to-head comparisons of Pravachol and Zocor with Lipitor were carried out. In an 18-month study involving 502 patients with an average LDL of 150, Lipitor (which lowered LDL to 79) was shown to be more efficacious than Pravachol (down to 110). In the PROVE-IT trial, Bristol-Myers Squibb, the group comarketing Sankyo's Pravachol, conducted a seven-year, head-to-head trial comparing Pravachol and Lipitor. The results were not what BMS expected. The clinical result was that Lipitor beat Pravachol. In another comparison, treating heart attack patients with Lipitor significantly lowered their risk of another heart attack, whereas Zocor failed to exert the same benefits. This comparison suggested that all statin drugs are not the same.

Statins are generally very safe. A running joke among cardiologists is that statins should be added to drinking water. An estimated 11 million Americans currently take statins, which greatly reduce the level of cholesterol and mortality in risk for heart attacks. Ironically, both Bruce Roth and Robert L. Smith (the inventor of Merck's Zocor) both take their own respective statins to lower their cholesterol levels. When former President Bill Clinton was diagnosed with hypercholesterolaemia, he was put on a statin. In May 2004, the U.K. Department of Health approved statins for over-the-counter (OTC) sale, with Zocor the first to be available at the pharmacies. Zocor lost its patent in 2003 in the United Kingdom.

One very important, though rare, side effect of statins is rhabdomyolysis, muscle-tissue breakdown that can lead to kidney failure. More than 50 people who were taking Baycol died of rhabdomyolysis. It seemed that the risk of rhabdomyolysis was much higher with Baycol than with other statins, and it was withdrawn from the market in 2001.

It is often said that with the advent of statins and antihypertensive drugs, there is little more for the pharmaceutical industry to do in the cardiovascular area. The claim is far from reality: although statins have proven to be extremely effective in decreasing low-density lipoproteins (LDL), increasing high-density lipoproteins (HDL) can significantly affect coronary heart disease beneficially.

Unlike LDL (the so-called bad cholesterol), which is usually too high, HDL (the "good cholesterol") is frequently too low. In 2004, Pfizer commenced the most expensive Phase III clinical trial in history for its HDL elevation drug, torcetrapib. The Phase III clinical trial is projected to cost $800 million. Torcetrapib works by inhibiting cholesterol-ester transfer protein (CETP), which shuttles cholesterol esters between LDL and HDL. In 1990, it was reported that Japanese patients who lack CETP had a low risk of cardiovascular disease, leading to the hypothesis that blocking the action of the CETP would increase the level of HDL. In 2004, *Forbes* magazine named torcetrapib one of five molecules that would change the world. The other four molecules were RNAi, carbon nanotube, nanowires, and polyhydroxyalkanoate (PHA). Although it is too early to tell, it is possible that agents that raise HDL may actually prevent (not just treat) cardiovascular disease.

CHAPTER 4

Sex and Drugs

Life is not to live, but to be well.
—Martial, first-century Roman poet

Aphrodisiacs

The word *aphrodisiac* comes from the name of Aphrodite, the Greek goddess of sexual love, fertility, and beauty. An aphrodisiac is any drug that arouses the sexual instinct.[1–9] Throughout recorded history, humans have gone to great lengths in pursuing enhancement of sexual activity and desire. In Shakespeare's *Midsummer Night's Dream* (II, i), Oberon begged for the love potion:

Family Planning © Chinese Postal Bureau

> *Fetch me that flower; the herb I show'd thee once:*
> *The juice of it on sleeping eyelid laid*
> *Will make a man or woman madly dote*
> *Upon the next live creature that it sees.*

Perhaps the best-known aphrodisiac is alcohol, which was recognized thousands of years ago for its possible aphrodisiac properties. Shakespeare described the effect of liquor through the porter in *Macbeth* (II, iii):

Lechery, Sir, it provokes, and unprovokes;
It provokes the desire,
But it takes away the performance.
Therefore much drink may be said to be an equivocator with lechery;
It makes him, and it mars him;
It sets him on, and it takes him off;
It persuades him, and disheartens him;
Makes him stand to, and not stand to;
In conclusion, equivocates him in a sleep, and,
Giving him the lie, leaves him.

Aphrodisiacs are not a mere recreational curiosity in medicine; they may genuinely help some patients on antidepressants who suffer decreased libido as a side effect. According to the mechanisms of action, libido lifters can be divided into seven categories:

1. Serotonin antagonists: cyproheptadine and granisetron
2. Adrenergic antagonists: yohimbine and trazodone
3. Cholinergic agonists: bethanechol chloride
4. Dopamine-enhancing drugs: bupropion, amantadine, and bromocriptine
5. Autoreceptor agonists: buspirone and pindolol
6. Stimulants: amphetamine, methylphenidate, and ephedrine
7. Herbals: ginkgo biloba and *L*-arginine

Aphrodisiacs can be obtained from plants and animals or made through chemical synthesis. Although many aphrodisiacs have not been rigorously proven effective in clinical trials, in reality, a psychological boost is likely enough to help, because many sexual problems are "in the mind" anyway.

Plant Natural Products

Yohimbine, an extract from the brown bark of the African tree *Corynanthe (Pausinystalia) yohimbe,* is perhaps the only drug that has lived up to its reputation as an aphrodisiac. Compared with Taxol from the Pacific yew tree (0.014%) and vincristine from the dry periwinkle plant (0.00025%), the yield of yohimbine from *Pausinystalia yohimbe* (6%) is extremely high.

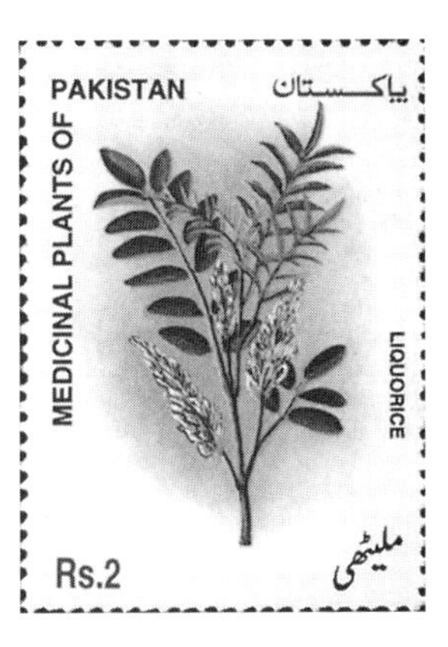

Ginseng © Chinese Postal Bureau; belladonna © Tanzania Post Office; and liquorice © Pakistan Post Office

In Spain, yohimbine is used to promote copulation of prized horses during mating season. Yohimbine has also been used as a sexual stimulant for cattle, sheep, and dogs. Yohimbine works as an α-2 adrenergic antagonist, similar to trazodone (trade name Desyrel), a synthetic antidepressant. Yohimbine's side effects are in large part associated with its central nervous system effects, which include anxiety, weakness, overstimulation, paralysis, and hallucination.

The reason that both ginseng and mandrake are touted as aphrodisiacs is likely that their roots form the shape of a baby. In controlled trials ginseng has been shown to enhance stamina for men, possibly through stimulation of metabolism. Mandrake (*Mandragon officinarum*), indigenous to countries bordering the Mediterranean Sea, contains hyoscine, which may account for some of its central nervous system effects. Much folklore was associated with its magic effects and with human fertility. In the Old Testament, the childless Rachel prayed: "Give me, I pray thee, of thy son's mandrakes" (Genesis 30:14). According to ancient superstition, it could only be harvested under moonshine, and a dog was needed to pull the roots out with a long rope. Terrible things would happen if one pulled the roots out by hand.

Belladonna flower, licorice root, nutmeg, vanilla beans, and pepper have all acquired reputations as aphrodisiacs in folklore.

Nutmeg was exceedingly important to the preservation of food before the age of refrigeration. One little-known reason for its popularity was its psychotropic properties. It was believed by some that nutmeg was imbued with aphrodisiac, soporific (anesthetic), and abortifacient properties. Clinical

Nutmeg, vanilla, and pepper

trials showed that consumption of large quantities of nutmeg elicited side effects that included headache, severe nausea, and dizziness lasting up to 1 hour, followed by hallucination and other sensory disturbances, which lasted up to 2 days. There was nothing erotic about those side effects.

Iboga (*Tabernanthe iboga*) is a plant indigenous to West Africa. The main ingredient of iboga extract is ibogaine, an alkaloid known as the "cocaine of Africa." Ibogaine's mode of action bears a striking resemblance to that of such tricyclic antidepressants as amitriptyline; therefore, ibogaine may modulate the level of serotonin in the central nervous system. Ibogaine, a controlled substance because it is a hallucinogen, gives the user relief from hunger and fatigue—African natives routinely chew the iboga root, which enables them to remain motionless for as long as 2 days while retaining alertness during hunting. Iboga is also believed to be an aphrodisiac possibly because yohimbine is another ingredient.

Ayahuasca (the vine of the soul) contains two alkaloids, harmine and harmaline, both of which cause sexual excitation in laboratory rats. In South American countries, Amazon Indians gave the *ayahuasca* brew to adolescent boys during their initiation rituals, which involve whipping and other, more sexually graphic activities.

If yohimbe is the king of aphrodisiacs for men, damiana is undoubtedly the queen of aphrodisiacs for women. Damiana, the shrub of *Turnera diffusa,* is the key ingredient in a Swedish blue drink called "Niagra," a play on the notoriety of Viagra. Studies showed that damiana contains β-sitosterol and aromatic oils, which may be stimulating and provide beneficial effects on sexual debility and nervous tension. It is estimated that in 2005, about 11 million women in the United States will report having "low desire." Some endocrinologists blame low levels of testosterone, produced in small amounts by the ovaries, for the symptom. Although it seems strange that testosterone, a male hormone, could be associated with female sexual dysfunction, random observations seem to corroborate the linkage between testosterone and female sexuality. For instance, after injecting a copious amount of testosterone as part of undergoing sex change, some females reported a significant surge of sexual desire. Taking advantage of those observations, Proctor and Gamble developed testosterone patches and tested them in patients with female sexual dysfunction who were also going through estrogen hormone replacement therapy. In December 2004, the FDA rejected the testosterone patches as the "female Viagra," citing safety concerns.

Animal Natural Products

For many years, rhinoceroses were hunted and killed to harvest their horns, which were thought to be an aphrodisiac. The reason men used rhinoceros horns to enhance sex drive could simply be the horn's phallic shape. Rhinoceros horns are useless as aphrodisiacs, but the false belief still helped cause the rhinoceros to become an endangered species. In order to solve the thousand-year-old myth, a group of Japanese chemists pulverized rhinoceros horns and extracted the chemicals with solvents. Detailed chemical analysis revealed that the horns merely contained polypeptide, sugars, ethanolamine, and free amino acids—not a completely unexpected result, because the rhinoceros horn is, after all, only modified epidermis. As a consequence, whatever aphrodisiac effects may have occurred were most likely only placebo effects. But then again, any man who slaughtered a rhinoceros and harvested the horns must have felt pretty confident. The only aphrodisiac effect of rhinoceros horn is probably its contribution to the English word *horny.*

The aphrodisiac effect of rhinoceros horns is merely a myth © Portugal Philatelic Department

Spanish fly (cantharide) is not a real fly but a small emerald green beetle found in southern Europe, mostly in Spain and France. For centuries, the Spanish fly love potion, made from pulverized dry beetles, has been considered an aphrodisiac. It has also been used as a method of inducing farm animals to mate. The active principle of Spanish fly, cantharidin (hexahydro-3,7-dimethyl-4,4-epoxyisobenzofuran-1,3-dione), is an extreme irritant and has been used as a blistering agent to remove warts. The irritation could provide prolonged stimulation of the erectile tissue in male and female genitalia, which may have contributed to the myth of its aphrodisiac effects. In reality, side effects from oral or topical administration of cantharidin can include pain, nausea, vomiting, and even death, with a lethal dose being a mere 32 milligrams in humans. Spanish fly is therefore not a genuine aphrodisiac, for it poorly mimics arousal by irritating the genital tract and water passages in the bladder, causing the genitals to burn, swell, and itch. At the end of the nineteenth century, Parke, Davis and Company in Detroit sold a popular aphrodisiac containing cantharides, strychnine, damiana, and zinc phosphate.

Probably nobody else contributed more to promulgating the myth of Spanish fly than Roald Dahl, the renowned writer whose credits include *James and the Giant Peach* and *Charlie and the Chocolate Factory.* In his humorous novel *My Uncle Oswald,* not only did Uncle Oswald make a fortune from the Sudanese fly (*cantharis vescatoria suddanii,* a fictional beetle), but he also experimented with its profound aphrodisiac effect on geniuses such as Albert Einstein, Sigmund Freud, George Bernard Shaw, Claude Monet, Auguste Renoir, and Pablo Picasso.

Synthetic Aphrodisiacs

Although decreased libido is a common side effect of many antidepressants, some drugs are known to possess aphrodisiac side effects. When *L*-dopa was first prescribed in the 1960s as a treatment for Parkinson's disease, hypersexuality was reported as a side effect. Some patients seemed to have excessive sexual arousal with the aide of *L*-dopa. In the movie *Awakenings,* the patient played by Robert DeNiro saw temporary respite from his Parkinson's disease and regained his sexual drive while on *L*-dopa. There were reports that 80-year-old men returned to their youth with nocturnal emissions and erotic dreams. Although the media touted *L*-dopa as a miracle aphrodisiac, it seems to have little impact on libido in people without Parkinson's disease.

There have been scattered reports of some drugs' aphrodisiac side effects, although most of them are anecdotal. A doctor described how he successfully treated a depressed man with a tricyclic antidepressant, clomipramine, who insisted on staying on the drug even after his depressive symptoms were long gone. It turned out that his wife wanted to him to stay on it, for he was able to maintain an erection much longer with the drug than without. Interest piqued, the doctor carried out a study and discovered that clomipramine indeed had a highly successful rate of treating premature ejaculation. In a Japanese study, imipramine and trimipramine showed similar side effects, as did both dopamine and serotonin.[7,9]

Bromocryptine, a synthetic derivative of the ergot alkaloids, was prepared by Sandoz chemists in Basel, Switzerland, in 1972. Andrea Genazzani of the University of Siena, Italy, treated women depressed from amenorrhea (lack of menstruation) with bromocryptine and reported, "In some women who had not had an erotic feeling in their entire lives, bromocryptine restored sexual desire and led to normal sexual activity."[7,8] Bromocryptine, a dopamine-enhancing drug, works by inhibiting the release of prolactin, which is known to inhibit synthesis of male hormones. Furthermore, some hypertension drugs such as hydralazine, endralazin, prazosin, and labetalol have also been reported to manifest some aphrodisiac side effects. *p*-Chlorophenylalanine, one of the halogenated amino acids, increased the sexual performance of male rats in the laboratories.

What constitutes a genuine aphrodisiac effect or just a placebo effect is difficult to discern. After all, sex, lust, and love are complicated phenomena that involve numerous factors such as vision, smell, and many cerebral factors, making definitive explanation of aphrodisiac effects nearly impossible. Therefore, drugs alone are not good enough, for humans are unique animals whose love lives are not merely physiological but also psychological—the only true aphrodisiac is the human mind.

Viagra and Erectile Dysfunction Drugs

It is estimated that 10 to 20 million American men suffer from impotence.[10–15] One of the old treatments was simply inserting an internal splint into the penis, a desperate measure indeed. British neurophysiologist Giles Brindley investigated means of treating impotence by injections of drugs into the penis. He first found success by papaverine injection. He then began to experiment with phenoxybenzamine after a colleague of his told him about its effect in hypertension. In 1983, in the annual conference of the American Urological Association, Brindley literally "showcased" his success. A few minutes before his lecture, he went to the bathroom and injected phenoxybenzamine into his penis. During the lecture the 57-year-old British gentleman pulled down his pants to display his achievement and invited the audience to physically examine his erection!

Before the emergence of Viagra, the Food and Drug Administration approved Upjohn Company's prescription drug alprostadil (Caverject) on July 6, 1995. Caverject, a member of the prostaglandin family, prostaglandin E_1, was expected to successfully treat 70–80% of impotence patients. Inconveniently, Caverject has to be injected directly into the penis before sex, not the most romantic move. On the other hand, Viagra, an oral drug, achieves the same purpose without cumbersome injections. Serendipity played an important role in the discovery of Viagra. As Louis Pasteur pointed out, "*Dans les champs de l'observation, le hazard ne favorise que les exprits préparés*" ("In the field of experimentation, chance favors the prepared mind").[10] Recounting the story of Viagra may help us to better appreciate Pasteur's point.

The saga began in 1985 when Simon Campbell and David Roberts, two chemists at Pfizer in Sandwich, England, put together a proposal to look for hypertension and angina drugs. They proposed to find compounds that would inhibit enzymes called phosphodiesterases (PDEs), which break down cyclic guanosine monophosphate (cGMP). At that time, little was known in the field of PDEs. One of the very few known PDE inhibitors was zaprinast, an antiallergy compound developed by May and Baker Laboratories (now part of Aventis). Zaprinast, a vasodilator, is not a clean drug—it inhibits PDE, as well as quite a few other enzymes. Theophylline is a PDE III and VI inhibitor, too. In order to make a better drug, one would need to find a more potent, more selective inhibitor that could be patented. One year later, the cGMP PDE approach received enough attention from management that five chemists were assigned to work on it, with Nicholas K. Terrett as the head of chemistry. Using zaprinast as the starting point, the team did what medicinal chemists do best—structural-activity relationship (SAR) investigation. By removing one nitrogen atom on the triazole ring, adding a sulfonamide group to reduce the molecule's lipophilicity (greasiness), and adorning the molecule with a couple of substituents, they created UK-92480 in 1989. In all, the team made 1,600 compounds for this program in 3 years, quite a feat considering that combinatorial chemistry was not reported until the early 1990s. Despite doubts in the company that achieving selectivity among the various PDEs could have been achieved, UK-92480 was shown to be selective in inhibiting one of the PDEs, PDE5, with weak activities against other PDE enzymes. It is only tenfold selective for PDE5 against PDE6, an enzyme involved in visual transduction, which may account for some of the visual side effects that have been occasionally observed with Viagra—some men who take Viagra will experience temporary color changes in their vision and see things as blue or green. In order to improve solubility, Terrett prepared the citrate salt of UK-92480, which would become sildenafil citrate and later Viagra.[15]

Nothing was unique about the story of Viagra thus far; it was just like any other drug discovery and development program. Things started to become interesting after the clinical trials began. Pfizer started Phase I clinical trials of sildenafil citrate in 1991 on healthy male volunteers and later commenced limited Phase II trials for indication of severe coronary heart disease (angina). Unfortunately, the drug was not efficacious for treating angina, and it was logical to terminate the trials. There is an urban legend that the healthy volunteers in Phase I trials refused to return tablets. In fact, the volunteers were in a clinical unit and did not have possession of the drugs. In further inquiries, some clever clinicians led by Ian Osterloh learned that the men "suffered" an unanticipated side effect, sometimes referred to in clinical trials as "unexpected benefits": the drug catalyzed their erections. The effect was especially striking for patients on high doses; 88% of them reported improved erections. However, especially at this time, a drug firm's ambition is to treat disease rather than to help healthy men achieve erections. Male erectile dysfunction is a serious disease, afflicting 10% of men under the age 40 but an astonishing 52% of men over 40 years old. In 1994, Pfizer initiated limited Phase II trials for 12 patients with male erectile dysfunction; 10 of them showed dramatic improvement in their erections, and the rest is history.[15]

However, one problem still remained. Pfizer had been expecting sildenafil to be a small product for angina. It now had a major blockbuster for male erectile dysfunction on its hands. Manufacture of the drug by the original medicinal chemistry route could not keep up with the rapidly expanding clinical program. A new route of synthesis was urgently required. By April of 1995 the Chemical Research and Development team at Sandwich in the United Kingdom, led by Peter Dunn, had invented a new synthesis of sildenfil. Dunn had left school at the age of 16 to work in a local chemistry factory. In 1984 he quit his job to study with Charles Rees at Imperial College, London, who helped him in his chemistry career. During the summer of 1995 the new synthesis was scaled up from 10 g to 1000 Kg in 13 hectic weeks to keep the clinical program on track. In 1996 Pfizer filed for a patent covering the new manufacturing method, with Peter Dunn and Albert Wood as inventors. The new synthesis gave an overall yield of 66% from 2-pentanone, which was an order of magnitude greater than the previous method.

On March 27, 1998, the FDA approved sildenafil citrate (trade name Viagra) for the treatment of male erectile dysfunction, and 4 million prescriptions were filled within the first 6 months. The sales reached $411 million in the first 12 weeks. Since its emergence, Viagra has received unprecedented publicity. The inventors found themselves being invited to do TV interviews and unexpectedly finding their pictures on the front page of national newspapers. Viagra, the little blue pill, has penetrated the cells of pop culture. It is one of the most commonly recognized names worldwide.

Viagra's mechanism of action is closely related to that of nitric oxide (see chapter 3). Sexual stimulation leads to the release of nitric oxide within the blood vessels of the penis, where it stimulates guanylate cyclase to increase cGMP levels in the corpus cavernosum. There are high levels

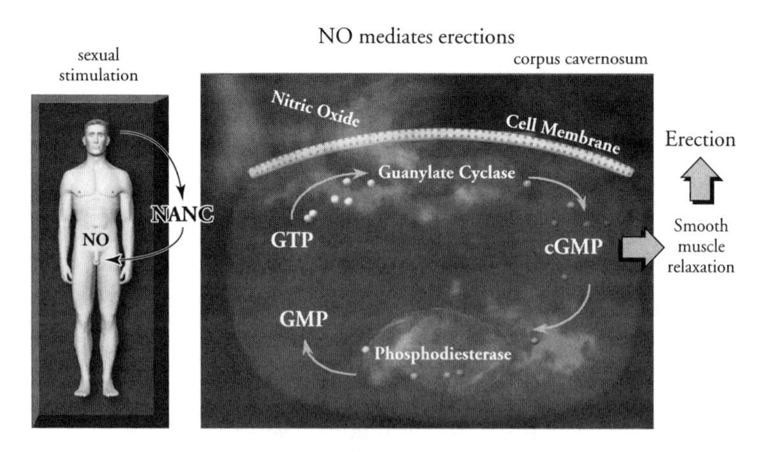

NO mediates erections © Pfizer Sandwich Laboratories

of PDE5 in the corpus cavernosum of the penis, and PDE5 is able to degrade cGMP and cause termination of erection. Nitric oxide production may be impaired in patients suffering from erectile dysfunction, leading to low levels of cGMP, which can be quickly degraded by PDE5. Inhibition of PDE5 by Viagra slows the breakdown of cGMP, allowing for higher concentrations of cGMP to build up in the corpus cavernosum, leading to an erection. Here is where it gets tricky: because nitroglycerin taken for heart conditions works via the nitric oxide production pathway as well, Viagra is contraindicated to nitroglycerin. Therefore, patients taking nitroglycerin or similar drugs should not take Viagra because the peripheral vasodilation from nitrates plus Viagra will divert blood away from a heart that is already compromised and will lead to further damage from lack of coronary blood flow. To overcome such a caveat, sexual dysfunction medicines using other mechanisms are desired. A cyclic peptide nasal spray, PT-141, stimulates melacortin receptors and was successful in Phase II clinical trials for erectile dysfunction and female sexual dysfunction.

On the heels of Viagra, two additional male erectile dysfunction drugs, Levitra and Cialis, emerged in 2003. Similar to Viagra, Levitra has a half-life of about 4 hours. Cialis's half-life is much longer, at 17.5 hours.

Riding on the commercial success and immense publicity of Viagra for men's erectile dysfunction, Pfizer invested millions of dollars and 8 years of clinical trials that involved 3,000 women in studying female sexual dysfunction. In the end, the results were equivocal. Not surprisingly, men and women display fundamentally different relationships between arousal and desire. For men, arousal invariably leads to desire, whereas arousal in many women has little effect on a woman's willingness, or desire, to have sex. In one trial, clinicians used a pelvic probe to measure any change in genital blood flow while women sat in front of erotic movies. Sure enough, the sex organs of women given Viagra were more engorged than those given placebo. The only problem was that, although Viagra was associated with greater pelvic blood flow, women experiencing this effect did not feel any more aroused. Mitra Boolell, leader of Pfizer's sex research team, pointed out that there is a disconnect in many women between genital changes and mental changes. This disconnect does not exist in men. With women, things depend on a myriad of factors. The bottom line is that the brain is the primary sex organ for women, and the genital area is only a secondary sex organ. In addition, hormone level seems play a more important role than blood flow.

The Pill

One of the most significant developments of the twentieth century was the discovery of oral contraceptives, popularly known as the Pill.[16–32]

The pursuit of fertility control began in antiquity. Men and women went to great lengths to avoid pregnancy. One ancient Egyptian contraceptive recipe used crocodile dung as the main ingredient! In the time of Hippocrates, recorded fertility control drugs included crocus, laurel, nettle seeds, peony root, and some mineral sources, as well as multiple barrier methods, all of which had little effect. Before the pill, couples often practiced withdrawal, otherwise known as coitus interruptus. Most other attempts at reducing fertility were largely ineffective as well until condoms became widely available in the early twentieth century. Much success was achieved, and the birth rate in the United States dropped by half.

Hormone means "to incite activity" in Greek; and sex hormones, in turn, are chemicals that modulate sexual activities. In 1930, a minute quantity of estrogen was isolated from the ovaries of 80,000 sows. In 1931, German biochemist Adolf Buttenandt collected 25,000 liters of men's urine and isolated 50 mg of androsterone, one of the male hormones.[16] In the late 1930s, an accidental discovery by Sir Charles Dodds revealed that the dimer of the plant phenol (anole aniseed) possessed high estrogenic activities. Supposedly, Dodds made the discovery because he was not a good organic chemist and could not purify his compounds. This led to Robert Robinson's synthesizing of a nonsteroidal estrogen, diethyl stilboestrol dimethyl ether, which Robinson cited as one of his chief achievements that won him the Nobel Prize in 1947. Interestingly, the United Kingdom listed diethyl stilboestrol dimethyl ether as a poison in 1939 out of fear that it would be used as an abortion pill.

In 1921, Austrian physiologist Ludwig Haberlandt proposed that progesterone, a female hormone, might be useful in fertility control because ovulation might be suppressed by modulating the levels of estrogen and progesterone. During pregnancy, progesterone maintains proper uterine environment and inhibits further ovulation. Although progesterone can therefore be considered nature's contraceptive, it is practically inactive when administered orally. When given intramuscularly, it often causes a severe reaction at the site of injection. A more ideal oral contraceptive needed to be more potent and more orally bioavailable than progesterone.

In 1951, Carl Djerassi at Syntex in Mexico City, Mexico, discovered the first oral contraceptive pill, norethindrone. As Isaac Newton stated that the reason why he could see far was because he stood on the shoulders of giants. At least two significant developments provided the foundation of Djerassi's success: one was Russell Marker's discovery of the synthesis of hormones from sapogenins, a class of steroids abundant in Mexican yams; the other was a novel synthetic methodology called the "Birch reduction," developed by Arthur Birch at Oxford University in 1944.

Russell E. Marker was an extraordinary organic chemist. He graduated from the University of Maryland in 1930. In order to get his Ph.D. degree, he was required to take some physical chemistry courses as part of the curriculum. He refused to do so because he already had a master's degree in physical chemistry and felt that it would be a waste of his time. He left without a degree, even though he had already completed his doctoral thesis and published it in the prestigious *Journal of the American Chemical Society.* He initially found employment at the Rockefeller Institute of Medical Research and later secured a professorship in organic chemistry at Pennsylvania State College in 1935. There, inspired by a Japanese chemist who isolated sapogenin (a steroid of botanical origin with cholesterol-like structure) from a yam-like plant of the genus *Dioscorea,* Marker developed a process (later known as *Marker degradation*) of transforming those abundant natural steroids into progesterone. During his 8 years of tenure at Penn State, he published 147 papers and secured 70 patents, which were assigned to Parke-Davis, the company that financed his research. In 1942, after having learned that sapogenins were abundant in certain types of wild Mexican yams (*Dioscorea*), he tried hard to convince Penn State and Parke-Davis to develop his process but was not successful. Typical of Marker, he quit his job again and moved to a cottage in Mexico, where he collected 10 tons of the yams. Moving back to the United States, he isolated diosgenin from the yams and transformed it into 2,000 grams of progesterone in a rented laboratory. One day in 1943, he showed up in the office of a local drug company in Mexico City, Laboratorios Hormona, with a jar under his arm wrapped in old newspaper. Inside the jar was 2 kilograms of progesterone, more than half of the world's supply, with a market value of $160,000. The two owners of the company were so impressed by Marker that they invited him to join them in forming a new company, Syntex (from Synthesis in Mexico), in 1944 to capitalize on his enterprise in progesterone. Although Marker left Syntex a year later because of disagreements with the owners over payments, profits, and patents, Syntex eventually was able to synthesize all four important steroid hormones, progesterone, testosterone, estrogen, and cortisone, from inexpensive starting materials found in the Mexican wild yams.

Arthur J. Birch was born in Australia and went to England to pursue a Ph.D. in chemistry, for none was granted in his home country at that time.[19,20] He received his Ph.D. in 1941 from Robert Robinson at Oxford University. Although most men of Birch's age joined the army when World War II broke out, Robinson insisted that Birch continue his research on steroids, stating, "The Americans tell us to put all our scientists into factories or the forces. But, if we ever win this war, we will need a few people to set up research again. I want you to be one of them."[19,20] During the Battle of Britain, rumors from the Polish Underground indicated that Luftwaffe fighter pilots were able to fly to unusually high altitudes because they were being treated with cortisone that allegedly had been made available by German chemists. Robinson wanted Birch to find the "fighter-pilot" hormone so that they could supply Royal Air Force (RAF) with it. In retrospect, it was fortunate that Birch never succeeded; otherwise, RAF pilots who would have taken it would have grown breasts and become sexually impotent—for all the steroid hormones that Birch synthesized were all markedly estrogenic.

However, Birch discovered a novel synthetic methodology that proved to be indispensable for the creation of the first oral birth control pill. In recounting his discovery, Birch claimed that part of the incentive was his being a lazy chemist, not wanting to take the long approach using conventional chemistry to synthesize the coveted "fighter-pilot" hormone. Intrigued by the steroid work of his fellow Australian graduate students John and Rita Cornforth, and inspired by the work of C. B. Wooster, Birch carried out the reduction of anisole using sodium in liquid ammonia and obtained β,γ-cyclohexenone. In essence, Birch succeeded in converting an aromatic compound into a functionalized cyclohexenone stereoselectively, which coincidentally correlated to the difference between the estrogenic steroids and androgenic steroids. Birch immediately realized the importance of his discovery, for he gave his paper a blanket title, "Reduction by Dissolving Metals." Djerassi was the one who first used the term "Birch reduction," which could very well have been "Robinson reduction" or "Robinson-Birch reduction" save for a complicated circumstance. At the beginning of the war, Robert Robinson arranged for Birch to receive some governmental funding channeled through ICI. As a consequence, theoretically, Birch became an ICI employee; thus ICI had the priority of patenting his discoveries. However, ICI pointed out that Birch should have been working on simple molecule "fighter-pilot" hormones rather than playing with reduction by dissolving metals in liquid ammonia. Robinson, being a consultant to ICI, was reprimanded for allowing his student to go astray from the sex-hormone field. In reality, Robinson, who sat on 37 committees during wartime, hardly knew what Birch was doing. As Birch recalled years later, the conversations they had usually went something like: "Where is Cornforth?" "In the lab vertically above."[19] And that was it. When Birch showed Robinson his manuscript "Reduction by Dissolving Metals," Robinson was furious and literally threw it out of his office, along with its author. Because Robinson refused to be associated with the paper, Birch became the sole author, and thus the process became the "Birch reduction."

Djerassi, a penniless Jewish refugee, fled Nazi Austria to immigrate to the United States in 1939, when he was 16.[22–27] In 1945, he completed his Ph.D. studies, focusing on the chemical conversion of testosterone to estrogen, in a record 2 years under Alfred L. Wilds at the University of Wisconsin. As a newly minted Ph.D., Djerassi went back to Ciba, where he had worked for a few years before graduate school. Fiercely ambitious, he was eager to continue pursuing steroid research, to establish a publishing record, and to become a professor, which was not really in line with what Ciba's management had in mind for an industrial chemist. In 1949, Syntex lured Djerassi away from Ciba to Mexico City and hired him as an associate director of chemistry in charge of their efforts in synthesizing cortisone. He was only 26 years old.

It was known in 1944 that elimination of the methyl group at C-19 position would greatly enhance the steroid's progestational activity. But

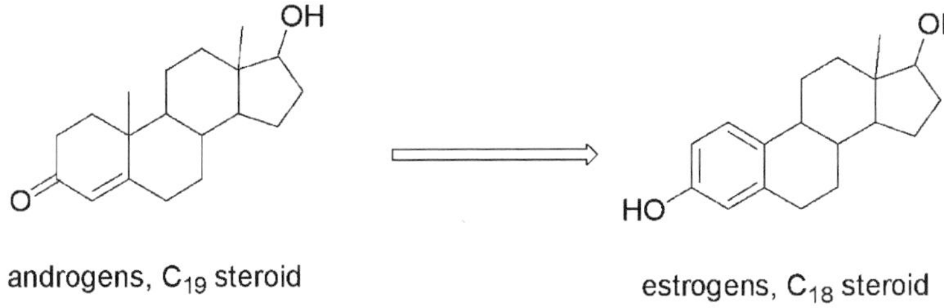

Chemical structures, androgens and estrogens

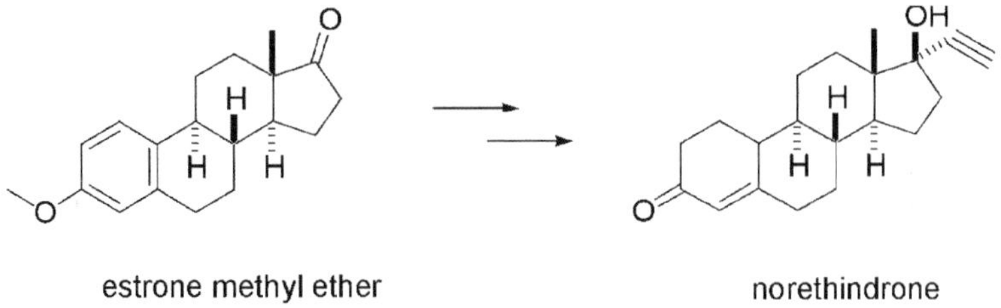

estrone methyl ether → norethindrone

Chemical structures, estrone methyl ether and norethindrone

nothing was done in the field until the Birch reduction was discovered, which enabled transformation of a readily available phenolic precursor to the 19-nor unsaturated ketone. Djerassi also immediately realized the utility of the Birch reduction in steroid chemistry and used it as the cornerstone of his design of the first oral contraceptive drug, norethindrone, which was synthesized by Luis Miramontes, a young Mexican undergraduate student working at Syntex, on October 15, 1951. Syntex alone used "kilos and kilos of metal and liquid ammonia,"[20] because the Birch reduction was indispensable in transforming estrone methyl ether into the corresponding cyclohexenone derivative. In early 1958, Djerassi hired Birch as a consultant to Syntex. When people asked "Why bring a fellow all the way from Australia to Mexico City?" Djerassi replied: "It's worth the few more dollars if you have hired a real brain."[25]

With norethindrone in hand, Syntex sent it for biological testing and found it to be the most potent orally active progestational agent at the time. Syntex promptly filed a patent in early November 1951, and Djerassi then published the chemical synthesis in the *Journal of the American Chemical Society* in March 1952. Lacking biological laboratories, drug development experience, and marketing outlets, Syntex licensed norethindrone to Parke-Davis to pursue the FDA registration and market the drug in the United States. In the middle of the trials, Parke-Davis suddenly chose to exit the contraceptive field for fear of religious backlash and returned the license to Syntex. As a consequence, Syntex lost a precious 2 years of time, aggravated by the fact that Parke-Davis was unwilling to hand over the data on safety in monkeys it had accrued for norethindrone to the Ortho Division of Johnson and Johnson, Syntex's new marketing partner.

Meanwhile, the world did not sit still. On August 31, 1953, more than 20 months after Syntex's patent was filed, G. D. Searle in Chicago, founded by Gedeon D. Searle in the late nineteenth century, filed a patent norethynodrel —acid→ norethindrone

Norethynodrel and acid gives norethindrone

for the synthesis of norethynodrel, discovered by Frank B. Colton, the chief chemist and an immigrant from Poland.[21] Norethynodrel is active orally but not so active when given subcutaneously because it is a prodrug, converted to norethindrone in the stomach. Djerassi believed that synthesis of a patented compound in the stomach constituted an infringement of Syntex's valid patent (similar situations have emerged a few times since then, as exemplified by BMS's Buspar and Wyeth-Ayerst's norgestrel, although mixed signals have been given by the courts). He pushed Parke-Davis to pursue a legal resolution, but Parke-Davis did not concur.[23]

Gregory Pinus, a reproductive biologist at the Worcester Foundation for Experimental Research in Massachusetts, was a long-time consultant to Searle. He scrutinized Colton's norethynodrel and Djerassi's norethindrone, along with many other drug candidates, in animal models. The renowned Harvard obstetrician-gynecologist, John Rock, led the clinical trials. Serendipitously, they discovered that addition of estrogen was beneficial in preventing breakthrough bleeding. When Searle's 10-mg Enovid was first approved in 1957 for the treatment of a variety of disorders associated with the menstrual cycle, it was a combination of the progestin norethynodrel and 1.5% of the estrogen mestranol. For fear of religious objection, Searle marketed Enovid as a treatment for gynecological disorders first. Syntex's norethindrone (as Norlutin, marketed by Parke-Davis) was approved at the same time and for precisely the same indication. The era of oral contraception began in May 1960, when Enovid was approved by the FDA for ovulation inhibition and was immediately introduced for such use. Therefore, G. D. Searle became the first pharmaceutical company to market the Pill.

The discovery of oral contraceptives was a great triumph for the pharmaceutical industry. More important, oral contraceptives contributed toward the emancipation of women from unwanted pregnancy. The social ramification was enormous, because the discovery of the Pill definitely helped fuel the sexual revolution in the 1960s and 1970s.

AIDS Drugs

HIV

In the early 1980s, the human immunodeficiency virus (HIV) was identified as the etiologic agent of acquired immune deficiency syndrome (AIDS).[33–47] More than 3 million people worldwide died from HIV/AIDS in 2003, according to a July 2004 United Nations report. During the same period, about 5 million people contracted the human immunodeficiency virus, bringing the total number of people living with HIV worldwide to 38 million. Although AIDS was called the "gay men's disease" at the beginning of the outbreak, it was soon discovered that sexual intercourse was not the only way of transmission. Blood transfusions and mother-to-baby transmission also spread the virus.

In comparison to the scourges caused by other viruses in history, we were more prepared and have achieved astonishing milestones against AIDS, thanks to our accumulated knowledge and efforts around the globe. HIV was identified and shown to be the cause of AIDS in less than 2½ years. It took only another 2 years for blood tests to become commercially available. In 1987, the first anti-HIV drug, AZT, was introduced. With the arrival of the HIV protease inhibitors and triple drug therapy (the cocktail therapy) in 1995, many patients who would otherwise have died are still alive. In 1996, *Time* magazine named AIDS researcher David Ho "Man of the Year" for his revolutionary idea of the cocktail therapy.

Who discovered HIV was such a contentious issue that it took the President of the United States and the Premier of France to settle the dispute.

The world against AIDS © Chinese Postal Bureau

In 1983 Françoise Barré-Sinoussi and Luc Montagnier, in the laboratory led by Montagnier at the Institut Pasteur de Paris, first detected and later isolated a retrovirus, lymphadenopathy-associated virus (LAV), which they believed was the cause of AIDS. During their research on the virus, Montagnier's laboratory collaborated with Robert C. Gallo, a renowned virologist at the National Cancer Institute (NCI), who was one of the most widely referenced scientists in the world in the 1980s and 1990s. Montagnier and Gallo frequently exchanged virus samples and information. In April 1984, Gallo held a press conference announcing that his laboratory had isolated a retrovirus, human T-lymphotrophic virus (HTLV-III), that he believed to be the cause of AIDS. Gallo was basking in scientific glory and was widely considered a leading contender for the Nobel Prize. Soon it was confirmed that Gallo's HTLV-III and Montagnier's LAV were identical. In 1986, a nomenclature committee was set up, chaired by Harold Varmus, an expert in avian retrovirus and then director of the NIH. The NIH committee settled on the name of human immunodeficiency virus (HIV).

In April 1984, Gallo's laboratory filed a patent on an HIV blood test kit using his HTLV-IIIB-ELISA (enzyme-linked immunosorbent assay), which was issued in a record 13 months via a special category involving national security. Although Institut Pasteur had filed a patent in the United States much earlier, in December 1983, it was not granted until a later date. Gallo's HIV test kit was approved by the FDA in 1985. An acrimonious legal battle ensued for the priority of the discovery of the HIV between the French and American teams. The contentious scientific and legal controversies came to an end in March 1987 when a historic agreement was signed by the directors of the NIH and the Institut Pasteur and ratified by Ronald Reagan and Jacques Chirac. The patents would become the joint properties of the two institutions, which would share the royalties. The three inventors from the NIH, including Gallo, would receive $100,000 annually from the royalties earned.

Even the intervention by two heads of state did not put the matter to rest. In November 1989, a Pulitzer Prize-winning investigative reporter, John Crewdson, published a 50,000-word article in the *Chicago Tribune* on the Montagnier-Gallo priority dispute. He concluded that Gallo had either stolen or allowed his samples to be contaminated with Montagnier's virus. The controversy generated resulted in congressional investigations. In the end, it was found that Mikulas (Mika) Popovic from Czechoslovakia, a cell biologist in Gallo's laboratory, had isolated HTLV-III from a pool by mixing several blood samples from different sources, including Montagnier's sample, which contained LAV. Pooling blood samples was an unusual practice in virology. In 1991, Gallo admitted in *Nature* that he had not discovered the new virus. In 1996, he left the NCI, where he had worked for 30 years, to become the director of the Institute of Human Virology at the University of Maryland Biotechnology Institute in Baltimore.

In 1987, the first anti-AIDS drug, AZT, was introduced by Burroughs Wellcome.[42,43] AZT, which blocks HIV reverse transcriptase activity, stands for azidothymidine, with the generic name of zidovudine and the trade name of Retrovir. Popular media often give the credit to Gertrude Elion of Burroughs Wellcome for having discovered AZT. In fact, although Elion and George Hitchings (see chapter 1, page 19) developed the concept of using nucleotides as antimetabolites in treating cancers, AZT itself was synthesized by a group led by Jerome Horowitz of the Detroit Institute of Cancer Research in 1964 as a possible anticancer drug. Horowitz, now a professor at Wayne State University, published his synthesis as a note in the *Journal of Organic Chemistry* in 1964.[43]

Since its birth, AZT had a checkered life as a drug looking for a disease to treat. AZT did not show efficacy in treating cancers; the drug also failed to prolong the lives of leukemic animals. In 1974, a German laboratory found it effective against viral infection in mice—Wolfram Ostertag of the Max Planck Institute for Experimental Medicine showed that leukemia helper virus (LLV-F) replication by AZT occurred via phosphorylation of AZT to the corresponding triphosphate, which cannot be incorporated into the growing strand of DNA. Ostertag correctly concluded that AZT-triphosphate worked by binding to the growing strand of DNA. Burroughs Wellcome acquired AZT and explored the possibility of using it to treat the herpes virus under the guidance of Gertrude Elion, although it did not make it to the market.

In 1984, shortly after Gallo announced his discovery of the retrovirus, HTLV-III, the head of the NCI, Samuel Broder, organized a team to screen antiviral agents as possible treatments for AIDS. In all, more than 50 pharmaceutical companies submitted their possible antiviral drug candidates to Broder's team for screening. Together with Dani Bolognesi, an AIDS researcher at Duke University, Broder obtained some of the potential antiviral compounds from Burroughs Wellcome. In February 1985, using an assay developed by Hiroaki "Mitch" Mitsuya, AZT was found to be active in vitro in the NCI laboratories in Bethesda. Wellcome patented AZT as an antiviral drug in June 1985 and promptly commenced the clinical trials. As with cancer drugs, the Phase I trials for AIDS drugs are done with patients rather than with healthy volunteers. The first trials to test AZT in patients with HIV showed dramatic efficacy. For ethical reasons, the company terminated the trials and switched patients on placebo to AZT immediately. The FDA approved the use of AZT on March 19, 1987, within 22 weeks. The recommended dose was one 100-mg capsule every 4 hours around the clock. Thus AZT established itself as the first antiviral drug in the arsenal against HIV. The mechanism of action of AZT is the blockade of the HIV reverse transcriptase activity. Reverse transcriptase, first isolated by David Baltimore and Howard Termin in 1970, is the enzyme that transcribes RNA into DNA. The success of AZT incited the development of many nucleotide anti-HIV drugs in an effort to minimize the toxicities that AZT displayed.

Among the newer reverse transcriptase inhibitors, Ziagen represents a vast improvement over AZT, a nucleotide whose gycosidic core structure is metabolized rapidly. Whereas AZT has to be taken every 4 hours around the clock, Ziagen allows a twice-daily regimen. When the oxygen on AZT is replaced with a methylene group, carbocyclic nucleoside analogs such as Ziagen are metabolized much more slowly by the body. Ziagen was developed by Glaxo Wellcome (now part of GlaxoSmithKline) using a technology developed by Robert Vince of the University of Minnesota, who licensed the patent to Glaxo Wellcome in 1993.

Robert Vince is a professor of medicinal chemistry and director of the Center for Drug Design at the University of Minnesota. After completing his Ph.D. training in 1966, he began his independent research in the field of antiviral medicine. In the mid-1970s, he designed an antiviral compound, carbocyclic Ara-A (cyclaradine), that was more effective in combating herpes virus than acyclovir was. Because he did not patent his discovery, it was difficult to entice the pharmaceutical industry to develop it. That experience taught him a lesson on the importance of intellectual properties. In the mid-1980s, inspired by the success of AZT, Vince started to tinker with nucleosides as HIV reverse transcriptase inhibitors. In retrospect, it was logical for him to replace the oxygen on the nucleosides related to AZT with a methylene group in order to improve bioavailability. But at that time, it represented a significant improvement. Along with a visiting researcher from China, Mei Hua, he synthesized a group of carbocyclic nucleoside analogs, which they called carbovirs. The NIH tested the carbovirs and found them to be the most active compounds in their screen against HIV since AZT. In fact, the carbovirs were the first compounds found active against HIV that were specifically synthesized for that purpose. In 1987, the University of Minnesota patented their synthesis and a group of antiviral drugs, listing Vince and Hua as coinventors. The university subsequently licensed the patent to Glaxo Wellcome, which arrived at Ziagen by substituting a propyl cyclopropyl group for the purine ring using the synthetic route developed by Vince. Because of Ziagen's favorable pharmacokinetics profile, it allows a twice-daily regimen and has brought in hundreds of millions of dollars in sales for the company.

The credit for any important discovery often seems to be a contentious issue. In this case, the stakes were high, as both AIDS and a large sum of money were involved. Glaxo claimed that Ziagen was not covered by the Vince-Hua patent because the patent did not cover Ziagen per se, whereas Minnesota contested that alkyl surely included cyclopropyl. In October 1999, the University of Minnesota and Glaxo settled this dispute, and as part of the settlement Glaxo agreed that the University patents were valid and enforceable. The settlement brought a financial windfall for Minnesota and the inventors. With the Ziagen money, estimated at $250 million thus far, Minnesota established a Center for Drug Design, with Vince as its director. Vince is putting his share of the Ziagen money to work on potential new AIDS drugs and other potential antiviral and anticancer agents at the center.

In addition to AZT and Ziagen, many HIV reverse transcriptase inhibitors exist. An organic chemistry professor at Emory University, Dennis Liotta, and his virologist colleague, Raymond Schinazi, discovered another reverse transcriptase inhibitor 3TC (lamivudine, Epivir), which allows a once-daily regimen. BMS's d4T was licensed from Yale University. The drug gained international fame when activists at Yale persuaded the university to rewrite a license agreement with BMS so that generic d4T could be sold in South Africa. BMS's ddI was approved in mid-1991, and nevirapine (trade name Viramune) by Boehringer Ingelheim was approved by the FDA in June 1996.

HIV Protease Inhibitors

The emergence of AZT and other reverse transcriptase inhibitors saved many lives. But they had only limited efficacy largely because of toxicities caused by interference with human cell metabolism. In the mid-1990s, HIV protease inhibitors became available, creating a category of highly active antiretroviral therapy (HAART). The availability of protease inhibitors dropped the fatality rate for AIDS patients by 70%.

HIV-1 protease, the enzyme that HIV needs in order to make new virus, is the best characterized of the virus's proteins, functionally and structurally. It is an aspartyl protease similar to renin in humans in its catalytic mechanism. Therefore, it is not surprising that many drug firms screened their old renin inhibitors (for regulation of blood pressure) to look for HIV protease inhibitors. Other human aspartic proteases include pepsin, gastricsin, and cathepsins D and E.

Saquinavir (trade name Invirase) by Roche was the first HIV protease inhibitor on the U.S. market.[44] Back in 1986, Roche undertook an ambitious international collaboration to tackle the HIV protease. While Roche's U.S. and Switzerland sites worked on the necessary molecular biology, biochemistry, and X-ray crystallography, the U.K. site in Welwyn would carry out the inhibitor design, biochemistry, medicinal chemistry, and in vitro virology. After choosing colorimetric assay as their in vitro assay, the chemistry team in Welwyn, led by Ian B. Duncan and Sally Redshaw, designed some inhibitors using the "transition-state mimic" concept, which was highly successful in producing potent renin inhibitors. They soon achieved an important milestone by defining the smallest peptide mimetic with which they could achieve acceptable levels of inhibition. They found that a tripeptide was ideal considering both potency and bioavailability. As a consequence, tripeptides became a common theme in many protease inhibitors to follow.

The Roche team made a heroic effort in fine-tuning the tripeptide, exploring their lead compound systematically by modifying each amino acid residue in turn. Their hard work paid off in 1991, when the team arrived at Ro 31–8959, which would later become saquinavir. Because saquinavir was a large molecule, the initial synthesis was long and tedious—26 steps, with an overall yield of about 10%. The Roche team made only 10 grams for in vitro studies. With the biological profile becoming more and more favorable, Roche's process group devised a convergent 11-step synthesis with an overall yield of 50%.

With enough active principle ingredient (API) in hand, Roche carried out the clinical trials, led by Keith Bragman, Roche's top European virologist. Saquinavir became the first HIV protease inhibitor for the treatment of AIDS when it was approved by the FDA in December 1995. Although saquinavir, a peptomimetic, was metabolized easily, it was found more beneficial to coadministrate with another protease inhibitor, ritonavir (Norvir) by Abbott. The combination could increase the plasma level of saquinavir significantly because ritonavir inhibited the enzymes that could degrade saquinavir.

Abbott's ritonavir (Norvir) was the second protease inhibitor on the market.[45] The circumstances under which ritonavir was discovered were unique. In the mid-1980s, Abbott hired pharmacologist Ferid Murad from the University of Texas Medical School at Houston as their research chief. Murad, who carried out his postdoctoral training with Edward Scolnick at the NCI and would win the Nobel Prize in Medicine in 1998 (for his work in nitric oxide), brought his academic flair to Abbott's research. For their HIV protease inhibitor program, he encouraged his scientists to write an NIH grant, which enabled them to hire some postdoctoral fellows to supplement their meager resources. The Abbott team was led by an X-ray crystallographer, John Erickson, and a medicinal chemist, Dale Kempf. Whereas it was rumored that Merck had 30 chemists on their HIV protease inhibitor project, Kempf had 3. Instead of screening their renin inhibitors, as most drug firms did at the time, they took advantage of Erickson's X-ray crystallography work on the HIV protease, which would prove to be very beneficial to their drug design. Integrating structure-based drug design and traditional medicinal chemistry, they prepared a series of symmetry-based inhibitors to match the C_2-symmetric nature of the HIV protease. Kempf dubbed his inhibitors *molecular peanut butter,* for they bind to both sides of the enzyme. Using that approach, the researchers arrived at A-77003, a tetrapeptide. Although A-77003 was potent in binding and cellular assays, it was not bioavailable with extremely high human biliary clearance (62 L/h). By reducing the molecular weight and replacing the existing amino acids with more soluble ones, they achieved an increase in bioavailability. They finally hit the jackpot when they discovered that the pyridine termini were oxidized into *N*-oxide by hepatic cytochrome P450. Simply replacing the pyridines with thiazoles and a little fine-tuning gave rise to ritonavir, whose bioavailability was 78% in comparison with 26% for the pyridyl analog.

Abbott collaborated with the NCI in testing the antiviral activities of their protease inhibitors, including ritonavir. The NCI's Samuel Broder and Hiroaki Mitsuya, who took part in identifying AZT, initially tested Abbott's compounds in vitro. After safety evaluation and a battery of other studies, Abbott moved ritonavir into clinical trials and found that it produced a rapid and profound decline in plasma viral RNA in AIDS patients. Interestingly, whereas many other protease inhibitors are metabolized by hepatic cytochrome P450 3A4 (a major isozyme), ritonavir is a potent inhibitor of P450 3A4. As a result, dual protease inhibitor therapy has proven to be a powerful regimen in terms of efficacy and minimizing drug resistance. In March 1996, Abbott's ritonavir (Norvir) won approval by the FDA.

Merck's indinavir (trade name Crixivan) was the third on the market.[46] Merck began its research on protease inhibitor in 1986, with Irving Sigal, a senior director in the Department of Molecular Biology, as the project champion. Following in the footsteps of his father, once a director of research at Eli Lilly, Sigal started at Merck in 1978 after his Ph.D. training at Harvard University. In the mid-1980s, Sigal and colleagues designed several experiments to define the role that HIV-1 protease played in viral replication. His "proof-of-concept" experiments provided the necessary momentum for Merck to commit the resources needed to pursue HIV-1 protease as a viable target for drug discovery.

On December 21, 1988, over the skies of Lockerbie, Scotland, a terrorist bomb blew apart the airliner Pan Am 103, killing everyone on board. Among 259 passengers and crew was 35-year-old Irving Sigal, returning from London 4 days before Christmas.

Losing the project champion was a setback to Merck's protease inhibitor program, but the team carried Sigal's torch and moved the project forward. Merck's medicinal chemistry team was led by Joel Huff. They initially screened their renin inhibitors for HIV protease inhibition and then carried out rational drug design by taking advantage of the known crystal structure of HIV-1 protease. In 1990 Wayne Thompson arrived at L-689,502, which was active in inhibiting the HIV protease but devoid of renin activity. Unfortunately, it was not bioavailable and effective only by injection. By that time, Roche's saquinavir surfaced in literature as a viable oral drug. Inspired by saquinavir's success, Joseph Vacca at Merck successfully incorporated a fragment of saquinavir into L-689,502. Bruce Dorsey, a new hire in 1989 in Vacca's group, and his associate, Rhonda Levin, succeeded in synthesizing L-735,524, which would become indinavir (Crixivan). Although in the monotherapy trials around 40% of the patients were below 400 copies of RNA after 6 months on the drug, HIV developed resistance to indinavir in some patients. Fortunately, it was found that the combination of indinavir and AZT or 3TC was quite effective in substantially suppressing the virus levels. Merck's studies of combination therapy were the first to prove the efficacy of the cocktail approach and became the standard for the industry. After filing with the FDA on January 31, 1996, indinavir received approval on March 13, 1996, in an accelerated review process, just a few weeks after the approval of ritonavir.

Other important protease inhibitors include nelfinavir (Viracept, approved in March 1997) by Agouron (now Pfizer) and amprenavir (Agenerase, approved in April 1999) by Vertex.

With the availability of reverse transcriptase inhibitors and protease inhibitors, AIDS has become a disease that can be controlled using a cocktail approach.[47] The current therapies are not enough to sustain suppression of the HIV over the long term. Patients may also suffer from treatment-limiting side effects or develop resistance to current therapies. Roche's Fuzeon, a fusion inhibitor, opened the door to a new paradigm in AIDS treatment. And many other approaches are racing toward the finish line as well. When the AIDS vaccine finally becomes available, this modern-day scourge will be under control.

CHAPTER 5

Drugs of the Mind

The desire to take medicine is perhaps the greatest feature which distinguishes man from animals.
—Sir William Osler

Alcohol

Alcohol produces a range of central-nervous-system-related biological effects, including anxiety reduction, euphoria, sedation, disinhibition, aggression, blackouts, tolerance, addiction, and withdrawal. The Chinese have used alcoholic drinks since 5000 B.C. Presumably, man ventured to drink the liquid from fermented grain, liked the intoxicating effect, and started to make it on purpose. Alcohol has been used as an anesthetic for millennia (see chapter 7). Alcohol is indispensable in medicine as a solvent. Laudanum, a staple of the medicine chest in the nineteenth century, was simply an alcoholic solution of opium. NyQuil, a cough syrup, and Listerine, an oral antiseptic, all contain copious amounts of ethanol. Alcohol has beneficial effects when consumed in moderate amounts. Research strongly suggests that moderate consumption of alcohol, especially red wine and dark beer, seems to have protective effects on the heart. The hallmarks of the Mediterranean diet are olive oil and red wine, and people from such countries have fewer cardiovascular events. Flavonoids, the active principle in red wine, are thought to exert beneficial cardiovascular effects.

According to the Bible (Genesis 9:20–21), Noah was the first man who discovered wine: "Noah, a man of the soil, was the first to plant a vineyard.

When he drank some of its wine, he became drunk and lay uncovered inside his tent." The New Testament gives an account of Jesus performing his first miracle—turning water into wine.

Despite the beneficial effects of moderate alcohol consumption, excessive use of alcohol damages the brain, heart, and liver. Even mild drunkenness can cause temporary loss of memory. The liver metabolizes alcohol with an enzyme called alcohol dehydrogenase, which turns alcohol into acetaldehyde. Because acetaldehyde is acutely toxic, people—including many Asians—who lack alcohol dehydrogenase cannot tolerate much alcohol. This is the reason that their faces become flush when they drink alcohol and that there are fewer incidents of alcoholism in Asians. Alcoholism is known to cause psychosis and alcoholic dementia. To fight the "demon rum," on January 16, 1919, the U.S. Congress passed the Eighteenth Amendment, prohibiting "the manufacture, sale, or transportation of intoxicating liquors." It was repealed 14 years later, the only amendment to the U.S. Constitution that has been repealed.

One of the earliest drugs used to combat alcoholism was disulfiram (trade name Antabuse), discovered by Erik Jacobsen and his colleague, Jens Hald. In 1949 Jacobsen and Hald, at the Medicinalco Pharmaceutical Company in Copenhagen, Denmark, were studying new vermifuges, a parasite. They discovered that disulfiram, a drug with four sulfur atoms, was very toxic to parasites but not to humans. To confirm the safety of the drug, they both took it themselves, partially to kill the parasites that infested them during their experiments (today's scientists would be dumbfounded about the lack of safety precautions during research at that time). Afterward, both of them went to a party, and, after consuming a few drinks, they experienced bright flushing of the face and neck extending to the chest and arms, ringing ears, a rapid pulse, headaches, giddiness, and drowsiness. All the symptoms were identical to those of drunkenness, which is caused by alcohol accumulation and a lack of alcohol dehydrogenase. Having accidentally discovered disulfiram's countereffect on alcohol, they published their observations in the British journal *Lancet,* and disulfiram has been used to combat alcoholism ever since. Unfortunately, disulfiram triggered aversion to alcohol only by causing users to get sick when they drank. Two additional drugs for treating alcoholism are on the market—one is naltrexone, a morphine analog; the other is acamprostate calcium (trade name Campral). The two drugs have a weak market, with a monthly prescription of 20,000. In terms of mechanism of action, all three antialcoholism drugs—Antabuse, naltrexone, and Campral—work by interfering with the way seven neurotransmitters in the brain interact with cells. Considering that 18 million people in the United States have drinking problems, a more ideal and more effective drug for treating alcoholism is desired to fill this unmet medical need.

Patients with mental disorders have a tendency to use stimulants, such as cocaine, amphetamine, and alcohol, to relieve their depression. Alcohol is the favored choice because it is rapidly absorbed into the bloodstream. Alcoholism is also prevalent among patients with bipolar disorder. But because alcohol is also a depressant, in addition to being a stimulant, the initial blunt of psychic pain is eventually replaced by more intensified depression.

Caffeine

The major sources of caffeine are tea and coffee, although cacao trees, kola nuts, and 60 other plants all contain differing amounts of caffeine.[1,2]

Shén Nóng (meaning "Divine Farmer" in Chinese) was credited with the discovery of tea. Shén Nóng was the emperor when agriculture started in China around 2737 B.C. He invented the plow, taught his people husbandry, and discovered the curative value of plants. On one of his trips to visit his constituency, his servant boiled water for him using the branches of a tea tree to make the fire. Some of the tea leaves fell into the wok, and the emperor found the water refreshing and exhilarating. From then on, he always asked for those particular tea leaves to make his drinks.

One of Shén Nóng's most important contributions was pioneering Chinese herbal medicine. Legend has it that he tasted 100 plants to gauge their medicinal utility. Unavoidably, he ingested poisonous leaves on many occasions. Whenever he did, he would chew tea leaves to expel the poison, and he went on to live more than 100 years. From then on, tea drinking became an integral part of Chinese culture. It was revered almost as a panacea, a cure for headache, indigestion, kidney trouble, and ulcers. It was also viewed as a guard against the noxious gases of the body and lethargy. The art of tea drinking reached its pinnacle in the Táng Dynasty (618 A.D. to 907 A.D.) when Lù Yû, a well-known poet, wrote *Chá Jing* (*The Classics of Tea*) at the request of a group of tea merchants. *Chá Jing* detailed every aspect of tea growing, harvesting, manufacturing, brewing, and drinking. Because the Táng Dynasty boasted the most civilized and richest culture in the world, scholars and merchants from other countries flocked to China to

An ancient tea tree and Lù Yû, the author of Chá Jing © Chinese Postal Bureau

learn agriculture, military, trade, and language. Tea culture rapidly spread through Asia and the Middle East. By the 1600s, tea was being imported to the West in general and England in particular. Nowadays, it is hard to imagine British culture without tea. Most interestingly, tea helped initiate the American Revolution, when King George III imposed upon the 13 colonies exorbitant tariffs on tea and other goods through the 1765 Stamp Act. The revolution began with the Boston Tea Party, when colonists threw tea overboard in protest. The rest is history. In a way, tea was partially responsible for the birth of the United States of America.

According to legend, coffee was discovered by a goatherd. A long, long time ago, Kaldi, a young Ethiopian goatherd, noticed that some of his goats became exceedingly energetic after having chewed the red berries of a hillside shrub. Interest piqued, Kaldi picked some berries and chewed some himself. Soon enough he found his sleepiness and weariness gone and felt refreshed, even exhilarated. The monks in a nearby monastery learned of Kaldi's experience and began to boil the red berries in their drinks so that they could keep awake during long hours of praying. Word started to spread about the magic berries, and coffee became a favorite drink in Africa and the Middle East. In the 1600s and 1700s, coffee drinking was adopted in Europe and has become popular ever since.

Coffee © The United Nations Postal Administration

In 1820, German chemist Friedlieb Ferdinand Runge isolated an alkaloid from the coffee bean, which he dubbed "caffeine," indicating something found in coffee. Caffeine is capable of eliciting many central nervous system effects. Mechanistically, caffeine's major physiological effect is the result of inhibiting the adenosine receptor in the brain. Without caffeine, adenosine acts as a neuromodulator that slows down the movements of neurotransmitters by binding tightly to their receptors, thus inducing sleep. Caffeine competes with adenosine for binding to the receptors of other neurotransmitters and keeps us awake. Despite its universal use, caffeine is rarely overdosed, simply because people become too anxious after consuming too much caffeine. Like all chemicals, caffeine is toxic at high doses. It takes 10 grams of caffeine, which is the equivalent of about that in 100 cups of coffee consumed one after another without interruption, to kill a person. Consumption of too much caffeine has been linked to a higher rate of kidney, bladder, and pancreatic cancer, fibrocystic breast disease, and osteoporosis, although the direct linkage is hard to pin down. Both France and Denmark have banned high caffeine-content drinks such as Red Bull, citing health concerns about the elevated caffeine level.

Sternbach, Valium, and Tranquilizers

The "fight or flight" response was essential for the survival of *Homo sapiens*—moderate anxiety is credited with keeping man alive in the face of danger. Therefore, anxiety may be viewed as our anticipation of danger, real or imaginary. However, too much anxiety may be problematic. In the 1950s, anxiety became a newly defined mental disease, and with it came the emergence of tranquilizers.

Frank M. Berger's Miltown (meprobamate) was the first tranquilizer to gain wide acceptance. During World War II Frank Berger fled Czechoslovakia for London and found employment at the British Drug Houses, Ltd., as a pharmacologist. Berger later moved to America and became the president and head of research at the Wallace Laboratories in New Jersey. In 1945, working with chief chemist William Bradley, he investigated a variety of compounds to find an antibacterial to kill Gram-negative bacteria for which penicillin did not work. When he injected one of the drugs, mephenesin, into mice, he found that those animals became temporarily paralyzed and that their muscles relaxed. This observation led Berger to look into mephenesin's tranquilizing properties. Unfortunately, it had a very short duration of action, and the potency was so low that a large dose was required. Berger then set out to find a drug superior to mephenesin. His persistent search for better tranquilizers led to meprobamate, which was synthesized in 1950 and which was structurally unrelated to mephenesin. Its duration of action was about eight times longer than that of mephenesin. It soon proved to be an excellent minor tranquilizer with a preferable safety profile. Berger christened his meprobamate with the trade name of Miltown, after the New Jersey town that he lived in at the time. The FDA approved its marketing in 1955, and it rapidly achieved outstanding commercial success as an anxiolytic agent. Miltown, touted as the "miracle cure for anxiety," became a cultural icon. People resorted to Miltown as a panacea for all kinds of ailments. The January 1956 issue of *Cosmopolitan* even claimed that with the help of Miltown frigid women who abhorred marital relations reported they respond more readily to their husbands' advances.

In the early 1960s, Leo Henryk Sternbach's Librium and Valium overtook Miltown as the leading tranquilizers.[3] Born in Poland, Sternbach earned a master's degree in pharmacy in 1929 and a Ph.D. in organic chemistry in 1931, both from the University of Krakow. After serving as a research assistant and lecturer for 6 years at his alma mater, he obtained a scholarship to study in Vienna. In April 1937, he attended a seminar given by Leopold Ruzicka (who would win the Nobel Prize 2 years later) at the Eidgenössische Technische Hochschule (ETH; Federal Institute of Technology) in Zürich, Switzerland. That day, Ruzicka's seminar on "Male Sex Hormone" galvanized Sternbach's passion for organic synthesis. He applied and was accepted by Ruzicka to work at ETH in October 1937. Working at ETH proved to be one of the best decisions Sternbach made in his life. Professionally, his experience at ETH enabled him to find employment at Hoffmann–La Roche in Basel in 1940. Personally, he met and fell in love with his landlord's beautiful daughter, Herta Kreuzer, whom he married in 1940. Herta Kreuzer, a Christian and 12 years younger than Sternbach, was very courageous to marry a Jew at the height of anti-Semitism in Europe.

Under the cloud of possible Nazi invasion, Roche moved its headquarters, along with research and development, to Nutley, New Jersey, in the United States. Sternbach moved to Nutley in 1941, and within 2 years he accomplished the challenging task of completing the total synthesis of biotin (Vitamin H). His synthesis was so practical that his synthetic route is still in commercial use today.

In 1955, encouraged by Miltown's success, Roche initiated its own program to search for tranquilizers. When Sternbach was assigned the task of finding new tranquilizers, he, and the scientific field for that matter, had limited knowledge of how the brain worked. His approach was purely empirical—"trial and error." He simply tested compounds that he made in animal models and hoped for the best. At the early stage of the project, Sternbach recalled his experience with benzodiazepines, which he had studied for his Ph.D. thesis on azo dyes and dyestuff intermediates. Although the chemistry was fascinating and yields were very high, benzodiazepines were colorless, thus useless as dyes. Sternbach published a paper in a small Polish journal, which did not draw much attention from the chemical world.

Resuscitating his old chemistry in Nutley, he prepared 24 derivatives of the benzodiazepines. The tests came back showing that none of them had tranquilizing effects. Disappointed at the failure, his boss lost confidence in him and moved him to work on other projects. Sternbach had no choice but to put this project on hold. Therefore, it is not surprising that his general attitude toward his supervisors was "Those who were above me were not my favorites."[4]

A year later, in May 1956, Sternbach's laboratory bench was piling up with compounds, because he did not like to throw anything away. One of his colleagues teased him that it was time for him to get rid of some of his "rubbish." During cleanup, he found two additional benzodiazepine analogs, which had not been submitted for pharmacological testing at the time. He sent them to the pharmacologist for testing, thinking that he had nothing to lose; moreover, he thought he might be able to write a chemically interesting publication on the synthesis of benzodiazepines. The test results came back. One of them had done well in mice, showing strong hypnotic effects. The pharmacologist, Lowell O. Randall, reported on July 27, 1957, "The product has hypnotic, sedative, and strychnine antagonist properties in the mouse identical to that of meprobamate."[5,6] Randall then carried out experiments in additional animal models using cats, dogs, tigers, lions, and monkeys. The drug, which became Librium, did well in all species as a tranquilizing agent. It was found to be superior to meprobamate, chlorpromazine, reserpine, and even phenobarbital, indicating its phenomenal sedative, muscle-relaxant, and anticonvulsant properties. Later, it was shown to be remarkably safe—in fact, there is almost no known drug safer than benzodiazepines. During the clinical trials, Hoffmann-La Roche cleverly involved as many doctors as possible and even encouraged doctors themselves to take Librium to dispel their reluctance to acknowledge anxiety as a disease. After receiving FDA approval in 1960, Roche marketed the drug under the trade name Librium, which means "to be free" in Latin. Roche tried Librium on leopards, tigers, and lions and captured the dramatic tranquilizing effects in films to show them to doctors. Three years later, Valium, meaning "to be well and strong" in Latin, was another benzodiazepine discovered by Sternbach and marketed by Roche in 1963. It was five times more powerful than Librium. Valium was also very safe, and a running joke was that the only way of getting killed by Valium was to be run over by a truck full of it. However, that is not to minimize the abuse liability of the benzodiazepines, which is their main drawback. Additional safety concerns also exist, at least when given chronically.

Leo and Herta Sternbach have two sons. The older son, Michael, worked for Roche as a salesperson. The younger one, Daniel, followed in his father's footsteps by studying at ETH and is now a medicinal chemist at GlaxoSmithKline in North Carolina. Daniel, a very productive medicinal chemist, holds 2 dozen patents, whereas his father Leo had 240 patents. At one point, Leo Sternbach alone was responsible for 20% of all Roche's patents. He and Roche advanced eight benzodiazepines to the market, which brought billions in sales for Roche. Sternbach's reward was a mere $10,000, and he donated most of it. In general, Sternbach's passion was working in the laboratory and discovering drugs. He never took part in infighting, backstabbing, or office politics. So confident was he in his ability as a chemist that he often categorized his opponents as ignoramuses, even idiots. Sternbach was well liked, and there was no jealousy among his colleagues over his success. He loved meat with fat and enjoyed playing Santa at Christmas parties and being hugged by his female admirers. His hobbies were playing bridge and the stock market.

Nothing has replaced benzodiazepines for their efficacy in ameliorating anxiety and related disorders. In 1964, *Newsweek* called the 1960s the decade of tranquilizers, with Valium and Librium as the most prescribed drugs. In 1976, Valium accounted for 1 out of 20 prescriptions in the United Kingdom. A professor in the United Kingdom called Valium the "opium of the masses,"[4] analogous to Karl Marx's famous declaration, "Religion is the opium of the masses." The rock band the Rolling Stones recorded a song titled "Mother's Little Helper," in which explicit reference is made to "a little yellow pill" that is taken to calm one down and help one get through a busy day.

The "little yellow pill" was suspected to be Valium. Statistics showed that Valium was indeed taken by more women than men. It was suggested that it was the result of the male-dominated medical profession controlling and suppressing women.

Benzodiazepines, as a class of drugs, possess a wide range of pharmacological effects. They are the drugs of choice for insomnia, although Ambien, a GABA A receptor agonist, and Valium are far more likely than other tranquilizers to cause side effects such as hallucination and agitation. The latter is more common in children than in adults (this is discussed further in the next section). Benzodiazepines are used as anticonvulsants and muscle relaxants and in managing alcoholism and heroin addiction. In the 1960s, nobody knew how benzodiazepines worked, although now it is believed to be through facilitation of synaptic transmission at the gamma-aminobutyric acid (GABA) receptor. The benzodiazepine-GABA receptor complex has since been isolated and characterized.

Antidepressants

Depression was romanticized at times due to its association with poets and artists, but in reality depression can be devastating. According to the *Diagnostic and Statistical Manual of Mental Disorders* (DSM-IV),[7] there are two types of depression: major depressive disorder and bipolar, or manic-depressive, illness. Both disorders are characterized by severe changes in mood as the primary clinical manifestation. Major depression is demonstrated by feelings of intense sadness and despair, with little drive for socialization or communication. Physical changes, such as insomnia, anorexia, and sexual dysfunction, can also occur. Mania is manifested by excessive elation, irritability, insomnia, hyperactivity, and impaired judgment. It may afflict as much as 1% of the population. Major depressive disorder is among the most common psychiatric disorders, with an estimated 12-month prevalence of approximately 10% in the general population and a prevalence of 12.9% and 7.7% in women and men, respectively. About 19 million Americans suffer from depression per year. In terms of disease burden, as measured by disability adjusted life years (DALYS), major depressive disorder ranks as the fourth most costly illness in the world, with estimated annual costs of depression in the United States amounting to approximately $43.7 billion.

Rauwolfia serpentina is a plant indigenous to India and the Indian subcontinent. The genus name, *Rauwolfia,* was chosen in honor of Leohard

Rauwolf, a sixteenth-century German botanist, physician, and explorer. The species name *serpentina* refers to the long, tapering snakelike roots of the shrub. For the same reason, *Rauwolfia serpentina* is also known as snakeroot. The Indians drink tea made from the snakeroot for its tranquilizing effect. It was Mahatma Gandhi's favorite drink when he meditated. In India, the *Rauwolfia* has also been used as a remedy to calm crying babies. Reserpine, an alkaloid, was isolated from *Rauwolfia serpentina* in 1952 by E. Schlittler, who disclosed the chemical structure of reserpine as an indole alkaloid a year later.[8–10]

In 1954, Nathan S. Kline of Rockland State Hospital in Orangeburg, New York, reported that reserpine was a helpful treatment for psychotic patients, showing an especially favorable effect on mania. Among 400 patients with mental retardation who were treated by Kline, the incidence of violent accidents, restraints, and seclusions were reduced after the treatment with *Rauwolfia,* or reserpine. Around that time, Kline also observed that reserpine induced a depressive state in normal patients by depleting neurotransmitters such as norepinephrine and serotonin. His observations led to the hypothesis that the biological basis of major mood disorders may include abnormal monoamine neurotransmission, which may be mediated by substances such as norepinephrine, serotonin, and dopamine. These substances are released from presynaptic neurons, cross the synaptic gap, and interact with receptors on the postsynaptic cells. The synthesis, transmission, and processing of these neurotransmitters provide a number of points of intervention for a pharmacological agent; manipulation of neurotransmission has been the mainstay of antidepressant therapy for over half a century. Currently, reserpine-induced depression has become a classic animal model for depression.

One of nature's remedies for depression was opium, which appeared to have a specific effect on depression, admittedly only a symptomatic one. On the other hand, an herbal supplement, St. John's wort (*Hypericium perforatum*), has been widely used in treating minor and mild depression since the early 1990s. The effectiveness of St. John's wort for *major* depression, however, has been a contentious issue due to conflicting data. According to the National Institute of Mental Health (NIMH) in 2004, St. John's wort is no more effective for treating major depression of moderate severity than placebo. In contrast, on February 10, 2005, an article in the *British Medical Journal* reported that extract of St. John's wort was at least as effective as paroxetine (Paxil, an SSRI) for the treatment of moderate to severe depression, while being better tolerated.

Cade and Lithium

Manic depression, also known as bipolar disorder, is a debilitating mental illness first identified by German psychiatrist Emil Kraepelin in the early 1900s. It has been an intriguing disease, with its supposed link to artistic creativity and superior leadership, which are frequently displayed in the disinhibited active, expansive, manic phase. History is replete with examples of well-known people with manic depression, including composers Robert Schumann, George Handel, and Hector Berlioz; British leader Oliver Cromwell; painter Vincent van Gogh; and poet and novelist Virginia Woolf.

Currently, about 20 million Americans are afflicted with manic-depressive illness. Lithium, an effective treatment for manic depression, is actually a lithium *salt,* such as carbonate, citrate, or acetate, not lithium *metal,* which was discovered by Swedish chemist/mineralogist Johann August Arfvedson (apprentice to the great chemist Johan Jakob Berzelius) in 1817. A. B. Garrod pioneered the use of lithium salt as a medicine in 1859 for the treatment of gout. The discovery of lithium for the treatment of manic-depressive disorder by John F. Cade was another textbook example of serendipitous discovery.[10–13]

Cade was born in 1912 in Australia. After his medical training, he became a house officer at St. Vincent's Hospital and later at the Royal Children's Hospital in Melbourne. There he developed severe pneumonia and was fortunate to survive thanks partially to a group of special nurses. One of them was Jean Charles, whom he married in 1937. In 1941, Cade left his wife and two young sons to serve in Singapore during World War II as an army medical officer. In 1944, he was captured by the Japanese and became a prisoner of war in Changi Camp. On his return from the war, he was a walking skeleton of 80 pounds. He returned to medical practice at the Bundoora Repatriation Mental Hospital, a small government mental hospital in Melbourne.[13]

In 1948, Cade embarked on a study of the cause of manic-depressive disorder and discovered that urine samples of schizophrenics and melancholics killed guinea pigs. Cade injected a few chemicals that he had in his possession in an effort to locate the offending toxin. He found that urea, the most abundant constituent in urine, was indeed lethal to guinea pigs. However, all his mental patients seemed to have normal urea levels, thus ruling out urea as the culprit for causing manic-depression. Cade then developed a theory (incorrect) that the mania associated with manic-depressive illness might be caused by the abnormal metabolism of uric acid, another abundant chemical present in all carnivores' urine. To test his hypothesis, he tried to inject uric acid into guinea pigs and ran into a problem: uric acid is highly insoluble in water. To overcome the difficulty, Cade decided to use the most soluble lithium salt, lithium urate, for injection into guinea pigs. He was surprised to notice that lithium urate seemed to have a profound sedative effect on the guinea pigs, which became lethargic after 2 hours during which, at first, no effects were observed. This was a lucky occurrence, for the sedative effect took place rather quickly for guinea pigs, whereas it generally would take 10 days for its benefits to manifest in humans. Subsequently, he injected guinea pigs with lithium carbonate and observed the same outcome. In a brilliant flash of insight, Cade predicted (correctly) that lithium might have some use in treating mania. To his wife's dismay, he bravely swallowed some lithium carbonate himself and saw no harm. In 1970 Cade justified his audacious move: "As lithium salts had been in use in medical practice since the middle of the nineteenth century, albeit in a haphazard way with negligible therapeutic results, there seemed no ethical contraindications to using them in mania, especially as single and repeated doses of lithium citrate and lithium carbonate in the dose contemplated produced no discernable ill effects on the investigator."[10] He then administered lithium carbonate to 10 of his manic patients and was rewarded with an astonishing triumph: all the patients were cured, and some even went back to work.

Cade published his pioneering findings in the *Medical Journal of Australia* in 1949 with the title "Lithium Salts in the Treatment of Psychotic Excitement."[11] Very few people took notice except a few Australian psychiatrists—an article published in an obscure Australian journal by an unknown government medical officer hardly warranted a headline in the *New York Times*. Cade himself later said self-effacingly that a discovery "made by a [then] unknown psychiatrist with no research training, working in a small chronic hospital with primitive techniques and negligible equipments, was not likely to command attention."[13] To make matters worse, excessive use of lithium chloride as a salt substitute for hypertension patients had killed several people in the United States. Doctors were simply "gun-shy" in prescribing lithium. More than 20 years would elapse until Mogens Schou reintroduced Cade's discovery to the world. Starting in 1955, Schou, a young Danish doctor, would talk to anybody who would listen about the therapeutic power of lithium. He kept presenting his results on lithium at every conference, meeting, and congress he attended.

Even when he presented results on other drugs, he would use lithium as a comparison, so he could promote the drug. His insistence surprised and even annoyed many psychiatrists, who stopped inviting him for presentations altogether. But Schou would bring up lithium during discussion sessions. In the end, his perseverance paid off. Initially, psychiatrists in Denmark began to take notice, and later those in Europe and America caught on. The publications grew exponentially on the therapeutic benefits of lithium for treating mania. Schou's validation and extension of Cade's original observations resulted in wide acceptance of lithium as the gold standard for treating manic patients; it is now the first drug prescribed after a diagnosis of bipolar disease. As William Osler pointed out, "In science the credit goes to the man who convinced the world, not the man to whom the idea first occurs."[14] Therefore, Schou's contribution was just as significant as Cade's with regard to lithium. After lithium gained wide acceptance, Schou was invited to every conference related to psychiatry, partly as an apology from his peers for their long indifference toward his sage words.

Lithium, a simple inorganic ion, has a pronounced psychotropic effect, although its mechanism of action is still unknown. It may reverse a major psychotic reaction by boosting the level of serotonin. In America, several states have regions with higher than normal levels of lithium in their drinking water. These communities have lower than expected rates of psychiatric admissions, suicides, homicides, and arrests relating to drug addiction. On the other hand, regions with lower lithium levels in water seem to have higher mental illness and crime rates.

Lithium as an antimanic is not perfect. Many side effects, including cardiac problems and weight gain, limit its widespread use. Furthermore, it seems to dampen the edge of artists, writers, and musicians, whose creativity seems to be closely associated with mania. Nonetheless, as the gold standard for treating mania, lithium has literally saved hundreds of thousands of lives. A true wonder drug, its power restores normalcy for 60–70% of manic-depressive patients.

For his landmark discovery of lithium in the treatment of manic depression, John Cade joined the ranks of many illustrious Australian scientists, such as Howard W. Florey of penicillin fame and Arthur J. Birch, whose name was immortalized through the "Birch reduction" in organic chemistry. Sadly, Cade became addicted to tobacco during his days in the prisoner-of-war camps. Despite the urging of his family, he was never able to kick the habit and eventually died of lung cancer at the age of 68 in

1980. An Australian film on his discovery of lithium was released in 2004 with the title *Troubled Mind—The Lithium Revolution.*

Kline, Iproniazid, and Monoamine Oxide Antidepressants

Iproniazid was initially prepared as an antituberculosis drug, and mood lifting was observed as a side effect, which in turn catapulted its use as an antidepressant.[10,15,16] Hoffmann-La Roche discovered isoniazid in 1951 for the treatment of tuberculosis. It has been credited with dramatically reducing the incidence of tuberculosis in the United States. In 1952, riding on the success of isoniazid, Herbert H. Fox and John T. Gibas at Hoffmann-La Roche prepared many derivatives of isoniazid. Iproniazid, the isopropyl derivative, was found to be equally as potent as isoniazid in animals but more potent in humans. Hoffmann-La Roche received approval from the FDA and marketed iproniazid for tuberculosis with the trade name Marsilid. Evert Svenson, the assistant medical director at Hoffmann-La Roche, had suggested to his superiors that Marsilid might have some use in depressed patients, but his proposal was met with laughter. Indeed, Svenson's view was so unorthodox that he initially had difficulties convincing his employer to supply clinicians enough Marsilid to conduct clinical trials for depression.[15,17]

During the clinical trials of Marsilid, central nervous system stimulation was also listed among the side effects. Independently, three groups simultaneously observed improvement of mood in chronically depressed, hospitalized patients also suffering from tuberculosis. Nathan S. Kline of Rockland State Hospital in Orangeburg, New York, who led one of the three groups, reported that his patients were "energized"[10] and that some even experienced "euphoria." Ironically, a fairly unique situation ensued, as recounted by Kline: "A group of clinical investigators were trying to convince a pharmaceutical house that they had a valuable product rather than the other way around."[10] Coincidentally, all three groups presented their respective results at a regional meeting of the American Psychiatric Association held in Syracuse on April 6, 1957. After that meeting, an article in the *New York Times* touted Marsilid's miraculous mood-elevation effects. Almost overnight, Marsilid found widespread use in treating depression as an off-label prescription (meaning that physicians may prescribe drugs that are on the market for indications other than the ones approved by the FDA). During the following year, about half a million patients were treated, thanks to an agreement between Hoffmann-La Roche and the FDA. Probably no drug in history was so widely used so soon after announcement of its application in the treatment of a specific disease. This circumstance arose not only because of overwhelming need for an effective antidepressant medication but also because the drug was already on the market, albeit for the treatment of tuberculosis.

However, some patients who took Marsilid developed jaundice due to liver toxicity. In all, 127 cases of hepatitis were reported. Statistically, there would normally have been about 100 cases of hepatitis even without Marsilid. Hoffmann-La Roche voluntarily withdrew Marsilid from the U.S. market in 1961 with the conviction that they would quickly find a safer drug, which never materialized. In retrospect, it might have been too hasty to take it off the market prematurely, robbing millions of an excellent antidepressant with limited side effects for a small number of patients. According to one of the pioneers in developing Marsilid, David M. Bosworth of St. Luke's Hospital and Polyclinic Hospital in New York, ". . . injudicious, undesired, and disastrous publicity . . ."[15] contributed to the removal of Marsilid from the medical armory. Interestingly, four patients still took Marsilid legally after its removal. One of them was Kline's patient, who did extremely well on the medication—even giving birth to her second child, having been on the drug all through pregnancy up to the time of delivery. Kline recounted: "Just at the time of the birth, the drug was withdrawn from the market and she was placed on an identical-appearing placebo. She promptly went downhill and the substitution of other antidepressant drugs was totally inadequate. Finally, in desperation, she wrote to the FDA with the before-and-after photographs."[10] The FDA granted Kline an exemption to reinstitute the treatment in her case. She immediately responded and remained on the drug for many years. From time to time, Hoffmann-La Roche produced a new batch of Marsilid, in part to keep Kline's one patient happy. In acknowledging Nathan Kline's numerous important contributions in psychiatry, the place at which he used to work in Orangeburg, New York, is now named the Nathan S. Kline Institute for Psychiatric Research.

In 1951, E. Albert Zeller at Northwestern University found that Marsilid inhibited monoamine oxidase (and at least 10 additional enzymes). The level of serotonin in the brain is closely associated with mood, and monoamine oxidase (MAO) is an enzyme that converts serotonin to 5-hydroxyindole acetic acid, lowering the level of serotonin. As a monoamine oxidase inhibitor, Marsilid is able to inhibit the oxidation of norepinephrine and serotonin in the central nervous system and lift the mood of depressed patients. Marsilid was the first of the monoamine oxidase inhibitors for antidepressant therapy. Obviously other factors are at work as well, because some of the most potent monoamine oxidase inhibitors in the laboratory have only weak or absent antidepressant activity in vivo. The discovery of monoamine oxidase (MAO) inhibitors and their use in the treatment of depressed patients was a major milestone in modern psychiatry. They have now been largely replaced by tricyclic antidepressants and selective serotonin reuptake inhibitors (SSRIs).

One of the side effects of MAO inhibitors is hypertension, also secondary to catecholamine excess, usually in association with the consumption of other catecholes or catechole precursors such as tyramine (the so-called cheese effect). In an effort to develop safer MAO inhibitors—at least from the hypertensive liability—two subtypes of the MAO receptor have been identified, MAO-A and MAO-B. Coming full circle starting with isoniazid and iproniazid, recent selective, reversible inhibitors of MAO-B have been developed—as antibiotics! The first is linezolid (Zyvox), a ketolide antibiotic, the first in a new class of antibiotics approved by the FDA in about 40 years (see chapter 2).

Imipramine and Tricyclic Antidepressants

In 1950, Robert Doenjoz, the director of pharmacological research at the drug firm J. R. Geigy in Basel, Switzerland, asked Roland Kuhn of the Cantonal Psychiatric Clinic in Munsterlingen, Switzerland, to test some of their antihistamines.[10] The initial intent was to learn whether they possessed hypnotic properties, which none of them did. Five years later, the tranquilizing properties of chlorpromazine (see later in this chapter) became known and heralded the first highly effective antischizophrenic drug. Its discovery ignited a flurry of research into the search for close analogs that could be used as treatment for a variety of psychiatric illnesses. Kuhn remembered that some of Geigy's antihistamines he tested produced effects similar to those of chlorpromazine. He then wrote several letters to Geigy and suggested that further studies of those antihistamines for central nervous system diseases were warranted. Geigy, initially hesitant, was

chlorpromazine (Thorazine) imipramine (Tofranil)

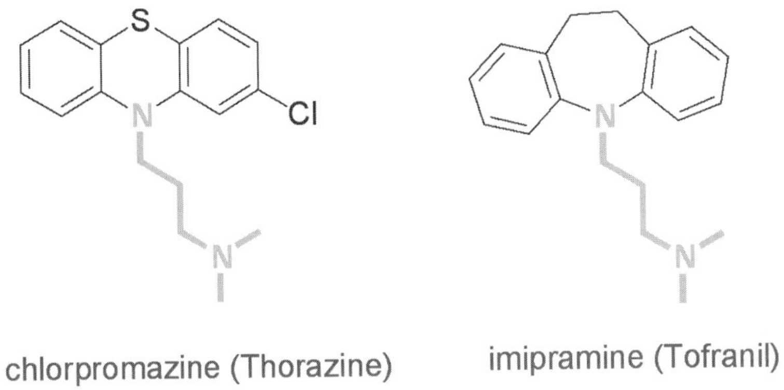

Imipramine and chlorpromazine have the same side chain

finally convinced and started a program to look for antischizophrenic drugs similar to chlorpromazine in collaboration with Kuhn.

The sulfur atom in the chlorpromazine molecule was perceived to be a liability because of its easy oxidation. Therefore, many replacements (called *isosteres* in medicinal chemistry terms) for the sulfur were synthesized. Because a sulfur atom is large, it took two carbon atoms to fill its space. Because the diamine side chain was important for the pharmacological activity (the side chain is known as the *pharmacophore* in medicinal chemistry terms), it was conserved in the molecule. The resultant drug, G22355, later known as imipramine, was synthesized and sent to Kuhn for testing. Unfortunately, after having administered imipramine to about 300 patients with schizophrenia, Kuhn was not able to detect much improvement.

At that juncture, most people would have given up. But Roland Kuhn was not "most people." He had already accumulated a tremendous amount of knowledge about imipramine from his careful observations. In his own words, "I examined each patient individually everyday, often on several occasions, and questioned him or her again and again. Many of the patients were also under the observation of my assistants and nursing staff and I always regarded their proposals and criticisms seriously."[10] He insisted that they do a thorough job with the drug. On his nudging, the clinical trials then moved to testing the effects of imipramine on severe depression, known as "endogenous depression." He later reported, "After treating our first three cases, it was already clear to us that the substance G22355, later known as imipramine, had an antidepressant action."[10] In 1958, Kuhn published his results in the *American Journal of Psychiatry* as an article titled "The Treatment of Depressive State with G22355 (imipramine HCl)," which ushered in the use of tricyclic antidepressants. These compounds are referred to as "tricyclic" because they all had three fused rings with a ubiquitous diamine side chain. Because the first effective antidepressant, Marsilid, was promptly withdrawn from the market, many consider imipramine the first antidepressant. It is ironic that Geigy was initially reluctant to develop imipramine because of their (wrong) perception that the market for antidepression drugs was too small.

Tricyclic antidepressants dominated the antidepression market for two decades. Strikingly, imipramine differed from chlorpromazine, an antischizophrenic drug, by only two atoms. Even though it is effective in the management of depression, it does have significant side effects and toxicities (flushing, sweating, orthostatic hypotension, constipation) due to its α-adrenergic blocking activity. The introduction of imipramine as an antidepressive agent by Geigy in the spring of 1958 brought disappointment in that these side effects were more troublesome than original studies had suggested. All the tricyclic antidepressants are especially toxic when overdosed, producing cardiac effects and seizures. These unwanted side effects limit compliance, with as few as 1 in 17 patients completing a therapeutic-dosing regimen.

In general, the mechanism of action of tricyclic antidepressants is inhibition of reuptake of the biogenic amines. When a neurotransmitter is released from a cell, it has only a short period of time to relay its signal before it is metabolized by monoamine oxidase and reabsorbed into the cell. All of the tricyclic antidepressants potentiate the actions of norepinephrine, serotonin, and, to a lesser extent, dopamine. However, the potency and selectivity for inhibition of the uptake of these amines vary greatly among the agents. Imipramine works by inhibiting noradrenaline reuptake at the adrenergic endings, as well as serotonin to a lesser extent. In 1970, Roland Kuhn, along with Frank Berger and John Cade, were given the Taylor Manor Hospital Psychiatric Award.

Prozac and Selective Serotonin Reuptake Inhibitors

The search for less toxic neurotransmitter reuptake inhibitors led to the development of second-generation antidepressants known as the selective serotonin reuptake inhibitors (SSRIs).[18–21] These agents differ from the older tricyclic antidepressants in that they selectively inhibit the reuptake (removal) of the neurotransmitter serotonin into the presynaptic nerve terminals, enhance synaptic concentrations of serotonin, and facilitate serotonergic transmission. Initially, AB Astra discovered and marketed an inhibitor for serotonin reuptake, Zimeldine, the prototype of the SSRIs. Unfortunately, a rare but serious side effect, Guillain-Barré syndrome, started to surface after Zimeldine was approved and administered in a large patient base. AB Astra pulled it off the market in the early 1980s.

The discovery of Prozac revolutionized the management of depression. Because it always takes a chemist to first make a drug, I shall begin the story of Prozac with a chemist, Bryan Molloy. In the 1960s, Molloy was a medicinal chemist working at the Eli Lilly Co. in Indianapolis. Born in 1939 in Broughty Ferry, Scotland, Molloy earned his B.S. and Ph.D. degrees in chemistry from the University of St. Andrews in Scotland in 1960 and 1963, respectively. Postdoctoral fellowships then brought him to the United States, where he carried out research at Columbia and Stanford Universities before he was employed by Eli Lilly as a senior organic chemist in 1966. At Lilly, Molloy's initial assignment was in the cardiac therapeutic area, where he looked into acetylcholine as a regulator of heart action. In 1969, pharmacologist Ray Fuller lured Molloy to the antidepressant project that he was running. Fuller wanted to take advantage of Molloy's experience with acetylcholine—the side effects of tricyclic antidepressants arose because they modulate both serotonin and acetylcholine, among others. Antidepressants without acetylcholine modulation would provide cleaner drugs with fewer side effects.

Sir James Black (see chapter 2) once stated that the most fruitful basis for the discovery of a new drug was to start from an old one. Prozac's journey began with Benadryl (diphenhydramine, a Parke-Davis over-the-counter drug), an old antihistamine well known for managing stuffy noses and allergies. Molloy began to tweak the molecule by replacing the original substituents on Benadryl with different functional groups. Through a process called structure-activity relationship (SAR) studies, he found out that the *N*-methyl ethylamine moiety was necessary to maintain the pharmacological activity (pharmacophore), so he kept that side chain constant. In time, he manipulated the diphenhydramine structure of Benadryl into the phenoxypropylamine series from which fluoxetine hydrochloride (Prozac) would later emerge. In 1970, Molloy made dozens of phenoxypropylamines, some of which blocked the effect of apomorphine-induced hypothermia (decrease of body temperature) in mice tested by his biological

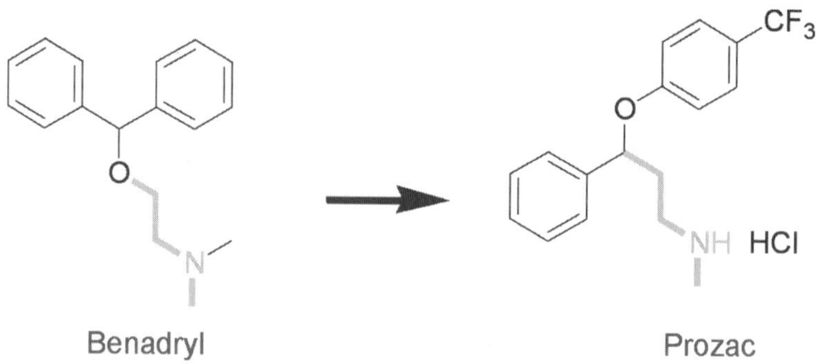

Benadryl

Prozac

The evolution from Benadryl to Prozac

collaborator, Robert Rathburn. Apomorphine, a morphine derivative, is a dopamine receptor inhibitor. The animal model using apomorphine-treated mice was thought to reflect the effect of acetylcholine.

In 1971, Molloy ran into David Wong, another important player in discovering Prozac. Wong, a Chinese immigrant, had joined Lilly 3 years earlier following his Ph.D. training in biochemistry at the University of Oregon. At the time, he was growing disenchanted with his research on antibiotics. By the early 1970s, the heyday of antibiotics was pretty much over. A couple of drugs that Wong was involved with did not go anywhere. In 1971, he began to shift his focus to neurochemistry. One day Wong and Molloy attended a seminar on neurotransmission given by Solomon H. Snyder from Johns Hopkins University (the author of a popular book titled *Drugs and the Brain*).[21] Snyder developed a "binding and grinding" method that enabled isolation of biogenic amines from rat brains. In essence, separating the ground-up products from rat brains would provide some fractions containing nerve endings that would still function chemically. With Snyder's consultation, Molloy and Wong began their quest to look for a more selective, and thus safer, antidepressant.

Using Snyder's "binding and grinding" method, Wong initially tested compounds that Molloy deemed most promising—the ones that obliterated the function of acetylcholine according to the apomorphine-treated mouse model. Unfortunately, they all turned out to be duds. They blocked the reuptake of norepinephrine but not serotonin. Wong's persistence paid off after he started testing the remainder of Molloy's compounds that had failed in the apomorphine-treated mouse model. One of them (fluoxetine, later to become Prozac) selectively blocked the removal of serotonin while sparing most other biogenic amines. Unlike some of the earlier compounds, it was relatively inactive in the reserpine-induced hypothermia test in mice. In clinical trials, Prozac was found to have a favorable side-effect profile and was much safer in overdose relative to the tricyclics. In 1988 the FDA approved Prozac, which rapidly revolutionized the treatment of depression thanks largely to its safety profile. Prozac transformed debilitating depression into a manageable disease for many patients. In fact, unlike the previous relatively toxic tricyclic antidepressants, which were prescribed primarily by psychiatrists, the much safer Prozac is frequently prescribed by nonpsychiatrists and general practitioners, taking the field of psychiatry more into the open. In 2000, it was the most widely prescribed antidepressant drug in the United States, with worldwide sales of $2.58 billion (which dropped significantly after its patent expired). However, Prozac, and all SSRIs, are generally *not* more efficacious than the tricyclic antidepressants and exhibit a marked delay in onset of action. The delay in onset of action is the reason that it takes 2–4 weeks for Prozac and other SSRIs to take effect. In addition, SSRIs also have their own set of side effects that result from the nonselective stimulation of serotonergic receptor sites.

Additional widely known SSRIs are sertraline hydrochloride (Zoloft; Pfizer) and paroxetine hydrochloride (Paxil; GlaxoSmithKline). Zoloft has been available in the United States since 1992 and had worldwide sales of $3.36 billion in 2004. Compared with Prozac, it has a shorter duration of action and fewer central-nervous-system-activating side effects such as nervousness and anxiety. Paxil generated worldwide sales of $3.08 billion in 2003, and its relatively benign side-effect profile favors its use in elderly patients.

It is well known that children and adults do not always respond to medications in the same way. The brains of children and adolescents are quite different from those of grown-ups. SSRIs that have demonstrated efficacy in treating adult depression have peculiar effects on children. In the early 2000s, data started to emerge that suggested that SSRIs may increase the incidents of suicide in children under 18. It underscored the intricacy of human brains and how modulation of serotonin levels could have other effects in different developmental stages. In June 2003, British authorities announced that Paxil should not be used to treat depression in anyone under 18. Shortly after, the FDA followed suit and recommended that pediatric use of Paxil in treating depression be avoided. For juvenile depression, Lilly's Prozac has the best data and evidence of benefit. In September 2004, the FDA required SSRI antidepressants to carry warnings about the risk of increased suicidal tendencies in young people. Currently, only Prozac is approved for adolescents 9–17 years old.

Due to our genetic disparities, each individual responds differently to different types of antidepressants. Many second- and third-generation antidepressants have been discovered and provide a wide variety of choice in managing depression. Behaving similarly to SSRIs, Effexor is one of the serotonin and norepinephrine reuptake inhibitors (SNRIs), whereas Wellbutrin is one of the norepinephrine and dopamine reuptake inhibitors (NDRIs), which are also used to help smokers quit. Effexor was Wyeth's largest selling drug, with over $3 billion in sales in 2004.

Antipsychotics

In ancient times, an insane person was often thought to be possessed by the devil or being punished by God for his sins. As a consequence, beating, bleeding, starvation, hot- and cold-water shock treatment, and incarceration were widely practiced on mental patients, which only worsened their conditions. In the eighteenth century, the Enlightenment heralded

Philippe Pinel unchained the insane © Warner-Lambert

the birth of psychiatry, among many other things. During the French Revolution, Philippe Pinel, a warden in an insane asylum in Paris, advocated unchaining the mental patients. He argued, "It is my conviction that these mentally ill are intractable only because they are deprived of fresh air and of their liberty."[22] He persisted in replacing cruelty and inhumanity with kindness, understanding, and rational therapy. His humanitarian and philanthropic conviction led to the cure and release of many mental patients and had an enduring impact throughout Europe. Furthermore, Pinel also carried out a systematic investigation and documentation of mental diseases. He is now considered the "father of psychiatry."

Although a certain stigma remains attached to mental illnesses, we have now amassed a tremendous amount of knowledge with regard to the impact of genetic, biochemical, and environmental factors on the human brain. Psychopharmacological drugs have significantly contributed to managing and understanding mental disease.

Schizophrenia is a mental condition associated with disordered thinking that is characterized by both *positive* symptoms, such as delusions, hallucinations, and disorganized speech and behavior, and *negative* symptoms, including apathy, withdrawal, lack of pleasure, and impaired attention. Other symptom dimensions include depressive/anxious symptoms and aggressive symptoms such as hostility, verbal and physical abusiveness, and impulsiveness. Older mental drugs included opiates, belladonna derivatives, bromides, barbiturates, antihistamines, and chloral hydrates. Before chlorpromazine became available in 1962, early treatments of schizophrenia included prolonged narcosis, known as "narcosis for psychosis." Meanwhile, history saw the emergence of excruciating treatments such as electroconvulsive therapy and shock introduced by fever, methiazole, and insulin. Austrian neurologist Julius Wagner von Jauregg invented the fever-shock treatment by introducing malaria in patients with psychosis and won the 1927 Nobel Prize in Medicine. Additionally, Egaz Moniz received the 1936 Nobel Prize in Medicine for his invention of lobotomy to introduce an organic syndrome for the treatment of schizophrenia.

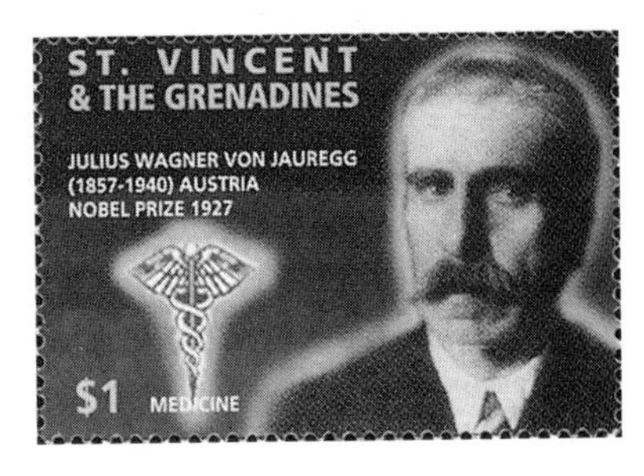

Julius Wagner von Juaregg © St. Vincent General Post Office

Chlorpromazine

The genesis of chlorpromazine is similar to that of imipramine and can be traced back to antihistamines, first discovered by Daniel Bovet at the Institut Pasteur de Paris in 1937. Unfortunately, Bovet's antihistamine proved to be too toxic.[10,23,24] In 1944, a group of scientists at Rhone-Poulenc Laboratories, led by chemist Paul Charpentier, began a program of systematically searching for safer antihistamines. Their starting point was older antihistamines: diphenhydramines in general and Benadryl in particular. In time, they successfully synthesized and marketed an antihistamine, promethazine. The molecule was an interesting hybrid consisting of phenothiazine, a moiety related to an antiparkinsonian agent, and a diamine side chain associated with the antihistamines. Similar to most antihistamines, promethazine had side effects in the central nervous system, which were mild antipsychotic properties. A surgeon in the French Navy, Henri Laborit, was looking for a compound with more central effects in his quest for a drug to treat surgical shock. Laborit found that promethazine was superior to other drugs, but its antishock effects were not pronounced enough. Intrigued by Laborit's proposition, Charpentier sought to enhance promethazine's "side effects" in the central nervous system. Structural activity relationship (SAR) investigations led to the synthesis of RP-3277 (chlorpromazine) in 1950. The structure of chlorpromazine differed only slightly from that of promethazine. Chlorpromazine had an extra chlorine atom and a slight difference in the diamine side chain.

By the end of 1950 a sample of chlorpromazine was sent to Simone Courvoisier, the head of pharmacology at Rhone-Poulenc, for testing. She noted that rats dosed with chlorpromazine became "indifferent": rats conditioned to climb a rope at the sound of a bell ignored the bell. In addition

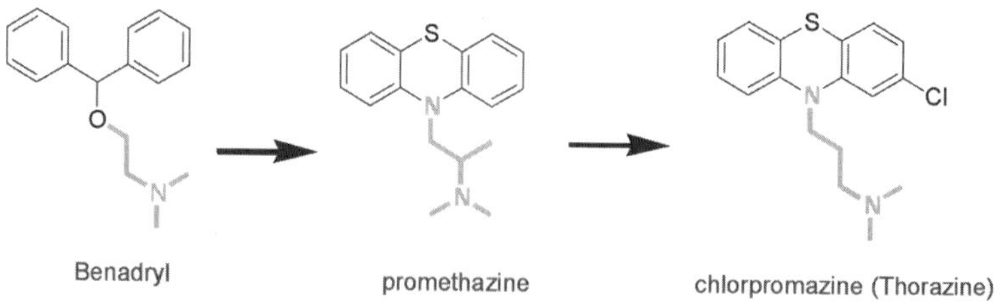

Chemical diagrams of Benadryl, promethazine, and chlorpromazine

to its outstanding "calming" activities, Courvoisier also determined that chlorpromazine had low toxicity. Further clinical trials were then carried out under the direction of Jean Delay and Pierre Deniker, two psychiatrists at the L'Hôpital Sainte-Anne de Paris. Under the influence of chlorpromazine, their patients became "disinterested" as well. More important, chlorpromazine subdued the hallucinations and delusions of psychotic patients. Chlorpromazine was introduced in December 1952 in France under the trademark Largactil. Largactil, meaning "large activity," was chosen to reflect a wide range of central nervous system activities that chlorpromazine elicited. It was the first conventional antipsychotic discovered that was superior to opium. With Smith, Kline, and French Pharmaceuticals as their comarketing partner, Rhone-Poulenc introduced chlorpromazine in the United States in 1954 under the trademark Thorazine. In the first 8 months, more than 2 million patients were administered the drug. It contributed to an 80% reduction of the resident population in mental hospitals. When mathematician John Nash (of the movie *A Beautiful Mind*), who suffered violent delusions, was diagnosed with schizophrenia, he was treated with Thorazine, which sedated him easily. Thorazine added a great impetus to the beginning of the psychopharmacological revolution.

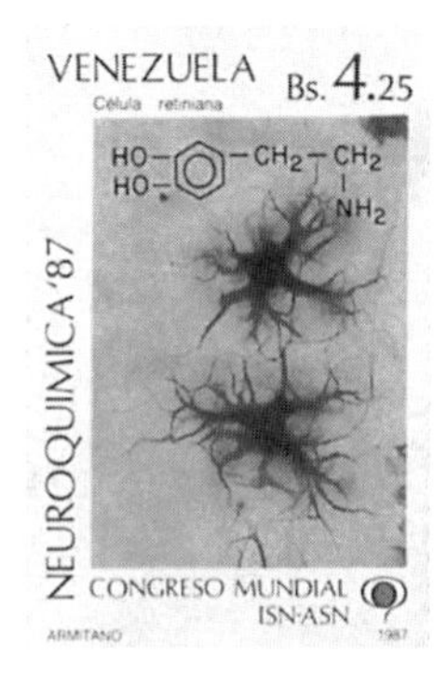

Dopamine © Oficina Filatelica Nacional, Venezuela

Subsequently, chlorpromazine was shown to be a potent dopamine D_2 antagonist with other pharmacological properties that were thought to cause unwanted side effects. Thus the D_2-receptor antagonism of the conventional antipsychotics mediates not only their therapeutic effects but also some of their side effects. With the discovery of newer atypical antipsychotics, older conventional antipsychotics are no longer used for first-line therapy, but they can still be effective as second-line or add-on treatments. In 1957, the American Public Health Association awarded the Lasker Prize for medicine to Pierre Deniker, Henri Laborit, Heinz Lehmann (the German-born Canadian psychiatrist who followed the lead of the French researchers in introducing chlorpromazine as a major tranquilizer), and Nathan Kline; the first three received the prize for their work on chlorpromazine, and Kline for his discovery of the antipsychotic actions of reserpine (see earlier in this chapter).

Haloperidol

Chlorpromazine revolutionized the specialty of psychiatry and brought legitimacy to biological psychiatry.[10,25–27] More important, chlorpromazine sparked a tremendous amount of research activity in searching for antipsychotic drugs. One of the fruits of the ensuing research was haloperidol, which was 50–100 times more potent than chlorpromazine with fewer side effects. Haloperidol was developed as a more potent and selective D_2 antagonist because the D_2-receptor blockade in the mesolimbic pathway is believed to reduce the positive symptoms of schizophrenia. Indeed, haloperidol is quite effective against the positive symptoms; however, it is ineffective in treating the negative symptoms and neurocognitive deficits of schizophrenia. In addition, administration of the drug typically causes extrapyramidal symptoms (EPS), including parkinsonian symptoms, akathisia, dyskinesia, and dystonia.

In 1935, after 12 years of medical practice, Constant Janssen in Belgium founded a pharmaceutical company called Janssen Pharmaceuticals. Like most ethical drug firms at the time, Janssen was churning out an astonishing number of drugs, but it lacked drugs that were covered by the company's own patents. The firm had no original research team, no patents, and therefore little hope of expanding its market. In 1951, Constant's son, Paul A. J. Janssen, graduated from the University of Ghent. After spending 18 months of mandatory military service as a physician stationed in Germany, Paul Janssen joined his father's company and became the head of research. Paul Janssen later recalled:

> The odds against success were apparently enormous. The available laboratory space was a small section of the existing analytical quality control laboratory. Trained personnel were virtually nonexistent and so was the budget. The only way out was somehow to concentrate on making new chemicals that could be synthesized and purified with simple methods and equipment, using the cheapest possible intermediates, and to efficiently investigate the pharmacology at minimal expense. The fact that the oldest member of our very small research group was 27 years of age, that we were all willing to work very long hours, seven days a week, and, being inexperienced, had no idea of the difficulties along the road but blind faith in ultimate success, were of course decisive factors in our favor.[10]

Their first success came when they discovered R951, a piperidine substituted with a propiophenone side chain. R951 was a narcotic analgesic, and it brought in money to fuel Janssen's further research. By eliminating R951's narcotic analgesic properties through simple chemical modifications—including adding one carbon to elongate the side chain and replacing the ester with a hydroxyl group—they arrived at a series of promising butyrophenones. On February 11, 1958, Bert Hermans, a young chemist in their laboratory in Beere, synthesized R1625, which later became haloperidol. Distinctive from chlorpromazine, haloperidol was a butyrophenone derivative. Therefore, haloperidol's chlorpromazine-like activity came as a surprise to Janssen and his colleagues. It was many times more potent than chlorpromazine. As a matter of fact, it was the most potent antipsychotic at the time. Haloperidol was both faster and longer acting. It was potent orally as well as parentally. More important, it was almost devoid of the antiadrenergic and other autonomic effects of chlorpromazine. Haloperidol also had a more favorable safety ratio and was surprisingly well tolerated when given chronically to laboratory animals.

Clinical trials, some of which took place at the Hôpital Sainte-Anne in Paris, where Jean Delay and Pierre Deniker tested chlorpromazine, confirmed that haloperidol belonged to the pharmacological family of neuroleptics. It became valuable in the treatment of agitation, delusions, and hallucinations in mental patients. Frank J. Ayd, Jr., a well-known psychiatrist, summed it up concisely: "Excited persons are candidates for haloperidol. Depressed persons generally are not."[10] Numerous inpatients with chronic disorders were able to leave the hospital and live at home thanks to haloperidol. Haloperidol (Janssen sold it in the United States using the trade name Haldol) remained one of the more prescribed neuroleptics 40 years after its discovery, until the emergence of atypical antipsychotics.

The discoverer of haloperidol, Paul A. J. Janssen, was a most extraordinary scientist. Over the years, he and his colleagues introduced approximately 80 drugs, including fentanyl and risperidone, an unprecedented number by an individual. He built the small drug firm founded by his father into a pharmaceutical giant, which merged with Johnson and Johnson in 1961. He published more than 850 papers, held more than 100 patents, and delivered more than 500 lectures all over the world in Dutch, English, French, German, and Spanish. Janssen once stated that "a good scientist is someone who succeeds in getting the different scientific disciplines to work in harmony with one another."[26,27] and he was the epitome of his

Paul A. J. Janssen. Photo by Peter Carmen © Janssen Pharmaceuticals

statement. In history, six people have been awarded the Nobel Prize for their contributions to drug discovery: Paul Ehrlich (1908), Gerhard Domagk (1939), Daniel Bovet (1957), George Hitchings (1988), Gertrude Elion (1988), and James Black (1988). According to Sir James Black, Paul Janssen was the greatest among them all. Although he was nominated several times, he never won, possibly because he had so many achievements that it was hard to summarize them in a sentence or two. According to Alfred Nobel's standard, Janssen was certainly the one who "shall have conferred the greatest benefit on mankind." Sadly, Janssen died suddenly on November 11, 2003, while attending a conference in Rome, Italy.

Atypical Antipsychotics

Atypical antipsychotics, also known as serotonin-dopamine antagonists, have reduced extrapyramidal symptoms (EPS) compared with conventional antipsychotics. They are also believed to reduce the negative, cognitive, and affective symptoms of schizophrenia more effectively. All atypical antipsychotics are potent antagonists of serotonin 5-HT_{2A} and dopamine D_2 receptors; however, they also act on many other receptors, including multiple serotonin receptors. The challenge remains to determine which of these secondary pharmacological properties may lead to improved efficacy and which are undesired and account for the side effects.

Clozapine, the first atypical antipsychotic, was developed in 1959 by the small Swiss company Wander AG (which also invented Ovaltine). During the clinical trials, clozapine showed strong sedating effects and proved to be efficacious for schizophrenia, but it also showed some liver toxicity. Wander planned to withdraw the drug because of difficulty in receiving regulatory approval. But the clinicians who participated in clinical trials urged Wander to provide more samples, because their patients fared better on clozapine than on typical antipsychotics. For a population of patients who responded poorly to standard therapy, clozapine was especially effective. Wander AG reluctantly proceeded with more trials and received approval to market clozapine in a few European countries in 1971, although liver toxicity limited its widespread use. Unfortunately, clozapine was removed from the market in 1975 due to rare but potentially fatal drug-associated agranulocytosis, a blood disorder that resulted in lowered white-cell counts. In Finland, eight patients died from subsequent infections. Additional side effects of clozapine therapy include sedation, weight gain, and orthostatic hypotension. Clozapine was reintroduced in 1990 by Sandoz, and it is now used as a second-line treatment that requires extensive monitoring of the patient's blood cell count. Over the years it has demonstrated efficacy against treatment-resistant schizophrenia, and some still consider it the gold standard for treatment-refractory patients despite the inconvenience of a weekly check of white blood cell counts.

The second atypical antipsychotic was risperidone (trade name Risperdal), introduced by Janssen Pharmaceuticals in Belgium in 1993. With haloperidol successfully on the market, Janssen systematically explored its clinical utilities. In one trial, they found that a combination therapy of haloperidol and the serotonin antagonist ritanserin improved negative symptoms, depression, and anxiety in patients with schizophrenia. Intrigued, Janssen initiated a medicinal chemistry program led by chemist Anton Megens. Based on the neuroleptics lenprone and benperidol, they ultimately discovered risperidone. Within a series of benzisoxazole derivatives, risperidone showed a desired combination of very potent serotonin and potent dopamine antagonism. In essence, risperidone possessed the attributes of both ritanserin and haloperidol, a typical antipsychotic. In patients with schizophrenia, risperidone is effective against both positive and negative symptoms with reduced EPS liability and has become a first-line therapy.

The third atypical antipsychotic was Eli Lilly's olanzapine (Zyprexa), launched in 1996, which brought in $4.28 billion in 2004 (about a third of Lilly's total sales). Zyprexa, an antagonist against several receptors, including dopamine, serotonin, histamine, adrenergic, and muscarinic receptors, was selected from a large series of chemical analogs based on behavioral tests. Like many atypical antipsychotics, one of the side effects of Zyprexa is weight gain. Ironically, the weight-gain side effects of some atypical antipsychotics may be used as an advantage when they are prescribed off-label to patients with anorexia. Because those patients normally would not eat enough food, the side effect of inducing hunger actually helps them to start eating.

Additional atypical antipsychotics are AstraZeneca's Seroquel (quetiapine; 1997), Pfizer's Geodon (ziprasidone; 2001), and Bristol-Myers Squibb/Otsuka's Abilify. One of the great advantages of Geodon and Abilify is that they do not have the weight-gain side effect.

Illegal Drugs of Abuse

The difference between legal and illegal drugs is in the eyes of the law. For instance, King Charles II of England in 1675 issued a short-lived ban on coffee drinking, stating that coffee caused men to become infertile. Alcohol was illegal during Prohibition in America. On the other hand, Ecstasy (MDMA), an illegal drug, is now approved by the FDA to be used in a small clinical trial to gauge its utility in terminally ill cancer patients, and cannabis (the active ingredient in marijuana) is used to stimulate appetite and possibly treat glaucoma. Although opium is a controlled substance now, it used to be in every physician's medicine chest. Laudanum, used as a "panacea" a century ago, was simply an alcoholic solution of opium. Opium, extracted from poppy seeds, was the initiator of the Opium War, which had a lasting impact in China politically and economically in the late nineteenth century.

Papaver somniferum, from which opium is extracted. Photo by author

LSD

LSD, *D*-lysergic acid diethylamide, is undoubtedly the most notorious psychedelic drug.[28,29] Albert Hofmann of Sandoz Laboratories in Basel, Switzerland, synthesized LSD in 1943 from lysergic acid, an ergot alkaloid of the family of alkaloids produced by a parasitic fungus, *Claviceps purpurea.* When grains, especially rye, became wet, the fungus would infest on the spoiled rye. Ingestion of ergot-containing rye could cause gangrene, known as "St. Anthony's fire" in the Middle Ages because the fingers and toes came to look as though they had been charred by fire. It was also believed (incorrectly) that a visit to the saint's shrine would cure the malady. Ingestion of ergot may have also caused the disorders of thinking and hallucinations that resulted in the "witchcraft" made famous in the Salem (Massachusetts) witch trials. In 1918, Arthur Stoll, also of Sandoz and a future collaborator of Hofmann's, isolated ergotamine, the key active principle of ergot alkaloids. In 1938, Stoll and Hofmann reported the synthesis of the diethylamide of lysergic acid, but they received little attention. In 1943, Hofmann was carrying out a systematic chemical and pharmacological investigation of partially synthetic amides of lysergic acid. During the process, he again synthesized *D*-lysergic acid diethylamide (LSD), among other compounds, with the intention of obtaining an analeptic (a circulatory stimulant). On April 16, 1943, Hofmann prepared LSD from lysergic acid and diethylamine using an amide formation reaction. He then proceeded to run column chromatography to remove the inactive isomer, *D*-isolysergic acid diethylamide. During the experiment, he experienced "a remarkable but not unpleasant state of intoxication which lasted 2–3 hours and was characterized by extraordinarily intense stimulation of imagination and an altered awareness of the world around me. On closing my eyes I saw a succession of a striking reality and depth, alternating with a vivid, kaleidoscopic play of colors."[29] Initially, Hofmann thought that he was intoxicated by chloroform, then he realized that he might have accidentally ingested a minute amount of LSD (which explains why protective gloves are worn today). Three days later, he decided to ingest a small (he thought) amount of LSD himself, 250 milligrams, which turned out to be five times the efficacious dose. He felt ill, and his assistant had to help him home. But after he came out of the "acid trip," he had a wonderful feeling about his experience. In the 1950s, Sandoz marketed LSD as a psychiatric panacea, useful in treating schizophrenia, alcoholism, sexual deviancy, and an assortment of other mental problems. Sandoz also suggested that psychiatrists try LSD themselves. The drug was a smashing success, so much so that people started to use it recreationally. They often reported that LSD use was a life-transforming experience. A Harvard psychologist, Timothy Leary, even organized an LSD cult, with himself as the spiritual guru. In the 1960s, the hippie culture and the sexual revolution were closely intertwined with the use of

LSD. The U.S. government classified LSD as a Schedule I drug (the most stringent category) in 1967.

The mechanism of action of LSD is still largely unknown, despite the serious scientific attention it has received. Initial suggestions that LSD was a serotonin antagonist have lost their credence on further scrutiny.

Amphetamine

German chemist. Lazar Edeleanu first synthesized amphetamine in 1887.[30] Little was done with amphetamine until 1927, when Gordon Alles at the University of California, Los Angeles, resynthesized it as a cheap substitute for ephedrine and investigated its therapeutic potentials.

Ephedra is a leafless bush indigenous to the desert regions of Asia and North America. Known as "Ma Huang" in Chinese, *Ephedra* has been used in Chinese herbal medicine for hundreds of years. The active alkaloid principle, ephedrine, was isolated by a team of Chinese and Japanese scientists in 1892. Smith, Kline, and French in Philadelphia developed ephedrine production in the 1920s and sold it as a popular remedy for asthma. In the mid-1920s, the raging civil war in China resulted in a severe shortage of Ma Huang. Gordon Alles (1901–1963), a chemist at a UCLA clinical office led by George Pines, synthesized amphetamine as a cheap substitute for ephedrine. Alles tasted his own product and reported, "Within minutes, I realized a notable subjective response was going to result."[30] He also discovered that the carbonate of amphetamine was volatile and thus could be inhaled, which afforded rapid relief of acute asthma attack. Smith, Kline, and French developed it and made it available in 1935.

The hallmark of amphetamine's pharmacological effect is sleep prevention. During World War II, both the Allies and the Axis liberally distributed amphetamine to their troops to keep soldiers alert during long battles. Its use was even more frequent for the pilots who flew long hours. At the end of the war, Hitler had his personal physician inject him with "vitamins" (most likely amphetamine) as often as eight times a day. Apparently, the vitamins did Hitler a lot of good, because he became very alert and could work through the night without being tired. He also constantly bit his nails, a sign of amphetamine toxicity. Hitler's Luftwaffe chief, Hermann Goering, was known to be addicted to methamphetamine, a methylated version of amphetamine. Winston Churchill also used amphetamine to help him through the long dark nights during the blackouts. After the war, the Japanese released their vast military stocks of amphetamine. Life in postwar Japan was tough, and many struggling young workers had to hold two jobs to make ends meet. Gulping down amphetamine kept them going during the long nights. Throughout the Vietnam War, U.S. soldiers consumed 200 million doses of amphetamine.

The pharmacology of amphetamine is similar to that of cocaine. Its neurochemical actions are exerted by blocking the reuptake of norepinephrine, dopamine, and serotonin. Amphetamine also inhibits monoamine oxidase, the same enzyme that the antidepressant Marsilid acts on. Amphetamine suppresses appetite, inhibits random eye movement (REM) sleep, and elevates mood, yielding elation and euphoria.

Methamphetamine's street names include *meth, speed, ice, crank,* and *quartz.* Its central nervous system effects manifest when it is smoked, swallowed, snorted, or injected. In recent years, the number of small, clandestine laboratories that make methamphetamine in the United States has mushroomed. Those illicit kitchen chemists make methamphetamine by reducing pseudoephedrine, an active ingredient in Sudafed, NyQuil, and several other over-the-counter decongestants. Because illegal meth makers often use sodium or lithium metal and anhydrous ammonia, they leave behind severe pollution in the environment. To keep cough syrup out of meth production, Pfizer introduced Sudafed PE in 2004, whose active ingredient is phenylephedrine. Although clandestine chemists can reduce the hydroxyl group easily, adding an extra methyl entails multiple sophisticated chemical manipulations, which is beyond the capacity of the improvising chemists. Because the extra methyl is key to methamphetamine's pharmacological effects, Sudafed PE cannot be used to make methamphetamine. For now, clever science is one step ahead of the illegal meth makers.

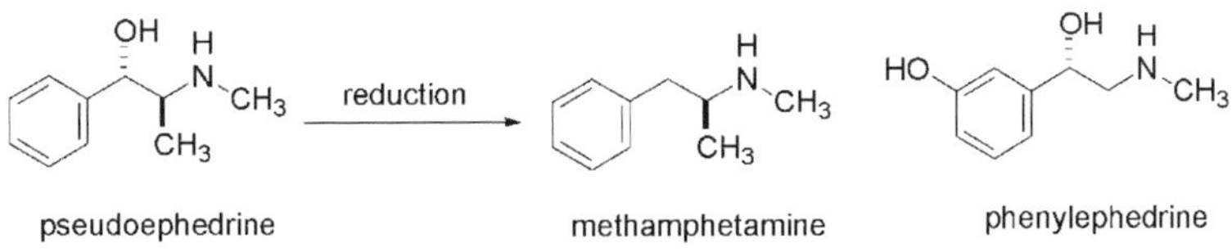

Pseudoephedrine, methamphetamine, phenylephedrine

Ecstasy

In 1912, the German drug firm Merck was looking for vasoconstrictive and styptic drugs (drugs that stop bleeding). In the process of synthesizing hydrastinin, a chemist isolated a by-product called 3,4-methylene-dioxy-methamphetamine (MDMA; street name Ecstasy). At the time, it was just a by-product of the purification, but Merck nonetheless filed a patent for this series of compounds in 1914; the patent has long since expired. Not much happened with MDMA until 1976, when Alexander Shulgin resynthesized it and reported its psychoactive properties. Since then, many psychiatrists used it as an adjunct to psychotherapy and saw tremendous benefits. It has also been reported that its use condensed a year of therapy into 2 hours. Unfortunately, abuse of MDMA began to emerge in the 1980s, quickly replacing amphetamine and LSD to become one of the most abused drugs. It was estimated that nearly 1 million people took Ecstasy in a weekend. MDMA, a potent hallucinogenic amphetamine analog, has been abused with greater degree because of its extraordinary central nervous system effects. If a person hit his or her finger after taking the drug, he or she would see the strike at the finger but not feel it until later. MDMA causes a disturbance in the correlation between sight and touching that occurs in the brain itself. MDMA is also known to increase the desire for sexual activity, possibly because it increases physical and emotional sensitivity. Young partygoers tend to use it so they can go all night with little mental and physical fatigue. Common acute adverse effects of MDMA are muscle tension and bruxism (grinding of the teeth), whereas chronic use can cause neurotoxicity, rhabdomyolysis (muscle wasting), and both cardiovascular and renal failure. In 1985, the Drug Enforcement Administration (DEA) labeled it a Schedule I drug. Recently, the pendulum has swung to the opposite direction, as MDMA has been approved by the FDA for use in a small clinical trial to gauge its utility in treating terminally ill cancer patients.

Heroin

British chemist C. R. Alder Wright first synthesized diacetylmorphine (Heroin) in 1874.[31] While Wright was experimenting with opium derivatives at St. Mary's Hospital in London, he obtained a white crystalline solid from morphine and acetic anhydride. In order to find out how the compound would behave, he fed it to his dogs, who did not like it at all and spat it out. Two decades later, a chemist at Bayer AG in Germany, Felix Hoffmann, replicated Wright's experiment and transformed morphine to diacetylmorphine in 1897 using a process similar to the one he had used in converting salicylic acid to acetyl salicylic acid (aspirin). Ironically, the commercial aspect of aspirin was seriously doubted by Bayer's management, whereas that of diacetylmorphine was viewed as a possible jackpot. This uninformed opinion was based on the belief that diacetylmorphine would be a nonaddictive painkiller superior to both morphine and codeine. The head of Bayer's pharmacology department, Heinrich Dreser, strongly believed (incorrectly) that the acetylation would eliminate morphine's addictive effect. The more clinical trials that Dreser carried out, the more convinced he was about diacetylmorphine's commercial potential. He tested the drug on frogs and rabbits, then tried it on himself and volunteers from the adjacent Bayer dye factory. Every test gave him resounding positive feedback. The workers felt that the drugs made them feel so "heroic" that Dreser named it *Heroin*. A year later, Dreser presented his astounding data to the Congress of German Naturalists and Physicians. He proclaimed that Heroin was 10 times more effective as a cough medicine than codeine but had only one-tenth its toxic effects. Bayer promoted Heroin as a safe substitute for morphine, addiction to which was becoming a growing plague. From then on, Heroin received widespread acceptance and was sold over the counter for all imaginable maladies across Germany, followed by Europe and America. In contrast, aspirin required a doctor's prescription for anyone who needed it. The rationale was that aspirin might have severe cardiac side effects. One would never have guessed that baby aspirin (81 mg dose) is used as a protective measure for heart conditions a century later.

In the early 1900s, Heroin began being abused in Europe and subsequently in America. In New York City, homeless Heroin addicts would gather scrap metals and sell them to pay for their habit. This is the origin of the nickname "junkie" for drug addicts.

CHAPTER 6

Diabetes Drugs

History is not history unless it is true.

—William Herndon

Diabetes mellitus is a multisystem disease associated with the loss of control of physiological glucose concentrations in the blood. The disease is broadly broken down into two types based on factors that include age, acuteness of onset, underlying glucose-handling deficit, and therapy.

Type 1 diabetes usually manifests acutely in the young, secondary to some underlying insult (possibly infectious) to the islet cells of the pancreas, resulting in an absolute lack of insulin. Type 2 diabetes is more frequently associated with maturity, obesity, and gradually increasing blood glucose concentrations; it may be asymptomatic for some time and discovered on routine glucose screening. In fact, as weight increases among the general population of the developed world, type 2 diabetes is becoming an epidemic. Type 1 diabetes always requires insulin replacement therapy, whereas type 2 can frequently be controlled with diet, weight loss, and oral medications that enhance residual pancreatic function.

Understanding Diabetes

Diabetes has been known since antiquity. In fact, the term *diabetes mellitus* comes from the Greek meaning "siphon and honey" due to the excess excretion (siphon or faucet) of hyperglycemic (sweetened, or honeyed) urine. In ancient times, most cases of diabetes were of type 1, with acute onset in the young, which was often fatal. Type 2 diabetes was extremely rare when sources of nutrition were scarce and obesity was not prevalent.

Diabetes was also known as "wasting" because diabetics were not able to metabolize the sugar content of food and eventually died from wasting away. Because of the effect of excess blood glucose, the blood of the diabetic is hyperosmolar (concentrated), and this triggers compensatory thirst (in an attempt to dilute the hyperglycemia and return the blood to a normal concentration). This excess thirst results in the common diabetic symptom of polydipsia (excessive drinking secondary to thirst, resulting in the urge to drink frequently) and polyuria (excess urination).

Even before many modern diagnosis tools became available, savvy doctors could diagnose diabetic men just by looking at their shoes for the telltale white spots from urine with high sugar content. In fact, tasting urine samples of diabetics was a routine diagnostic tool for diabetes. Even the breath of a severe diabetic was sweet—a sickly smell as a result of acidosis. In addition, it has been mentioned that ants would track to the urine of diabetics.

Serious complications of diabetes include nephropathy (kidney diseases), neuropathy (nerve damage), and retinopathy (blindness). Diabetes is the most common cause of blindness and amputation in the elderly in the United States.

The earliest treatment for diabetes was documented in the Ebers papyrus, unearthed in Egypt, which dates from 1500 B.C. It contains a prescription: "to eliminate urine which is too plentiful: a measuring glass filled with water from the bird pond, elderberry, fibers of the asit plant, fresh milk, beer-swill, flower of the cucumber, green dates, make into one, strain, and take for four days."[1]

Claude Bernard (1813–1878), a French physician and physiologist, made significant contributions to the understanding of diabetes by elucidating the functions of the pancreas gland. As a young man from a village in the Beaujolais region, Bernard aspired to be a playwright. One of his plays debuted in Paris and received scathing criticism from renowned artistic critic Saint-Marc Girardin, who found his literary work unacceptable. Instead, Girardin encouraged Bernard to pursue medicine, a profession in which it was easier to make a living. Bernard went to college and then medical school. He was not an illustrious student, graduating toward the bottom of his class in 1839. Undaunted, he became a protégé of the brilliant experimentalist

Claude Bernard © La Poste, France; François Magendie © Transkei Post Office

François Magendie. After 4 years in Magendie's laboratory, Bernard began his independent research in a private laboratory, funded by his wife's large dowry. He rapidly rose to eminence in the field of medical research and illuminated many aspects of human life. The four most significant areas were: (1) the digestive functions of the secretions of the pancreas gland, (2) the glycogenic function of the liver, in which Bernard isolated an "animal starch" that he termed *glycogen* (the stored form of sugar), (3) his discovery of the role of vasoconstrictor nerves in regulating blood flow in blood vessels, and (4) his foray into poisonous alkaloids, such as curare, shedding light on their central nervous system (CNS) effects.[2–5]

Bernard's mentor, Magendie, had isolated pancreatic juice from live animals but failed to understand its physiological function. Earlier in his career, Bernard explored this field with a series of brilliantly designed experiments. He observed that no secretion of pancreatic juice occurred in a fasting dog. But with a fed dog, the acid chime would enter the duodenum to stimulate the flow of pancreatic juice, which, in turn, transformed starch into sugar. In 1849, he published a landmark paper on the role of the exocrine pancreas, showing its crucial contribution to the metabolism of fats. Bernard's glycosuria (sugar in the urine) experiments demonstrated that sugar was produced in the liver with the aid of an enzyme. In other words, the stomach had to turn food into sugar for the pancreatic juice to metabolize. His observations shed light on the functions of the pancreatic juice, which was essential in solving the jigsaw puzzle of diabetes. His exploration with endocrinology created a paradigm for the future studies of metabolism. Bernard was also the first to employ the term "internal secretion" in a lecture in 1855 at the Collège de France. He is considered the "father of experimental physiology." Bernard once predicted with an extraordinary insight: "One day we will be able to follow the journey of a molecule in the body from its entrance all the way through to its exit."[3] Although we are not there yet, we are certainly a lot closer.

Bernard sacrificed his family life for medicine. Although he was able to fund his research using his wife's dowry, his marriage of convenience (he had married his wife for her money) came back to haunt him. Resenting his use of live animals to study experimental physiology, his wife and two daughters were so repelled that they became antivivisectionists and deserted him (even today, antivivisectionists still consider Bernard their greatest enemy). They would have much preferred that he give up experimental medicine involving animals and move into a more lucrative medical practice that involved human patients. Bernard found solace through his intimate communications with Madame Raffalovich, a young, beautiful, and intellectual woman. Bernard's letters to her were published in 1950 as *Lettres Beaujolaises*.[2] In defense of vivisection, Bernard was unequivocal: "I think one has this right, completely and absolutely . . . it is only possible to rescue living beings from death after sacrificing others. I do not accept that it would be moral to try more or less dangerous or active remedies on patients without having tried them initially on dogs."[3]

Bernard lived to see his discoveries and achievements recognized and rewarded. He became the chairman of the Sorbonne and was elected president of the French Academy. Emperor Louis Napoleon, an admirer of

Claude Bernard elucidated the functions of the pancreas gland © Warner-Lambert

his, invited him to the senate, an honorary position without much power, where he served as a "rubber stamp" in the Bonaparte regime. Bernard passed away on February 10, 1878, in Paris. France held a national funeral for him, the first scientist so honored. Streets in France have been named after him.

In the early 1900s, German scientist Joseph von Mering (1849–1908) worked with his Strasbourg colleague Oscar Minkowski (1858–1931) to study the function of the pancreas gland in digestion. By chance, they discovered that removal of the pancreas gland resulted in diabetes. At one point they removed the pancreas from a dog; one day later they noticed a swarm of flies hovering over a pool of the dog's urine. Intrigued, they analyzed the urine sample and found that the sugar level was extraordinarily high. They suddenly realized that they had done what many had attempted to do without success—created an animal model for diabetes. They later demonstrated by a series of definitive experiments in dogs that diabetes mellitus indeed involved the pancreas gland.[1]

There is another version of the story detailing the discovery of the linkage between pancreas and diabetes. Minkowski had removed the pancreas of a dog for an experiment. The next day, he noticed that the dog was lying in a pool of urine. He chastised the cleaner for allowing the dog to soil the floor. But the cleaner replied that it was impossible to stop it. Interest piqued, Minkowski drew some urine using a pipette and tasted it. The canine urine tasted similar to those Minkowski found in diabetic patients. He then made the connection between the pancreas and diabetes. The findings by Minkowski and von Mering laid the foundation for future generations looking for a treatment for diabetes.

There are two types of cells in the pancreas that are involved with the production and excretion of chemicals to aid in handling foods, fats, and sugars. One group of cells, called acinous cells, produces powerful enzymes that are excreted into the intestine for the digestion of food. The other cells are the endocrine "islet of Langerhans." Paul Langerhans, a German medical student, studied the anatomy of the pancreas through dissection. In 1869, he "discovered" an islet attached to the pancreas, now known as the islet (the islands of Langerhans cells). However, Langerhans did not appreciate the function of the "islet of Langerhans," which we now know is to produce insulin. Before 1921, as many as 400 attempts to isolate insulin failed because the proteolytic enzymes of the crude pancreatic extracts destroyed insulin, a peptide.

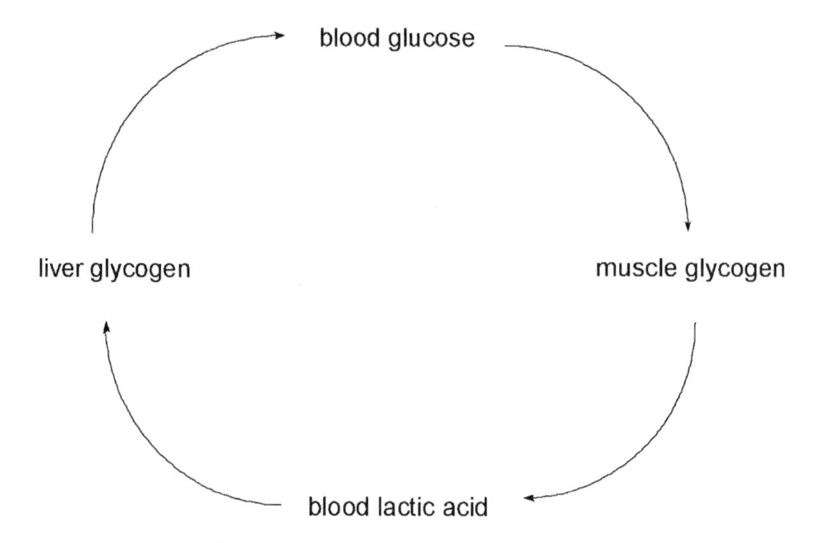

The Cori glucose and glycogen cycle

Carl Ferdinand Cori (1896–1984) and Gerty Theresa Radnitz Cori (1896–1957) also contributed significantly to the understanding of diabetes. They attended medical school at the German University of Prague. In 1922, they immigrated to the United States to escape the escalating anti-Semitism in Europe.[6,7] They focused their investigations on carbohydrate metabolism at Washington University in St. Louis, Missouri. They elucidated the landmark "Cori cycle" with regard to the metabolism of sugar. During muscle contraction, glucose is converted to lactic acid when oxygen is in short supply. The buildup of lactic acid explains why muscles are sore after severe physical exercise. Some lactic acid, in turn, is stored as glycogen until needed again. The couple won the 1947 Nobel Prize in Medicine or Physiology for their discovery of the course of the catalytic conversion of glycogen. The Coris were also extraordinary mentors. Six of their students went on to win Nobel Prizes.

The Discovery of Insulin

Insulin was discovered by Fredrick G. Banting and Charles H. Best at the University of Toronto in 1921.[8–26]

Banting was born on November 14, 1891, in Alliston, a small farm town 40 miles north of Toronto. He studied medicine for 4 years at the

University of Toronto, went on to serve as an army surgeon during World War I for 1 year in the European theater, and won a medal for bravery. As a war hero, he came back to Toronto but found that his dream of working in a hospital was beyond reach because work was scarce after the war. He bought a house in London, a small city about 110 miles west of Toronto, and set up a practice there in the summer of 1920. Business was slow, and he was barely able to make a living. In order to supplement his income, Banting obtained a part-time job at nearby Western Ontario University as a lecturer in physiology and a demonstrator in surgery and anatomy.

On October 31, 1920, while preparing his class on carbohydrate metabolism, Banting studied an article by Moses Barron in the *Journal of Surgery, Gynecology, and Obstetrics.* Barron's article on the relationship between diabetes and the pancreas kept him up late into the night. Around 2 A.M., he woke up and wrote down an idea that he conceived during his fitful sleep. It read: "Diabetus Ligate pancreatic ducts of dog. Keep dogs alive till acini degenerate leaving Islets. Try to isolate the internal secretion of these to relieve glycosurea."[13]

In addition to the misspelling of two words, *diabetes* and *glycosuria*, the idea was later proven to be a misconception. With his hallmark stubbornness, Banting was determined to carry it out and look for the internal secretion from the pancreas extract as a means to treat diabetes.

Banting closed his practice in the summer of 1921 and moved to Toronto to pursue further research on that idea. At the University of Toronto, he obtained support from J. J. R. Macleod, a renowned professor of physiology and world expert in carbohydrate metabolism. Macleod initially thought Banting's idea was naïve because many scientists with better training and better facilities had had similar ideas but had never succeeded in the laboratory. Macleod estimated that there had been more than 400 attempts within 30 years to isolate the active hormone from the pancreas before Banting tried. Some scientists had spent their entire careers doing so without success. Moreover, Macleod was underwhelmed by Banting's superficial textbook knowledge of research in general, physiology in particular. He nonetheless provided Banting with some laboratory space, several dogs for experiments, and a student assistant, Charles Best, telling him "We must leave no sod unturned."[20] After a brief discussion of the research plan, Macleod left Toronto to spend his summer vacation in his native Scotland.

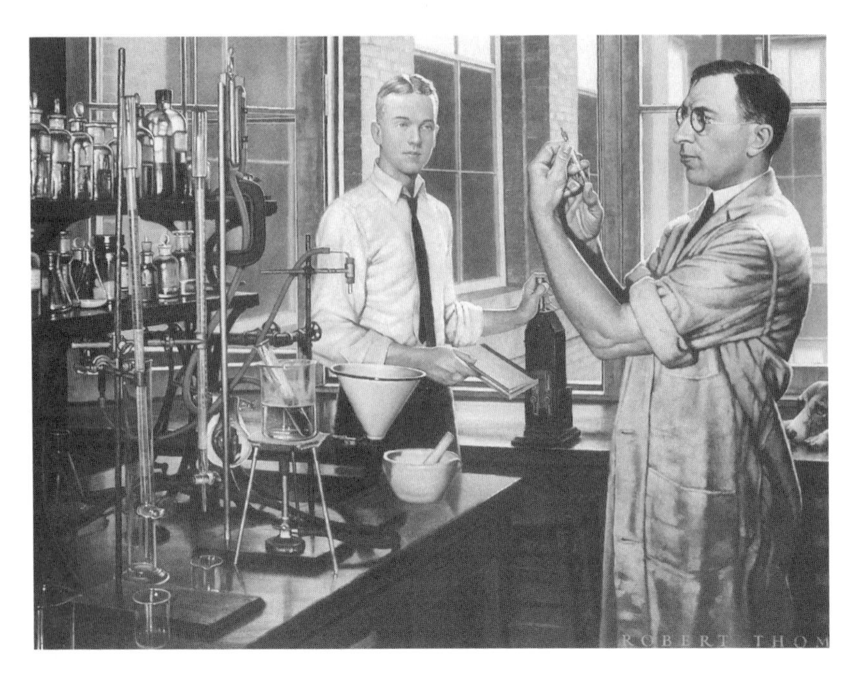

Fredrick G. Banting and Charles Best © Warner-Lambert

In the smothering heat of the summer, Banting and Best toiled in the basement laboratory carrying out surgical removal of the pancreas on dogs and artificially creating diabetes. Due to lack of experience, they ran into many frustrations and failures. Several dogs died, and Banting had to buy more dogs with his own money from the street. They also tied the pancreas ducts of some dogs and prepared crude extracts from the degenerated pancreases 7 weeks later. Being a surgeon, Banting took on the surgical part of the work, whereas Best took advantage of his biochemistry knowledge and carried out determination of the sugar levels and the glucose-to-nitrogen ratios (the G:N ratios), in both blood and urine. In August, injection of their precious extracts, termed *insletin* by Banting, significantly lowered the sugar levels of the diabetic dogs in both blood and urine. According to Best, "This was a dramatic result and marked the point at which we were sure that we could prepare the internal secretion of the pancreas in very potent form."[22]

In September, Macleod came back from Scotland. Although he lacked the same level of enthusiasm as Banting and Best, he was nonetheless impressed on seeing several energetic dogs that had been living comfortably for weeks without pancreases. Meanwhile, Banting was confident enough to request additional resources from Macleod, who at first refused, claiming Banting's research was "no more important than any other research in the department."[20] When Banting threatened to ask the University of Toronto how important his research was, the aristocratic Macleod simply replied, "As far as you are concerned, I am the University of Toronto."[20] This exchange, in addition to Macleod's initial "snub" and discouragements, had sowed the seeds for a long and acrimonious feud between Banting and Macleod. When Banting threatened to go to either the Rockefeller Institute in New York or the Mayo Clinic in Minnesota to do the research, Macleod yielded and provided the resources Banting had asked.

In October, Banting found a good source of pancreases from fetal calves, a significant development because the supply of degenerated dog pancreases became the bottleneck. Having been born a farm boy, Banting was familiar with stock breeding and was well aware of the availability of fetal calves, which were plentiful in local slaughterhouses.

In December, Banting, Best, and Macleod decided to try fresh whole beef pancreas. When the extraction was done in alcohol rather than water, it afforded them very effective batches of pancreas extracts. The old method using water required the later removal of the water, which was done at high temperature and partially destroyed the insulin. Meanwhile, Macleod assigned James Bertram Collip to join the team to help with the pancreatic extraction process. Collip, a professor of biochemistry at the University of Alberta in Edmonton, was on sabbatical leave working in the Macleod group at that time.

At end of the year, at the conference of the American Physiology Society held at Yale University in New Haven, Banting presented a lecture titled "The Beneficial Influences of Certain Pancreatic Extracts on Pancreatic Diabetes," listing Macleod, Banting, and Best, in that sequence, as authors. Never an eloquent public speaker, Banting "did not present well," according to his own account. Things became even worse during the question-and-answer session because he was not familiar with the literature and background. Macleod came to the rescue. He handled the questions masterfully, although he always used "we" as the subject. This obviously unnerved Banting. Unable to sleep all night on the train back to Canada, he began to suspect that Macleod was trying to steal his credit.

The first test in humans using the extract prepared by Banting and Best was done on a 14-year-old diabetic boy. It did not work well and caused sterile abscesses, most likely because of the low potency of the crude extract and the presence of many impurities. It was a "crushing defeat for Banting," according to his biographer, Michael Bliss.[14]

On January 19, 1922, Collip finally came up with a procedure that yielded more pure, and thus more potent, extracts of the hormone. When Banting and Best inquired about the experimental details, Collip replied that, in consultation with Macleod, he had decided not to tell B2 (as he referred to the duo, Banting and Best) and that he was thinking of taking out a patent under his own name. Banting was so outraged that he almost punched Collip in the face. History can only speculate what the six-foot-tall Banting would have done to Collip, a small-statured fellow, if Best had not been present.

After having solved the interpersonal conflicts and technical problems of the extraction process, the team produced enough purified extract to try on patients. The first few recipients of insulin were emaciated and on the verge of death. They were literally brought back to life through insulin injection, becoming living miracles. The news first appeared in the local newspaper, the *Toronto Star*, and then spread around the world. Banting was inundated with patients. Insulin actually resurrected dying diabetics. Diabetes, a disease known to be invariably fatal, was controlled for the first time, so that diabetics could live normal lives. Meanwhile, Toronto scientists changed Banting's *insletin* to *insulin. Insulin* has a Latin root; *insula* means "island." To ensure the quality of insulin production, Banting, Best, and Collip took out a patent and sold their rights to the University of Toronto for $1 each.

Despite the splendid success of the clinical trials, the University of Toronto was not equipped for the mass production of insulin. The University of Toronto licensed the technique to Eli Lilly and Company in Indianapolis, which began to extract the insulin hormone in 1922 from the pancreas glands of slaughterhouse animals. George Walden, a bench chemist, made an astute observation that pH is more important to the stability of insulin than temperature. At the isoelectric point (the pH at which peptides precipitate because they are charged as zwitterions), insulin was more pure and more potent. This was not surprising, theoretically, because insulin is a protein made of 51 amino acids, behaving as a polypeptide. However, the practical ramifications were enormous. Lilly was able to extract potent insulin consistently from pork and beef pancreases, making insulin available to a large number of patients. When insulin supply was still scarce, J. K. Lilly, then chief executive officer of Lilly, gave a batch of insulin to Banting while he was visiting. Banting broke down, buried his head in Lilly's arm, and wept. Lilly was so deeply moved by Banting's sincere care for diabetes patients that he decided to back him "to the limit."[20] Insulin catapulted Lilly from a small backwater drug firm to a powerhouse in the pharmaceutical industry.

Although Macleod initially gave full credit for the discovery of insulin to Banting and Best under pressure, he somewhat deviated from that view when he met with August Krogh, the Nobel laureate from the previous year. On learning that the Nobel Prize was at stake for the insulin discovery, he commented to Krogh to the effect that "without my advice and guidance, Banting and Best would have gone off the wrong track."[21] In October 1923, largely based on such "hearsay" from Krogh, the Karolinska Institute awarded the Nobel Prize for Physiology or Medicine jointly to Banting and Macleod. Banting was outraged by the inclusion of the unworthy (to him) Macleod and the exclusion of Best. He challenged his friends to name one idea in the research from the beginning to the end that had originated in Macleod's brain or one experiment that he had done with his own hands. His immediate reaction was to decline the prize. He later changed his mind and announced that he would share half of his prize money with Best. Following suit, Macleod declared that he would split his share of the prize with Collip. In 1972, the official history of the Nobel Foundation admitted the unwise choice made in 1923: "Although it would have been right to include Best among the prize-winners, this was not formally possible, since no one had nominated him—a circumstance which probably gave the committee a wrong impression of the importance of Best's share of the discovery."[26]

The former chairman of the Nobel Prize selection committee for Physiology or Medicine, Rolf Luft, flatly stated in 1981 that the choice of the Nobel Prize in 1923 was the worst mistake that the commission has ever made. He dismissed Macleod as a manager and promoter who "put Collip and the Lilly Company into business."[26]

J. J. R. Macleod
© Guyana Post Office Corporation

Banting was instrumental to the discovery of the miracle drug insulin, which has saved millions of lives. His hard work, determination, and persistence were the right attributes for transforming his original idea into reality. No doubt teamwork was essential to the success of the endeavor, for without the skills that Collip brought as a biochemist, a more purified extract would not have been possible. Looking back at history from a grander vantage point, Banting came along at the right time and was in the right place, in the right discipline, and working with the right team. Indeed, the idea of isolating an active hormone extract from the pancreas was not new. But only at that time did more reliable methods and techniques become available for determination of the sugar level and the glucose-to-nitrogen ratio in blood and urine. For instance, the Benedict test of glucose level was a recent development that obtained good results by using just a small amount of blood sample. With regard to timing, Banting's decision to move to Toronto to pursue his idea was partially due to his professional and personal problems. Giving up his practice and house in London provided an escape for him at the time. The University of Toronto surely was the right place. Not only did it have the right facilities and literatures, but it also boasted Macleod, the world's leading authority in carbohydrate metabolism, as the chair of biochemistry. Banting, trained as an orthopedic surgeon, was well equipped to use his surgical skills to carry out pancreatectomy and duct ligation. Having been born a farm boy, he was familiar with stock breeding and was well aware of the availability of fetal calves, which temporarily solved the pancreas supply issue.

Above all, the single defining factor for the discovery of insulin was Banting's persistence. The fact that he was not familiar with the 400 or so failures before him was a blessing. He had a strong and unshakeable belief in his idea. In this case, too much knowledge would actually have been a detriment. Macleod, like many scholars, could easily talk himself out of such an idea because he had mastered the existing literature on failed precedents. Although Banting's original idea was proven wrong, he believed in it until the last moment of his life.

After the discovery of insulin, honors showered Banting, deservedly, as a benefactor of mankind. He was soon appointed a professor of medical physiology at the University of Toronto and the director of the newly established Banting Research Institute. Determined not to become another Macleod, he would never add his name to any publications that he did not contribute to both intellectually and experimentally. His marriage to Marion Robertson in 1924 was not a happy one and ended in divorce in 1932. Marion had expected to have a prominent social life, but Banting spent most of his time in the laboratory, striving to make another discovery greater than insulin. He often resorted to alcohol to fill his emotional void.

Fiercely patriotic, he threw himself into war research after World War II began. He took part in the investigation of aviation medicine and biological warfare. He volunteered himself as a human guinea pig to test the effect of mustard gas and seriously damaged his leg, which was almost amputated. On February 29, 1941, he was on a plane to England on a war-related mission, one of the very first to attempt to fly across the Atlantic Ocean. The plane crashed in Gander, Newfoundland. Banting was mortally injured. He was only 50 years old. Today, the lake into which his plane crashed has been renamed Banting Lake. The house in which Banting used to live, in London, Ontario, has been converted to a museum about insulin discovery. In front of the very house in which Banting conceived his idea that led to the discovery of insulin, an eternal flame, the flame of hope, is burning. When a cure for diabetes is found, it will be extinguished.

After the discovery of insulin, Charles Best took the wise advice of Henry Dale (the 1936 Nobel laureate), went on to finish his M.D. training in England, and came back to the University of Toronto. He carried out high-caliber research, contributing to knowledge on choline, a dietary factor, and heparin, an anticoagulant. In all, he had a happy life and career, retiring as the head of the Best Research Institute at the University of Toronto. Best spent a great part of his career trying to rewrite medical history. According to historian Michael Bliss's research, Best overly exaggerated his role in the discovery of insulin while understating the contributions of Macleod, Collip, and others. In a letter of his, Best shows himself as human as anyone else:

> I have to confess that even after all these years the revival of the memory that Prof. Macleod and later Collip instead of being grateful for the privilege of helping to develop a great advance, used their superior experience and skill, with considerable success, in the attempt to appropriate some of the credit for a discovery which was not truly theirs, still makes me warm with resentment.[23]

Macleod's reputation was so tarnished by Banting that it was difficult for the only two Nobel laureates of the University of Toronto to work at the same university. Macleod, as a foreigner, was also cold-shouldered by many Canadians in the climate of intense nationalism going on at the time. In 1928, he went back to Scotland as the Regius Chair of Physiology at the University of Aberdeen, where he carried out illustrious research until his death in 1938. On leaving Toronto, he shuffled his shoes repetitively on a mat before boarding his train, stating, "I am wiping away the dirt of this city."[9] Macleod never returned to Canada.

Collip went back to the University of Alberta in Edmonton, where he did important work on parathyroid hormone. During the 1930s, he and Banting both mellowed, reconciled, and became close friends. The day Banting died, he was wearing the sheepskin gloves that Collip had given him the day before. After Banting's death, Collip wrote profusely about Banting: "Banting was a most unselfish individual. He was always mindful of helping others and it was almost a religion with him to encourage, stimulate and assist young research workers."[25]

The intriguing story of the discovery of insulin was depicted in a movie titled *Glory Enough for All,* adapted from Michael Bliss's book *The Discovery of Insulin.* Insulin has done so much for the human race that we are indebted to everybody who was involved in its discovery. As Banting says to Best in the movie, "There will be enough glory for all of us if we can get it right." Indeed, the discovery of insulin was one of the greatest achievements of modern medicine.

Insulin Aftermath

Insulin, one of the greatest discoveries in history, has saved millions of lives. Many interesting events have taken place since the discovery of insulin. For instance, Banting won the 1923 Nobel Prize for his discovery of insulin, which, in turn, fortuitously made another Nobel Prize possible 11 years later. George R. Minot (1885–1950) was a severe diabetic, kept alive only by insulin. At Harvard, he carried out important research in the pharmacology of the liver, unlocking the secret of pernicious anemia. Minot earned the Nobel Prize for Physiology or Medicine in 1934.

August Krogh played a decisive role in convincing the Karolinska Institute to award the Nobel Prize to Banting and Macleod. Right around the time of the emergence of insulin, Krogh's wife and collaborator developed diabetes. Krogh traveled to Toronto, brought back the expertise of the time, and convinced the Danish pharmaceutical industry to produce insulin, helping to found a drug firm, Novo Nordisk.

Impotence—erectile dysfunction—is a serious side effect of diabetes. About 50% of diabetic men develop erectile dysfunction within 10 years of diagnosis. Initial insulin treatment restored the sexual drive of some male diabetics. The added benefit helped to convince many men to accept insulin treatment in the early days, when there were still clouds of doubt.

Insulin served as an unwilling accomplice to a medical practice of dubious distinction: "insulin coma therapy," also known as "insulin shock therapy," for the treatment of mental illness in general and schizophrenia in particular. The movie *A Beautiful Mind* portrayed the torturous ordeal endured by John Nash during the treatment of his schizophrenia at the Trenton State Hospital in New Jersey. This aggressive treatment was invented by an ambitious Polish physician, Manfred J. Sakel (1900–1957), in Vienna in the 1920s. After insulin injection, the glucose level of a diabetic actress, who was also a morphine addict, dropped precipitously, and she slipped into a mild coma. But her addiction to morphine subsided after she woke up. Sakel realized that he had accidentally overdosed his patient on insulin, which lowered the glucose level in the brain even further. On further experimentation, Sakel extended the insulin coma therapy to schizophrenia and claimed a stunning 88% cure rate. The treatment was hyped and adopted throughout the world, not surprisingly considering that no other treatments were available for schizophrenia at the time. In the 1940s, after 20 years of extensive usage, it was found that only a very small percentage of patients with schizophrenia were helped by insulin coma therapy. It was simply too dangerous and too expensive; the benefit did not justify the torture and suffering for most mental patients. Insulin coma therapy was replaced by electroconvulsive therapy (ECT), which was beneficial for severe refractory schizophrenia and catatonic depression. Insulin coma therapy was abandoned in the 1960s when modern neuroleptics started to emerge. Manfred Sakel was known to be a quick-tempered and arrogant man with a vicious and dogmatic spirit. Deeply embittered by the discarding of his invention, Sakel became more aggressive and paranoid. He died of a heart attack in 1957.

John Jacob Abel was the first to isolate the crystalline form of insulin in 1926. Abel was a professor of pharmacology at Johns Hopkins University and a peer of medical luminaries such as William Halsted and William Osler, both of whom were also on the faculty. Abel began his foray into the crystallization of insulin at the age of 67. Abel started with Lilly insulin from beef pancreas via isoelectric precipitations. He carefully prepared exquisitely precise buffers by adjusting pH values with pyridine, brucine, and acetic acid. Using those buffers, he obtained several fractions,

Insulin crystals
© Postens Filateli, Denmark

one of which had the highest sulfur content and the highest physiological activity. In November 1925, Abel was rewarded with one of the most beautiful sights of his life—glistening crystals of insulin forming on the side of a test tube. The crystals, melted sharply at 213°C, were ten- to fifteenfold more potent than the insulin samples he started with.

Abel's crystalline insulin was an extraordinary scientific achievement, which was more impressive considering that scores of groups around the world were attempting to reach the same goal. Accompanying the publication of his results, the *Journal of Pharmacology* published an editorial titled "Crystalline Insulin—Another Chemical Triumph." It called Abel's work "an outstanding accomplishment in the life of a man already distinguished by his conquest in biochemistry."[27]

In early 1926, Abel suddenly became unable to make crystalline insulin. A cloud of suspicion gathered around his work. His greatest critics and competitors were his British colleagues, especially Henry Dale. They proposed the "absorption theory," in which the crystals were the active principle of insulin absorbed on an inert crystal protein, or even cysteine. After a year's tension, frustration, and hard work, Abel finally regained his ability to make insulin crystals in early 1927. On the tail of a new and exciting idea, Abel often worked himself into a state of collapse, obsessed with a desire to get to the bottom of the matter and impatient with all interruptions. As a matter of fact, insulin crystallization was not really magic. A swath of parameters affected the crystal formation. For one, beef insulin was much easier to crystallize than insulin from pig pancreas. Abel's initial failure could be traced back to the source of insulin: Lilly had mixed pork and beef insulin at one point. Another factor was crucial as well: crystalline insulin contains traces of zinc. Abel originally borrowed the brucine he used from a colleague in the department of chemistry who had kept it in a zinc-alloy container. Abel later purchased it from a commercial source in glass containers.

Insulin is a natural protein with 51 amino acids connected in two chains with two disulfide (-S-S-) bonds. The two chains are called the *alpha* chain (glycine chain), with 21 amino acids, and the *beta* chain (phenylalanine chain), with 30 amino acids. After 10 years of persistent pursuit,

Frederick Sanger of Cambridge University deciphered the amino-acid sequence of bovine insulin in 1955. The reason that Sanger chose insulin was not only that it is a protein but also that insulin was the only protein that could be bought in a pure form at the time. Insulin was the first protein whose amino-acid sequence was completely deciphered. Sanger also reported his laboratory techniques using partition chromatography for the determination of the order in which amino acids are linked in proteins. Sanger's work made it possible for other chemists to identify the exact amino-acid sequence and structure of other compounds. The 40-year-old Sanger was awarded the Nobel Prize in Chemistry in 1958 "for his work on the structure of proteins, especially that of insulin." Many Nobel laureates would take on big administrative or teaching jobs, but Sanger stuck with research and became the only chemist to win the Nobel Prize in Chemistry twice. The second one was in 1980, shared with Walter Gilbert "for their contributions concerning the determination of base sequences in nucleic acids."

After the publication of Sanger's amino-acid sequence of bovine insulin in 1953, the race was on, with organic chemists around the world attempting to make insulin by way of total synthesis. After a decade of fierce competition, an unlikely winner emerged: a group of chemists from Shanghai and Beijing in the fledgling People's Republic of China.[28]

In 1958, Chairman Mao Tse-Tung initiated the "Great Leap Forward" movement in China. The overzealous and unrealistic goal was to surpass Great Britain's gross domestic product (GDP) in 15 years. One of the blunders was that a national steel and iron campaign was waged, exhausting a tremendous amount of resources with little success. On the basic scientific research front, the Chinese launched a national campaign and successfully synthesized bovine insulin in 1965. The insulin from their synthesis gave exactly the same crystalline form and biological activities as natural insulin. Considering that the People's Republic was founded only in 1949, it was indeed a gigantic scientific feat. Curiously, the article announcing the achievement began with the following statement:

> The successful total synthesis of a protein was accomplished in 1965 in the People's Republic of China. Holding aloft the great red banner of Chairman Mao Tse-Tung thinking and manifesting the superiority of the socialist system, we have achieved, under the correct leadership of our party, the total synthesis of bovine insulin.[28,29]

The article went on to criticize the work of P. G. Katsoyannis at Brookhaven National Laboratory:

> Since 1963, Katsoyannis and his coworkers in the United States of America have announced on several occasions the total synthesis of insulin, but have so far not supplied the necessary data for their claims. Their published notes and preliminary communications provide neither information about the experimental conditions chosen nor quantitative data of the activity of their final products.[29]

Today, 40 years after its publication, even the distance of history does not obscure the article's strong and frank political statement and the direct attack on Katsoyannis, deservedly or not.

The crystalline form of insulin made it possible to determine the 3-dimensional structure using X-ray crystallography. Whereas it had taken Frederick Sanger 10 years to figure out the amino-acid sequence of insulin, the 3-dimensional structure of insulin took 34 years and was accomplished by Dorothy Crowfoot Hodgkin at Oxford University. It was a daunting task, considering that she had to determine the relative positions of 777 atoms in insulin, whose molecular weight is 5800. Hodgkin, the 1964 Nobel laureate in chemistry, started her research adventure with insulin in 1934 and completed the deciphering of the 3-dimensional structure of insulin on August 14, 1969. Hodgkin later reckoned that it was the most exciting event of her life. She always reminisced about walking around the town of Oxford in pure delight the day she developed the first diffraction photograph of insulin. In addition to being a great research scientist, Hodgkin was also an excellent teacher. One of her more famous students was Margaret Thatcher, the former British prime minister, who made a career move from chemistry to politics. The teacher, a liberal, and the student, a conservative party chairwoman, seemed to get along. They were spotted having tea together from time to time at 10 Downing Street.[30]

In 1983, B. H. Frank and R. E. Chance produced human insulin by cloning recombinant DNA in *E. coli,* instead of the traditional extraction from the pancreases of pigs and horses. Insulin was the first protein hormone that was isolated, crystallized, sequenced, and synthesized by genetic engineering. It is also the first hormone for which a tetrameric receptor, equipped with tyrosine kinase activity, was detected and defined.

Oral Diabetes Drugs

Type 1 diabetes is insulin dependent. With the wondrous efficacy that insulin bestows, type 1 diabetes is largely controllable with insulin injection. However, another, more prevalent, form of diabetes, type 2 diabetes, is not insulin-dependent. Oral diabetes drugs are generally required for most patients with type 2 diabetes.

Type 2 diabetes is closely associated with age, lifestyle, obesity, lack of exercise, and genetic predisposition. The more obese a person is, the more likely he or she will be inflicted with type 2 diabetes. With lifestyle changes in industrialized countries, type 2 diabetes is rapidly becoming an epidemic. In the United States alone, more than 16 million people are stricken by type 2 diabetes. The vast majority of patients with diabetes in America have type 2, up to 90–95%. Those patients slowly lose the ability to produce insulin in their pancreases, which lose sensitivity to the insulin that is produced. Type 2 diabetes is also closely related to diet. After World War II, the incidence of diabetes among Native Americans increased twenty-fivefold within a decade. During evolution, the human body became acclimated to its habitat. The mostly vegetarian dietary intakes by Native Americans had significant protective effects until they changed to a high-sugar/high-fat diet after the prosperity of the mid-1940s.

Oral diabetes drugs may be classified into two categories: (1) agents that augment the supply of insulin such as sulfonylureas and (2) agents that enhance the effectiveness of insulin, such as biguanides and thiazolidinediones.

Sulfonylureas

More than 20 million patients worldwide are treated with hypoglycemic sulfonylureas. It was realized that insulin contains a large amount of sulfur in the protein molecule. Therefore, sulfur was deemed essential for insulin's biological functions. Initial attempts to capitalize on this observation were naïve: scientists simply prepared colloidal sulfur (elemental sulfur) and gave it to animals and humans either orally or parenterally. Although a slight decrease of glycemia and glycosuria was observed, it did not amount to any significant benefit to insulin-independent diabetics.

Sulfonylureas, on the other hand, are effective in controlling type 2 diabetes. Marcel J. Janbon discovered the antidiabetic effects of sulfonylureas in 1942 by chance.

Gerhard Domagk had discovered Prontosil in 1932, and from then on sulfa (sulfanilamide) antibiotics flourished. In 1942, France was under Nazi occupation. Janbon, professor of pharmacology at the University of Montpellier, gave a sulfa antibiotic drug to soldiers to alleviate their typhoid fever. Some patients who took this drug felt tired and dizzy, an indication of a drop in blood glucose levels. Three patients subsequently died of hypoglycemia (low glucose). Janbon infused glucose into some of his patients, who promptly recovered. He made an astute hypothesis that some sulfa antibiotics, namely sulfonylureas, could be used in the treatment of diabetes. The particular sulfa antibiotic drug that lowered sugar levels as a side effect was isopropylthia-diazole, IPTD. To confirm his theory, Jabon assigned Auguste Loubatirères, a medical student in the university, to further investigate the correlation between the sulfa drug and hypoglycemia as part of his doctorate. Loubatirères carried out a thorough investigation into the effects of IPTD on animals and concluded that IPTD could be useful in treating diabetes.

Many additional sulfonylureas, including tolbutamide, chlorpropamide, and glibenclamide, were found to be effective in lowering glucose. Over time, sulfonylurea use strikingly increased. Tolbutamide, which possesses a toluene moiety, is metabolized readily and thus requires a twice-daily treatment. The advantage of tolbutamide was that it did not possess antibacterial properties, thus avoiding buildup of bacterial resistance. Unfortunately, tolbutamide (trade name Orinase), marketed since 1957, was found to be associated with increased cardiac mortality and was withdrawn from the market in 1997. Chlorpropamide, on the other hand, has a chlorine substituent in place of the methyl group in tolbutamide. Therefore, chlorpropamide is less prone to be metabolized and can be taken once daily; the same is true for glibenclamide.

Sulfonylureas are also known as insulin secretagogues because they stimulate the secretion of insulin from beta cells in the pancreas. In other words, they increase the sensitivity of beta cells toward glucose, enabling them to release insulin. The exact mechanism is unknown, but sulfonylureas probably promote depolarization of the beta-cell membrane by closing off ATP-gated potassium channels. Normally, these channels are closed when intracellular levels of glucose increase. Sulfonylureas are ineffective in completely insulin-deficient patients, for successful therapy likely requires about 30% of normal beta cell function.

The main complication of sulfonylurea treatment is the excessive pharmacological effect seen with most antidiabetic agents (including insulin)—hypoglycemia (the level of glucose in blood is too low) in normal patients, as well as in diabetics. Hypoglycemia may cause fainting or even potential brain damage. Thankfully, sugar intake, such as a glass of orange juice, can easily reverse hypoglycemia.

Biguanides

Sulfonylureas serve to increase insulin production by the pancreas. But insulin production in the pancreas is limited. In due time, the pancreas will not be able to generate insulin any longer, so there is a need for drugs to increase insulin sensitivity of the pancreas. Biguanides and thiazolidinediones belong to this category. Guanidine itself was too toxic, even though it lowered the blood sugar level of diabetics. On the other hand, biguanides were known as hypoglycemic agents (drugs that lower glucose levels) in animals as early as 1929. Biguanides work by enhancing the effectiveness of insulin. They can lower excessive blood glucose levels only if insulin is present in the body; biguanides do not stimulate insulin release by themselves.

Phenformin, an early biguanide, was one of the few available oral agents for some time. However, phenformin has limited utility due to its serious side effects. Phenformin was withdrawn from the market by the FDA in 1977 because it caused severe lactic acidosis. Metformin (Glucophage), an analog of phenformin, has 10 to 15 times fewer incidences of lactic acidosis and is thus safer than phenformin. The FDA approved metformin for use in the United States in 1957. Bristol-Myers Squibb still sells it under the brand name of Glucophage.

The highest concentrations of metformin are found in the gut and liver. It is not metabolized but is rapidly cleared from plasma by the kidney. Because of rapid clearance, metformin is usually taken two to three times daily.

Thiazolidinediones

Similar to biguanides, thiazolidinediones (TZDs) are insulin sensitizers.

A Japanese pharmaceutical company, Sankyo, discovered the first thiazolidinedione, troglitazone (Rezulin). Whereas Sankyo sold troglitazone itself in Japan, Parke-Davis Pharmaceuticals licensed its U.S. rights from Sankyo. Parke-Davis carried out its own clinical trials of troglitazone in the United States and proved that it was safe in clinical trial. The FDA granted approval of troglitazone for the treatment of type 2 diabetes mellitus in 1997. However, as is occasionally the case when a drug makes a leap from the thousands of patients in clinical trials to the public, severe liver side effects started to surface when the prescription population grew to millions. When the problems began to surface, the FDA sent all physicians a warning letter, asking them to monitor liver function of the patients who took troglitazone. Unfortunately, only a mere 5% of doctors complied. In time, some patients developed liver damage that required surgical operations and even liver transplants. On March 21, 2000, Parke-Davis voluntarily removed troglitazone from the U.S. market when two other glitazones, Actos and Avandia, became available.

Actos and Avandia are still on the market. Actos (rosiglitazone) is made by Takada, Japan's number one drug maker, with Lilly as their U.S. co-marketing partner. And Avandia is made by GlaxoSmithKline. Although they all possess the same TZD functional group as troglitazone, they are less toxic because they are more potent than troglitazone, thus requiring smaller doses. They were the fruits of empirical medicinal chemistry and rodent pharmacology used by the pharmaceutical industry over the past 40 years. Although the cellular targets were initially unknown, Actos and Avandia have been successfully employed in the treatment of hypertriglyceridemia, or type 2 diabetes, in humans for several years. It was only recently that they were all found to be peroxisome proliferator-activated receptor-γ (PPARγ) agonists.

CHAPTER 7

Anesthetics

In science the credit goes to the man who convinced the world, not the man to whom the idea first occurs.

—Sir William Osler (1849–1919)

History

The easiest pain to bear is someone else's.

In the preanesthesia era,[1–7] the prospect of surgery was so terrifying that it was not uncommon for a tough-hearted man to commit suicide rather than go through that unbearable, excruciating agony. It is hard to believe that there was a time when nothing was effective to alleviate surgical pain. The patients were simply strapped down and that was *it.* As a consequence, speed was the most important attribute of a surgeon in those days. A great English surgeon, Robert Liston at the University College Hospital, once boasted that he had amputated a leg in 29 seconds, along with a testicle of his patient and a finger of his assistant. The operation rooms were often strategically located at the tops of towers in the hospitals to keep fearful screams from being heard. During wartime, surgeries were even worse than battlefield injuries, because during the fight soldiers were temporarily "hypnotized" and became oblivious to pain.

Before anesthesia, surgeons resorted to whatever means were available to deaden the pain of their patients during operations. The three most popular methods were alcohol, ice, and narcotics. Legend has it that a surgeon first conceived the idea of operating during a patient's alcoholic coma

Chinese surgeon Bian Què © Chinese Postal Bureau

when he noticed that a drunkard had had parts of his face chewed away by a hog but was not aware of it during a drunken stupor. Chinese surgeon Bian Què (401–310 B.C.) was reported to have operated on a patient's brain using herbal extracts to render him unconscious more than 2,000 years ago. Hua Tuo (115–205 A.D.) made his patients take an effervescing powder (possibly cannabis) in wine that produced numbness and insensibility before surgical operations.

Cold deadens pain by slowing the rate impulse conduction by nerve fiber. Some surgeons used ice to numb limbs before amputations. This method was invented by Baron Dominique Jean Larrey (1766–1842), surgeon of Napoleon's Grande Armée.[4] Larrey was born in Baudean on July 8, 1766, and later studied medicine at Toulouse. He also served as the chief surgeon of the Royal Guard in all of Napoleon's military campaigns from 1805 onward, including the Battle of Eylau, the Battle of Wagram, the Russian Campaign, and Waterloo, where he was taken prisoner after being wounded. In 1807, immediately after the battle of Eylau, he noticed that soldiers suffered little pain during surgeries when they had developed gangrene of the extremities due to frostbite. From then on, he began to use ice before his surgeries, a practice rapidly taken up by many surgeons. Napoleon greatly admired Larrey and made him a baron, stating: "To the French Army's Surgeon General, Baron Larrey, I leave a sum of 100,000 francs. He is the worthiest man I ever met." Larrey died in Paris on August 1, 1842.

Other pain relievers included narcotics such as opium, laudanum, and henbane. As early as the seventeenth century, the eminent British chemist Robert Boyle was reported to have injected a warm solution of opium into a large dog. The hope was that the bloodstream would carry the drug to the brain and cause unconsciousness. The effects were quick to manifest, and the dog immediately began to nod and reel. In order to preserve the dog's life, Boyle ordered him to be kept awake by whipping.

Mandrake (*Mandragora officinarum*) is a plant related to the potato family. The either genuine or perceived insensibility-causing effect of mandrake was well known for centuries. In Shakespeare's *Antony and Cleopatra,* Cleopatra cried:

Give me to drink of mandragora . . .
That I might sleep out this great gap of time
My Antony is away.

Unfortunately, all aforementioned attempts to alleviate pain proved to be largely unsatisfactory. In 1839, the French surgeon Louis Velpeau lamented, "To obviate pain in operation is a chimera which we are not permitted to look for in our day."[1] Ironically, his pessimism became obsolete in a short decade.

In the nineteenth century, two of the most important discoveries in medicine were surgical antisepsis and surgical anesthesia. Oliver Wendell Holmes, a well-known physician, novelist, and poet, coined the word *anesthesia* in 1846. He derived the name from the same Greek word, meaning "lack of sensation." Whereas Joseph Lister was indisputably the sole discoverer of antisepsis using carbolic acid, the real discoverer of ether anesthesia is one of the most contentious issues in medical history. Three out of the four contenders succumbed to insanity in the end.

Wells and Laughing Gas—Nitrous Oxide

Joseph Priestley (1733–1804) synthesized nitrous oxide in 1775.[8] Priestley was a luminary in both theology and history in Shrewsbury and London. He made a short yet astonishingly fruitful foray into chemistry. Besides nitrous oxide, he also discovered oxygen (via his theory of phlogiston) and 20 other gases, including ammonia, hydrogen chloride, sulfur dioxide, carbon oxides, and silicon tetrafluoride. He was a nonconformist clergyman whose dissent from the Church of England and his support of the 1789 French Revolution led him to flee to America in 1794, settling in Northumberland, Pennsylvania. Today, the American Chemical Society maintains a memorial in his home in Northumberland.

Not much happened with nitrous oxide after Priestley's first synthesis until Humphrey Davy started experimenting with it. Davy (1778–1829) was a rare genius, although he did rather poorly at school, which he left at age 13. The old education methods seemed to stifle him, but he was known to have given scientific lectures to children on the street when he was only 8. Having been recognized by a slew of mentors during his youth, Davy quickly ascended to the position of superintendent of Thomas Beddoes's

Pneumatic Institute at the tender age of 21. Sometime during his tenure there he bravely inhaled nitrous oxide, despite the fact that all previous researchers, including Joseph Priestley, had announced that nitrous oxide was poisonous. As it turned out, impure nitrous oxide gas contained corrosive nitric acid, which damaged the mucous membrane in the mouth because it was prepared by treatment of copper or iron with diluted nitric acid. Later, Davy found out that only purified nitrous oxide was safe to inhale. Pure nitrous oxide gave him soaring euphoria and later unconsciousness, which he exclaimed to be an absolutely intoxicating sensation. Davy related that the sensations it gave were, in general, analogous to those associated with intoxication from fermented liquors. He called nitrous oxide "laughing gas" because many people involuntarily started laughing uncontrollably after inhalation. In 1800 Davy published a flamboyant and widely read book on nitrous oxide titled *Researches, Chemical and Philosophical; Chiefly Concerning Nitrous Oxide and Its Respiration.* Davy presciently suggested in his book: "As nitrous oxide in its extensive operation appears capable of destroying physical pain, it may probably be used with advantage during surgical operations." Unfortunately, it seemed as though nobody took notice and tested the idea until nitrous oxide was used as a general anesthesia in dental and surgical practices in 1844.

Thanks partially to the publicity generated by Davy's book, nitrous oxide rapidly became a fad as a salon amusement, known as "laughing gas frolic." The exhilarating sensation it caused earned laughing gas another name, "exhilarating gas." It was especially popular with wealthy young people seeking escape from boredom by getting high. Inhaling gases became so much in vogue that even carbon monoxide was fashionable for a short while (thankfully). The pink, rosy cheeks caused by carbon monoxide poisoning (asphyxia) were considered beneficial to one's health. In the early 1800s, "laughing-gas sideshows" became popular in America as well. Itinerant "doctors" traveled with bags of nitrous oxide, giving scientific lectures. They would gather a big crowd and charge a quarter for a ticket. They would then ask volunteers to take a sniff of the laughing gas and let them make fools out of themselves.

On December 10, 1844, an itinerant lecturer, Gardner Q. Colton, set up a laughing-gas sideshow at the Union Hall in Hartford, Connecticut. To demonstrate the effects of inhaling nitrous oxide, he asked for volunteers from the crowd to try it. A young man, Samuel Cooley, was one of those brave volunteers. He was quickly overcome by hilarity and started dancing and running around. After the show, a friend of his, a local dentist, Horace Wells (1815–1848), noticed that Cooley had banged his shin and bled quite a bit. When questioned, Cooley claimed that he had been completely unaware of the wound while under the influence of nitrous oxide.

It was at that moment that Wells made the monumental connection between nitrous oxide and pain relief. Wells later wrote: "Reasoning by analogy, I was led to believe that surgical operations might be performed without pain."[1] Without wasting any time, Wells tried his idea the very next day. With Gardner Colton on hand to administer nitrous oxide for him, Wells had his former student, John Riggs, extract a bothersome wisdom tooth of his. Wells woke up and did not feel any pain. What he exclaimed was possibly the understatement of the century: "A new era in tooth-pulling!"[1] Little did he know that that event would change medical history.

After a month and a half, having pulled four teeth for his patients with the aid of nitrous oxide, Wells was ready for a bigger audience and greater glory. In those days Boston was the center of politics, commerce, science, and medicine in New England. To gain more acceptance and notoriety, Wells traveled to Massachusetts General Hospital, which was and still is the teaching school of Harvard Medical School and the mecca of medicine.

In January 1845, Wells secured an invitation from John C. Warren, a renowned surgeon and the head of surgery at Massachusetts General. Wells was to address the senior class of Harvard Medical School on his theory of painless dental extractions. The lecture was to be followed by a demonstration. On that day, Wells chose a student volunteer from among a crowd of skeptical onlookers. After administering nitrous oxide that he had procured in Boston, Wells pulled the patient's tooth; the patient screamed loudly during the extraction. Although the patient later confessed to having experienced significantly less pain than he normally would have had, the damage was done. The audience perceived the outcry as a sign of total failure; they started laughing, hissing, and yelling "humbug!"[9]

We will never know what exactly made the patient cry out. It could have been the quality of nitrous oxide; it could have been the silk bag that Wells used leaked; or it could just have been that Wells withdrew the gas-bag too prematurely. In hindsight, the ill-fated demonstration might have underscored the low anesthetic potency of nitrous oxide. Now it is well known that nitrous oxide is less potent than ether for its anesthetic effects. And ether, in turn, is less potent than chloroform.

Wells's confidence was utterly shattered by the experience. Fearing that he had become a laughing stock of the medical community and deeply embittered by his lack of public recognition, Wells left the dental profession to pursue other strange and unusual ways to make a living. He spent many years peddling canaries, shower baths, and fake European masterpiece paintings and engravings. Unfortunately, he only made a pittance. Wells's personality was characterized as volatile, erratic, and wayward. After the failed demonstration at Massachusetts General, his mind was incessantly preyed on by the imaginary sound of "humbug!" After James Simpson's announcement of chloroform's anesthetic power in late 1847, Wells decided to reseize the glory of being the discoverer of inhalation anesthesia by promoting chloroform. In January 1848, he traveled from his Hartford home to New York City, where he began experimenting with the newest anesthetic, chloroform. Somehow the experiments went awry. He started to behave strangely under the influence of nitrous oxide, ether, and especially chloroform. Sadly, Wells became one of the very first chloroform addicts, and he was in a state of constant delirium. It is evident now that he suffered from a profound depression and hallucinations; historians agree that Wells was insane in his last days. In a delusional state, he was caught throwing sulfuric acid on two streetwalkers on Broadway and was promptly arrested and incarcerated in Tombs Prison. Wells somehow managed to slip in a bottle of chloroform and a razor. On January 24, 1848, he cut the femoral artery in his left thigh under the influence of chloroform. He died at the age of 33. His discovery, which benefited mankind, became an accomplice to his own destruction.

In his suicide note to his wife, Wells wrote:

> Oh! My dear wife and child, whom I have destitute of the means of support—I would still live and work for you, but I cannot—for I were to live on, I would become a maniac. I feel that I am but little better than one already, and now while I am scarcely able to hold my pen I must bid all farewell. May God forgive me.[1]

During his funeral, his widow said at the graveside: "My husband's great gift, which he devoted to the service of mankind, proved a curse to himself and his family."[9]

Redemption came shortly after Wells's death. Just a few days after his death, the Paris Medical Society officially credited Wells as the first to use general anesthesia to perform painless surgery. In 1873 the medical and dental professions of Britain passed a resolution officially awarding Wells credit for the discovery of anesthesia. In his hometown of Hartford, Connecticut, citizens erected a statue of Horace Wells in Bushnell Park in 1875. At the base of his statue are a walking stick, a book, a bag (for nitrous oxide gas), and a scroll. Even today, in a park called Place des Etats Unis at the heart of Paris, there is a statue of Wells that commemorates his discovery of nitrous oxide for general anesthesia. Adjacent to Wells's sculpture in the plaza stands a monument to Washington and Lafayette. At Wells's grave a monument was erected. On the left it says "I Sleep to Awaken"; on the right, "I Awaken to Glory."[8]

Long, Morton, Jackson, and the Great Ether Controversy

Ether, short for diethyl ether (Et_2O), is relatively easy to prepare by dehydration of ethanol, the main ingredient of alcoholic drinks, in the presence of acid. Because sulfuric acid was often used as the catalyst, ether was

Paracelsus © Warner-Lambert

also known as sulfuric ether. Ether was discovered in 1275 by Spanish chemist Raymundus Lullius, although another source gave the credit to German scientist Valerius Cordus. Swiss physician and alchemist Paracelsus (1493–1541) first discovered the hypnotic effects of ether. Among many significant contributions to science made by Humphrey Davy, one of the most important ones was his recognition and mentoring of another chemistry genius, Michael Faraday (1791–1867), who experimented with ether and proposed that it be used to induce a similar effect to that exerted by nitrous oxide.

One contender to claim the discovery of ether anesthesia was Crawford W. Long.[10,11] In 1841, Crawford Long (1815–1878) was a country doctor who introduced nitrous oxide and ether inhalation at his parties. By taking advantage of the intoxication of the partygoers, Long was rumored to have kissed the girls without getting himself into trouble. On March 30, 1842, Long successfully carried out painless removal of a neck tumor after dropping ether on a cloth that covered the site of the tumor. He charged the patient 25¢ for ether and $2 for the operation, but he did not publish his results. Later he had to resort to friends and patients to corroborate his claim. Publication would have established the validity of his claim of priority beyond any doubt. A lesson can be learned from the incident that publications are an important and integral part of discoveries, although today patents are a more prevalent form of publication for many discoveries.

After Long died of natural causes in 1878, the state of Georgia put up a statue of Long in the state capitol. Long's discovery of ether for surgical anesthesia is widely considered one of the 10 greatest discoveries in medicine. March 30, the anniversary of Long's great accomplishment, is celebrated in the United States as Doctor's Day, when all medical specialties are honored.

The second contender in the discovery of etherization was William T. G. Morton. William Morton (1819–1868) was born to a well-to-do family in Charlton, Massachusetts. When he was seventeen, his father's business collapsed, and he lost his wealth. After this, the younger Morton had to find employment as a bookseller in Boston to support his family for about a year. Between 1836 and 1841 he traveled around the country: Rochester, New York; Cincinnati, Ohio; and St. Louis, Missouri. In late 1842, Morton moved to Baltimore, Maryland, and then back home to Charlton, Massachusetts, after which he studied dentistry in Boston for the next couple of years. In 1843, Morton formed a partnership with Horace Wells and commenced the practice of dentistry in Boston. The short-lived partnership lasted only 3 weeks, at which point Wells sought to dissolve the partnership because he found Morton to be the ". . . most deceitful man . . . who uses his tongue in making random statements."[12]

In the spring of 1844, Morton visited his former creditor in the nearby town of Farmington, a lady who had kindly loaned him $1,000 to start his dental business a year earlier. There, Morton met the beautiful 16-year-old Elizabeth Witman, the creditor's niece. They quickly fell in love, and Morton asked her father, Edward Witman, for the lady's hand in marriage. Her father, a wealthy man, deemed Morton unworthy to be a son-in-law because he was a dentist and would not sanction the marriage until Morton pledged to go to school to study medicine. Later, Elizabeth wrote: "Dr. Morton had paid me attentions which were not well-received by my family, he being regarded as a poor young man with an undesirable profession. I thought him very handsome, however, and he was very much in love with me, coming regularly from Boston to visit me."[12] In the end, love prevailed. William and Elizabeth were married on May 29, 1844. During their honeymoon, Morton brought a skeleton with him to study human anatomy—talk about a "skeleton in the closet"! Morton did indeed go to Harvard Medical College for a period of time, although he never earned an M.D. degree. Sometime in 1845, the couple moved to Charles T. Jackson's house in Boston as tenants. Jackson was a preeminent professor in medicine and geology at Harvard University and a well-known physician, chemist, and geologist in Boston. Little did they know that their lives would be so intimately intertwined until their respective deaths.

Morton began experimenting with nitrous oxide to aid his dental practice after Wells's ill-fated demonstration at Massachusetts General. His scholar-landlord, Jackson, suggested that he try ether. Morton shut the door for secrecy and inhaled from a handkerchief saturated with ether, from which he was able to achieve insensibility for 7–8 minutes. In the 1944 movie *Great Moment,* Morton was depicted experimenting with ether on Elizabeth's pet spaniel and goldfish, to the chagrin of his wife. There was no doubt that Morton's motivation was to make his dental practice more lucrative, because many of his prospective patients turned away from the installation of dental plates for fear of the pain during the operations.

Regardless of Morton's motives, on Friday, October 16, 1846, in the surgical amphitheater (now the Ether Dome) of Massachusetts General Hospital, Morton successfully demonstrated to the public the anesthetic effect of ether and its utility in surgical operations. That day is now known as "Ether Day." On that day John C. Warren granted Morton the stage to try his new painless agent, despite Wells's initial failure with nitrous oxide. Before the operation to remove a tumor on the neck of a patient, Gilbert Abbott, Morton applied ether using an inhaler that he assembled at the last minute. Warren carried out the operation without the fearful screams that had always accompanied surgical operations previously. Afterward, Warren slowly and emphatically exclaimed: "Gentlemen! This is not humbug."[9] After having witnessed the first painless operation, Henry J. Bigelow said it the best: "I have seen something that will make its way around the world!"[11] That day was the best day in the 27-year-old Morton's life, just a decade after his father had suffered his financial debacle. Sadly, it was also the day that his demise began. From then on, his life was almost entirely consumed in the profit from ether and in defending his role in its discovery.

The news of triumph over pain spread through American hospitals like wildfire. New York, Chicago, St. Louis, and Buffalo rapidly adopted Morton's discovery. Henry Bigelow described the ether anesthesia in a letter to

William Morton applying ether at Massachusetts General Hospital © Warner-Lambert

Robert Liston, who performed the first major operation under ether in England on December 21st, 1846. The patient, a middle-aged man, lost consciousness during amputation of his right leg. He woke up and started sobbing when he saw his stump but felt no pain. The audience was deeply moved, and Liston was full of joy: "Hurrah, Rejoice! An American dentist has used the inhalation of ether to destroy sensation in his operations. In six months no operation will be performed without this precious preparation."[1] He was right.

The issue of who deserved the credit for discovering etherization after "Ether Day" caused a great controversy. The debate involving the Wells–Morton–Jackson triangle raged on for decades, until Jackson's death in 1880. Ether brought miserable ends to Wells, Morton, and Jackson.

Before Ether Day, Morton had applied for and been granted a U.S. patent (Letters Patent No. 4848) in November 1846, with Jackson and Morton listed as coinventors. In order to conceal the real chemical composition of ether, Morton camouflaged it with some fragrances and called it *Letheon* (from Greek mythology, in which a drink from the river Lethe could expunge all painful memories). Because physicians are bound by Hippocrates' oath, profiting from a life-saving discovery was unthinkable. As a consequence, Morton's patenting of Letheon was met with disdain and was called "nostrum mongering." During the war with Mexico, the U.S. Army and Navy took advantage of the anesthesia but did not honor the Morton–Jackson patent. Jackson promptly renounced the patent and relinquished his rights after having realized that the government had annulled the patent it had just granted.

Morton, on the other hand, behaved exactly the opposite way. In his book *Tarnished Idol,* Richard J. Wolfe wrote that Morton:

> engaged in the most scurrilous and unprincipled debate, resorted to vile character assassination, and committed other misdeeds in order to annihilate those who opposed him when he attempted to obtain substantial remuneration from Congress for a discovery in which he played only a partial role.[12]

The 1944 movie *Great Moment* concluded with the following reflections: "Dwarfed by the magnitude of his revelation, reviled, hated by his fellow men, forgotten before he was remembered, Morton seems to be very small indeed until he reined himself. . . ." Morton fervently lobbied Congress for pecuniary reward for his discovery. He sought $100,000 of federal compensation for his discovery of the anesthetic properties of sulfuric ether by invoking names of several inventors and discoverers who had received awards from their grateful governments. In particular, the British Parliament had awarded Edward Jenner (1749–1823) a combined £30,000 in 1802 and 1806 for his discovery of the vaccination to combat smallpox. Morton garnered the staunchest support of Daniel Webster (1782–1852), a Massachusetts senator (1823–1827) and U.S. Secretary of State during the last 2 years of his life.

Despite the fact that the House approved the pecuniary reward for Morton's discovery, the bill failed in the Senate. On Saturday, August 28, 1852, the U.S. Senate voted 28–17 to deny Morton's claim of priority in his discovery of the anesthetic properties of sulfuric ether. The patenting of Letheon was one of the reasons that the Senate ultimately denied the monetary reward. Morton was understandably "dispirited, and crushed by this disappointment, sick in body and mind from the reaction from the intense excitement of the previous nine months."[12] He fell terribly ill for 30 days and almost died. In 1853, Morton was expelled from the American Medical Society and disowned by his fellow dentists for his conduct. After relinquishing his profession, he moved to a cottage near Boston, which Morton christened "Etherton." In 1859, Nathan P. Rice published *Trial of a Public Benefactor, As Illustrated by the Discovery of Ether Anesthesia.* The complimentary book liberally quoted Morton, who embellished many details.

On July 15, 1868, Morton died in Central Park in New York. His death was initially attributed to sunstroke, or simple stroke. But Henry J. Bigelow at Massachusetts General, one of Morton's staunchest supporters, seemed to believe that "Morton fell to apoplexy, induced by the publication in behalf of Jackson, of a nature to prejudice a subscription then arranged in New York for his benefit."[12] An apoplexy is a stroke or a cerebral accident. In an ironic twist, the two attending physicians at St. Luke Hospital who took care of Morton were Charles F. Heywood and John C. Dalton. Both of them had been young medical students who were present at the surgical amphitheater of Massachusetts General on Ether Day, when Morton had successfully showed the world the power of ether anesthesia. It was tragic that money and fame became such a plague to Morton, who had a beautiful and loving wife, five children, and a successful dental practice. Without the need to pursue the evil of greed and recognition, he just might have had a happy and long life.

In Mount Auburn Cemetery, citizens of Boston erected the "Morton Monument." The inscription on his tombstone reads "Inventor and revealer of inhalation anesthesia: before whom, in all time, surgery was agony, by whom, pain in surgery was averted and annulled; since whom, science has control of pain."

The third contender to claim the "etherization" discovery was Charles T. Jackson, who was trained at Harvard College and in France to become a physician. After coming back to Boston in 1840, he found that his "hobby" as a geologist and chemist was much more in demand than was his training as a medical doctor. In 1850, Jackson engaged in a furious litigation for the inventorship of the telegraph with Samuel Morse, whose invention had been inspired by a conversation with Jackson aboard the trans-Atlantic steamer *Sully* in 1835. After 5 long years of legal wrangling, the U.S. Supreme Court ruled in favor of Morse in 1854.

Just a few days after Ether Day, Jackson claimed that he had started using ether to banish pain as early as 1845, when he accidentally inhaled chlorine gas, which nearly suffocated him. To relieve his pain and neutralize the acidity generated by chlorine, he alternately inhaled ether and ammonia. Later, Jackson experimented further with ether and used it as numbing drops applied to a patient's tooth during extraction with satisfactory result. Indeed, Morton admitted that it was Jackson who recommended that he use ether. The admission was possibly made to leverage Jackson's fame to help Morton's monetary pursuit—most people certainly would not trust Morton's reputation. Jackson wrote to the French Academy of Sciences via Élie de Beaumont, stating his pivotal role in the ether discovery without even mentioning Morton by name. He simply referred to him as a certain Boston dentist whom he induced to administer the ether vapor for tooth extraction. Jackson also claimed that it was he who had urged Morton to go to Massachusetts General to administer the ether vapor before a surgical operation. Jackson enlisted the support of his brother-in-law, Ralph Waldo Emerson, the renowned author, poet, and philosopher. The Jackson camp also helped to expose Morton's crimes of swindling money in the *National Police Gazette* on February 10, 1855.

However, Morton's reputation as the discoverer of etherization seemed to grow despite the U.S. Senate's refusal to satisfy his pecuniary requests. Dispirited, Jackson turned to an old-fashioned anesthetic, alcohol, to alleviate his mental anguish. Jealousy and resentment kept him awake at night. In the summer of 1873, Jackson's mind finally began to give in, and he started to speak a strange language that nobody else could understand. One rainy day, he was found crying at Morton's tombstone, oblivious to the downpour. Having become incapacitated, he was involuntarily committed to a mental institution, McLean Asylum in Belmont, near Boston. McLean Asylum was and still is affiliated with Massachusetts General, to which Jackson had often consulted with regard to etherization. He had frequently visited McLean Asylum around 1847 to administer ether to the mental patients to provide them temporary relief. Repaying Jackson's invaluable contribution to the discovery of ether anesthesia, Massachusetts General underwrote all expenses incurred during his stay of 7 years at McLean. On August 28, 1880, Jackson's anguish was mercifully ended with his death.

Existing pictures of Jackson all depict him wearing the four medals bestowed on him by foreign governments and organizations. In Boston Garden, a downtown public park, citizens of Boston erected an "Ether Monument" in 1868 for the discovery of ether anesthesia. In Massachusetts General Hospital, the Ether Dome, having witnessed Wells's heartbreak and Morton's triumph, is preserved as a national shrine.

In hindsight, one could conclude that Wells was the father of anesthesia who ushered in a whole new era in medical history. He should not be credited with ether, though, because nitrous oxide and ether are two distinctive chemicals. Long, Jackson, and Morton, especially Morton, all deserved the credit for the discovery of etherization. Indeed, despite Morton's being ignorant and greedy, he was the one who deserved the most credit, because without his demonstration of ether anesthesia at Massachusetts General, the medical profession and the public would not have become aware of it at that time. Despite their bitter ends, these men all were, willingly or unwillingly, benefactors of mankind.

Simpson and Chloroform

The genesis of chloroform can be traced back to three chemists: Samuel Gunthrie in the United States, Eugène Souberian in France, and Justus von Liebig in Germany, who almost simultaneously synthesized it in 1831.[13–15]

Samuel Gunthrie (1782–1848) was trained as a physician and served as an army surgeon in the War of 1812 against the British. But his real passion was chemical experiments, for which he gave up his medical practice and devoted most of his life to independent research. After having spent decades in his own laboratory in his home in Brimfield, Massachusetts, he bragged that he had experienced more than 100 explosions in his lifetime. It was a wonder that he lived as long as he did; his death was related to nerve damage inflicted from those explosions. In 1831, he was the first to make chloroform, albeit inadvertently. In order to find an inexpensive way to make the "Dutch liquid," he treated acetone with chlorinated lime (calcium hypochlorite, $Ca(OCl)_2$). Using distillation from a copper apparatus, he obtained a volatile (chloroform's boiling point is 65.7°C), sweet-tasting liquid. What Gunthrie actually obtained was an ethanol solution of chloroform, which he sold to locals as "Gunthrie's sweet whisky," believing it had beneficial effects on one's health. He later further distilled the ethanol solution and obtained pure chloroform.

French pharmacist Eugène Souberian also prepared chloroform in 1831 by mixing alcohol with bleach followed by distillation. Few records exist on the details of his discovery.

Justus von Liebig (1803–1873) pursued his Ph.D. in organic chemistry in Paris under the tutelage of Joseph Louis Gay-Lussac (1778–1850), whose marriage to his wife Josephine was an extraordinary story in the history of chemistry. During the 1789 French Revolution, Gay-Lussac entered a draper's shop in Paris by chance. The scene surprised him. Behind the counter, a charming girl of 17 was reading a book in the intervals between serving customers. Further inquiry revealed that she was intently studying a treatise on chemistry. It turned out that her formal education had been abandoned as a result of the Revolution. Impressed by Josephine's ardent desire for education, Gay-Lussac sent her back to school at his own expense. Therefore, it was not all that surprising that their marriage in 1808 was a remarkably happy and successful one. A romantic French savant, Gay-Lussac was known to indulge in a frisky waltz around his laboratory to celebrate progresses that he deemed worthy. Young Liebig was often astonished to find himself in the arms of his boss when he reported his successes with his experiments.[13]

In 1824, Liebig waltzed out of Gay-Lussac's laboratory as a newly minted Ph.D. at the age of 21. He was appointed the chairman of chemistry at Giessen University, which incited furious jealousy among several of the professors already working there because he was so young. Fortunately, time would prove that the choice was a wise one for the department. Liebig would soon transform Giessen from a sleepy university into the mecca of organic chemistry in Europe. Liebig is now considered the father of organic chemistry.

In 1831, Liebig treated chloral with caustic potash and distilled the reaction mixture to obtain chloroform. Eager to claim priority, Liebig rushed to publish his results without careful characterization, and he mistakenly assigned the structure as pentachlorodicarbon (C_2Cl_5). In 1840, the eminent French chemist, Jean-Baptiste André Dumas, correctly elucidated chloroform's chemical structure as $CHCl_3$.

Medically, not much happened with chloroform until James Young Simpson (1811–1870), a Scottish surgeon and obstetrician/gynecologist in Edinburgh, first employed chloroform as an anesthetic.

James Simpson was the seventh of eight children born to a poor baker's family. His genius was recognized early on in his childhood. He went to school at age 4 and then to Edinburgh University when he was 14. Because he was so often called "Young Simpson," he changed his name to James Young Simpson.[14]

In 1839 an ambitious Simpson waged a £500 campaign to secure the position of the chairman of midwifery at Edinburgh. At that time, a 28-year-old bachelor was not a good candidate for such a prestigious post. He married a woman he did not love so that he could be more qualified. Simpson was elected to the chairmanship by one vote more than his opponent. He quickly dissolved his short-lived marriage after the announcement of his appointment.

In 1846, Simpson started to use ether in obstetrics. As one would imagine, with ether being highly flammable, fire accidents were not uncommon for doctors administering ether under candlelight. In addition, slow induction of anesthesia and irritancy were also associated with etherization. In October 1847, Simpson started to look for alternatives to ether. Because he was liberally experimenting with so many inhalations, his neighbor was obligated to call in every morning to make sure that he was still alive. At one point, he invited a few friends to his house. After dinner and drinks, he served his friends a whiff of chloroform. They all felt an intoxicating sensation and then passed out after a laughing spree. Chloroform was found to be much more potent than ether as an anesthetic. In November 1847, Simpson revealed his discovery at the Edinburgh Medico-Chirurgical Society, after which chloroform quickly became widely used. At one point, Simpson sent William Morton a letter describing details of his work on chloroform and the formula for its preparation.

Because patients were made unconscious during childbirth, and because they often reported experiencing a delightful sensation, it was rumored that chloroform transformed the agony of childbirth into an orgasm for a few hours. Indeed, chloroform in moderate doses brought on erotic fantasies, which made some believe that chloroform was a sexual stimulant. Some even dubbed chloroform anesthesia as "chloroform orgasm."[14]

That caused an outcry from the church and laymen alike. Religious people believed that pain during childbirth was God's punishment for the sin committed by Adam and Eve. They called chloroform a "decoy of Satan" and anesthesia a violation of the laws of God, whom they believed inflicted pain to strengthen faith and to teach the new mother the need for self-sacrifice for her children. They considered anesthesia for childbirth a challenge to Divine Providence and cited Genesis 3:16: "in pain thou shalt bring forth children." Simpson rebuffed them by also quoting the Bible in regard to the creation of Eve in Genesis 2:21: "And the Lord God caused a deep sleep to fall upon the man, and he slept; and He took one of his ribs, and closed up the place with flesh instead thereof."

When Queen Victoria of England was to give birth to Prince Leopold on April 7, 1853, she requested John Snow, the first full-time anesthesiologist in history, to administer chloroform for anesthesia. Evidently she was happy with the result, because Snow was asked for an encore for the birth of Prince Beatrice in 1857, which was equivalent to today's celebrity endorsement—the approval by the queen was all it took for the general public to accept chloroform as an anesthetic. In 1858, Snow published a vastly popular book titled *Chloroform and Other Anaesthetics.* In contrast to Morton's avaricious money-grubbing and fame-snatching schemes, Snow freely published his results in medical journals, disseminating his discoveries with regard to apparatus, techniques, and applications. Snow's other important discovery was the mode of transmission of germs during the cholera epidemics of 1849 and 1854.[15]

In 1858, E. R. Squibb and Company (forebear of today's Bristol–Myers Squibb) in the United States was established by Edward R. Squibb. During the Civil War, their ether and chloroform production reaped sizable profits. Chloroform revolutionized battlefield surgery for its quick onset and convenience.

Unlike Wells, Long, Morton, and Jackson, Simpson was showered with honors from all over Europe and the world. He was so intoxicated by the laurels that he erroneously announced that chloroform was superior to ether and declared himself to be the *sole* inventor of anesthesia. Before his death on May 6, 1870, from angina pectoris, he wrote letters to all his adversaries in which he asked their pardon for wrongs he had committed. He was buried in St. Andrews Chapel, Westminster, next to Humphrey Davy's memorial. On a marble statue of Simpson, the inscription reads:

> *To whose genius and benevolence*
> *the world owes the blessing derived*
> *from the use of chloroform for*
> *the relief of suffering.*
> *LAUS DEO*[14]
> [*Laus Deo* means "Praise to God" in Latin.]

Unfortunately, despite its potency, the safety of chloroform proved to be less than ideal in comparison with nitrous oxide and ether. As chloroform usage became widespread, cardiac deaths and liver damage became more prevalent. Some people were killed when chloroform replaced all oxygen (asphyxia) in their bloodstreams. In 1937, the last straw was the discovery that chloroform sensitized the pacemaker and caused ventricular fibrillation, after which it was slowly eliminated from use.

Baeyer, Fischer, and Barbiturates

Adolf von Baeyer (1835–1917) synthesized barbituric acid from urea and malonic acid in 1864. At that time, he was "charmed by a young beauty Barbara X," thus he incorporated her name with the letters "ur" to designate the derivatization of the acid from urea. During his lectures, Baeyer used to say, "At the time I was in love with a Miss Barbara. So I named my urea derivative 'barbituric acid.' "[16]

Another version exists on the genesis of the name of barbituric acid—curiously, also told by Baeyer himself. "I frequently took my noon meal with a group of young artillery officers. I told them about the striking new compound that I had found and they at once insisted that I name it after their patron Saint Barbara because this was her day in the Catholic calendar. I was glad to give them the pleasure."[16]

Baeyer's real joy was in his laboratory, and he deplored any outside work that took him away from his bench. As a scientist, Baeyer was free of vanity. Unlike other scholastic masters of his time (Liebig, for instance), he was always ready to acknowledge ungrudgingly the merits of others. Baeyer's famous greenish-black hat was a part of his perpetual wardrobe, and he had a ritual of tipping his hat when he admired novel compounds. At the first glimpse of the fine white crystals of barbituric acid, the master ceremoniously raised his hat in silent admiration.[13]

Adolf von Baeyer
© Dominican Post Office

Barbituric acid, the first of all barbiturates, was the precursor of an extremely important class of therapeutics. However, barbituric acid itself is of no therapeutic significance; it is too polar to penetrate the blood-brain barrier and is unable to reach the brain. In order to accomplish that purpose, some hydrophobic attachments have to be adorned to the molecule to make it greasier.

Baeyer's student, Emil Fischer (1852–1919), was perhaps the greatest organic chemist ever. When he was young, his father, Lorenz Fischer, sent him to study chemistry because "the kid is too dumb to be a business man."[17] In 1903, Fischer prepared diethylbarbituric acid, the first derivative of barbituric acid, from diethylmalonic acid and urea, using sodium ethoxide as a condensing agent. During the latter part of the nineteenth century Fischer's friend Joseph von Mering (1849–1908), at the Clinical Laboratories in Halle, investigated the hypnotic properties of barbiturates in general, diethylbarbituric acid in particular. Von Mering christened diethylbarbituric acid *Veronal,* later known as Barbital in the United States. After the publication by von Mering and Fischer, barbiturates as a group of hypnotic drugs flourished, with more than 2,500 barbituric acid derivatives prepared and their hypnotic value tested. Barbiturates became the most versatile of all sleep drugs at the time.

Fischer was once approached by the novelist Hermann Sudermann, who complimented his Veronal: "You know it is so efficient, I don't even have to take it," he declared. "It's enough that I see it on my nightstand." "What a coincidence," Fischer exclaimed. "When I have problems falling asleep, I take one of your novels. As a matter of fact, it's enough that I see one of your wonderful books on my nightstand and I immediately fall asleep!"[18]

Adolf von Baeyer received the Nobel Prize in Chemistry in 1905 at age 70. His apprentice, Emil Fischer, won it in 1902 when he was 50, 3 years before his teacher. Sadly, World War I turned Fischer's life upside down. He lost two of his sons, medical officers in the German Army, and the Allies confiscated his patent right to Veronal.[17] That, in combination with inflation, reduced Fischer from a wealthy man to a penniless one. To make matters worse, he learned that he had inoperable bowel cancer. Afflicted with severe depression, Fischer, one of the greatest organic chemists of all time, committed suicide on July 15, 1919, at age 67.

The name change from Veronal to Barbital was associated with historical events. After World War I broke out, the United States enacted the Trading with the Enemy Act, embargoing German goods, including chemicals and medicines. There was a dire need to secure drugs from a domestic source. In 1919, Abbott Laboratories hired Ernst Henry Volwiler, the first graduate student of Roger Adams at the University of Illinois, as a research chemist for $35 a week. Volwiler plunged into work immediately and reproduced diethylbarbituric acid (Veronal).[19] Because Veronal was a German trade name, it was sold in America using the brand name of Barbital, which was used mostly as an anesthetic. Later, Volwiler and Donalee L. Tabern discovered pentobarbital, an oral hypnotic barbiturate, which Abbott sold under the trade name of Nembutal. Nembutal distinguished itself by being extremely cheap and having a long narcotic duration, making it a good candidate for abuse. On August 5, 1962, Marilyn Monroe, the famous Hollywood actress and cultural icon, was found dead on her bed. An autopsy revealed that her stomach contained 47 tablets of Nembutal. In 1997, 39 cult members of Heaven's Gate committed suicide by swallowing Nembutal.

From 1932 to 1933, Volwiler and Tabern discovered one of modern medicine's major anesthetics, Sodium Pentothal (also known as thiopenthal and Pentothal), an intravenous barbiturate. They prepared about 200 derivatives of barbiturates in search of short-acting and potent hypnotics. After compilation of the data, the very first one, a sulfur analog of Nembutal, turned out to possess the most striking potential. The presence of a sulfur atom enabled the molecule to be cleared quickly from the body, making it safer than longer acting barbiturates. As Tabern recalled later: "We injected it into a dog and he went to sleep fast and woke up in ten minutes feeling fine—and we were sure we had it."[20] Tabern later became the head of research at Abbott. Volwiler, on the other hand, was Abbott's CEO from 1950 to 1958.[21]

John Silas Lundy at the Mayo Clinic carried out many tests on Pentothal. Although it worked marvelously on most patients, somehow certain subgroups seemed to be less successful. Lundy noted a greater resistance to the drug among persons who had red hair, who were born in the South, or who were Jewish, which prompted Lundy to write to Tabern: "I don't yet know why this should be, but I'd hate to have to administer the stuff to a red-haired rabbi born in New Orleans."[20]

When Sodium Pentothal was used as the sole anesthetic, it would cause cardiodepressant effects. On December 7th, 1941, the Japanese bombed Pearl Harbor, and Sodium Pentothal anesthesia was extensively used in operations on the victims after the bombing. Stunningly, many patients died of Sodium Pentothal overdose. The mystery was solved by Bernard B. "Steve" Brodie at New York University.[22] It turned out that Pentothal, mostly absorbed by fat and muscle tissues, did not break down in the body rapidly. Not knowing the buildup, additional injections tended to cause overdose and even deaths.

Unlike nitrous oxide, ether, and chloroform, Pentothal was fireproof, nonexplosive, easy to transport, and comparatively easy to inject. It proved to be extremely useful during World War II and was administered in 78% of cases requiring anesthesia, with a mortality rate of mere 0.018%. The navy claimed that Pentothal was the most valuable anesthetic on all its hospital ships. Pentothal became the anesthetic of choice, just as Abbott had claimed: "To know intravenous anesthesia is to know Pentothal."[20] To some extent, they were right.

The most dramatic and publicized uses of Pentothal were in the healing of shattered minds, in peripheral maladies, and in police work. During World War II, many British medical doctors employed Pentothal on emotionally wracked civilians and soldiers as a way of treating acute combat neuroses.

Pentothal, along with scopolamine and Amytal (an oral barbiturate anesthetic), were dangerously given the label of "truth serum." They were erroneously thought to influence people so that they would only tell the truth when given those drugs. It was Lundy who first noticed that patients who received a dose of Pentothal would wake up during operations, have a conversation, and answer questions. The patients would have no recollections of the pain or of those conversations. Lundy wrote: "In each instance, the answers seemed truthful. Perhaps this suggests a method for obtaining information from insane persons or criminals."[20] Abbott Laboratories always denounced the use of Pentothal by the police for squeezing out information from suspects. The courts repeatedly threw out information obtained this way.

Pentothal is a fast-acting barbiturate that produces almost instantaneous short-term unconsciousness after a single dose. It is used in many state penitentiaries in the United States as part of the regimen of lethal injection, in which Sodium Pentothal works as an irreversible anesthesia. Jack Kevorkian used Pentothal to help his terminally ill patients to euthanize themselves.

Three types of barbiturates are classified according to their duration of action. Very short-acting barbiturates are mostly used in anesthesia, whereas intermediate- and long-acting barbiturates are administered for treating anxiety and insomnia.

Longer-acting barbiturates, or "downers," are used by substance abusers. They calm down stimulation and are popular in the entertainment industry for helping performing artists to cope with the emotional roller coaster that they have to deal with on a daily basis. Use of barbiturates tends to interfere with the ability of cells to utilize oxygen. In 1955, the United States produced 864,000 pounds of barbiturate (4 billion tablets) as a sleeping aid, enough to put 10 million adults to sleep every night of the year—26 pills for each American—and one out of every seven Americans took barbiturates.

Köller, Cocaine, and Local Anesthetics

Inhalation and general anesthesia were a great boon to surgery. However, when it came to eye surgery, they were not useful because eye movement is even faster during anesthetic sleep than random eye movement (REM). Therefore, local anesthesia is more appropriate for such purposes. In 1884, Carl Köller began to use cocaine as a local anesthetic for eye surgery. Nowadays, cocaine is a well-known illicit drug, validating the point that nothing of itself is good or evil; only the manner of its use makes it so. Cocaine has been both a blessing and a curse to mankind since Köller's discovery.

Indigenous Indians in South America, especially in Peru and Bolivia, have been known for centuries to chew on coca leaves of *Erythroxylon coca* to relieve fatigue and pain. Andes and Inca Indians consider it a sacred gift from the sun god. In 1860, Albert Niemann, working in the laboratory of Friedrich Wöhler in Göttingen, isolated the active principle from the leaves of the coca shrub. Wöhler, who first made urea from inorganic ingredients, named the white crystalline alkaloid *cocaine.* New compounds were routinely tasted, and Wöhler noticed that "Cocaine was a substance which had a somewhat bitter taste and exerted a numbing influence upon the gustatory nerve, so that they became almost completely insensitive."[23]

In 1884, Sigmund Freud experimented with cocaine as a cure for morphine addiction. After procuring some cocaine from E. Merck, the young neurologist swallowed a small quantity of the drug, which calmed his stomach and boosted his libido. He immediately wrote to his fiancée, Martha Bernays:

> Woe to you, my princess, when I come. I will kiss you quite hard and feed you until you are plump. And if you willfully resist, you shall see who is the stronger, a gentle little girl who doesn't eat enough or a big wild man with cocaine in his body. In my last depression I took coca again and a small dose lifted me to the heights in a wonderful fashion. I am busy collecting the literature for a song of praise to this wonderful substance.[23]

Meanwhile, Freud applied some cocaine locally to himself and found that it temporarily paralyzed the sensibility of a certain area without any marked effect on the central nervous system. He enlisted the service of his friend, Carl Köller (1857–1944), an ophthalmologist and house surgeon at Vienna General Hospital, to investigate the utility of cocaine. In the summer of 1884, they attempted to develop a novel and reliable type of local anethetic using cocaine. Unfortunately, Freud's personal affairs involving his future wife led him to travel, and he would not come back to the matter for 2 years.

While Freud was away from Vienna, Köller, being an ophthalmologist specializing in operations on the eye, tried dropping cocaine solution into the eyes of frogs, guinea pigs, himself, and later several of his colleagues. They all proved the effectiveness of the cocaine eyedrop as a local anesthetic. Because young Köller could not afford to travel to Heidelberg for the Heidelberg Ophthalmological Society meeting, he asked his friend, Joseph Brettaur, to deliver the preliminary communication for him. Köller is credited as the first to discover local anesthesia. Two years later, Freud came back to Vienna and lamented:

> When I got back to Vienna, I found my friend Köller, who had been working with me upon cocaine, had made decisive advances

in its use. He therefore can rightfully be considered the discoverer of local anesthesia with cocaine which has been so important in minor surgery. But I am not disposed to feel a grudge against my wife because that honor did not accrue to me.[23]

In reality, that was not such a bad thing for Freud. He is now a household name, known as the "father of psychoanalysis," whereas Köller is recognized by only a few specialized professionals.

During the same time in America, William Halsted, the renowned Baltimore surgeon, experimented with cocaine as well. Halsted and his two associates injected cocaine in each other numerous times to gauge its dosage, efficacy, and toxicity. Unfortunately, Halsted became extremely addicted to the drug. The soaring euphoria and subsequent plummeting depression incessantly haunted him. He once commented on his experience with cocaine: "Only those who have experienced the distress which follows so promptly the brief period of exhilaration can at all comprehend it."[23]

Cocaine is still used as a topical anesthetic by ear, nose, and throat surgeons because of its unique ability to combine local anesthesia and intense vasoconstriction. Despite cocaine's usefulness as a local anesthetic, its central nervous system and cardiovascular systemic toxic effects started to become apparent as its popularity increased. The pharmaceutical industry embarked on a pursuit for less toxic and shorter-acting replacements.

Over the years, medicinal chemists have tweaked the cocaine molecule and synthesized many derivatives with the aim of retaining the local anesthetic properties without the addictive effect of cocaine. Piece by piece, chemists removed fractions of the molecule and arrived at many local anesthetics devoid of the debilitating addictive effects. Benzocaine, the simplest of the "-caines," was identified by von E. Ritsert in 1890 as a topical anesthetic. Its utilities were limited, however, because of benzocaine's poor solubility in water. German chemist Alfred Einhorn began making synthetic analogs of cocaine in 1894 and first prepared procaine in 1904. Sold under the brand name Novocaine, it was the first man-made local anesthetic that was widely accepted in clinics. More important, Novocaine was not as rapid in its action as cocaine, and it did not present the peril of strong addition that cocaine engenders. The name change of Novocaine to procaine took place during and after World War I. The United States manufactured Novocaine using the German patent annulled during the war but sold it under the brand name Procaine. Currently, procaine is seldom used for peripheral nerve or epidural blocks because of its low potency, slow onset, short duration of action, and limited ability to penetrate tissue.

Procaine has an ester functional group, which is rapidly metabolized by esterase in plasma. As a consequence, procaine is a short-acting anesthetic, rendered ineffective in a matter of minutes. Steve Brodie of the National Institutes of Health (NIH), working with organic chemists at Squibb, transformed the ester group into a "stronger" group: an amide. The result was procainamide, which has a much longer duration of action and is still widely used. On at least one of the Apollo missions, procainamide was taken to the moon as a precautionary measure for the astronauts. Brodie joked in reference to the funding agency, the American Heart Association: "They certainly got their money's worth on that one."[22]

The utility and popularity of procaine incited a flurry of "me-toos". In all, more than a hundred "-caines" were synthesized.

John F. Kennedy, America's 34th president, injured his back in 1936 while playing football at Harvard. His back injury was exacerbated in 1943 when he was commanding a patrol torpedo boat, PT-109, which sank near the Solomon Islands. During his presidency, in addition to a dozen or so medicines, Kennedy had to have multiple procaine injections before his public appearances so that he would look healthier than he really was.

Anesthetics Today

It is amazing that in such a short period of time, anesthesia has become an extremely specialized medical profession. For general and local anesthesia, hundreds of anesthetics are now available. It is impossible to imagine modern medicine without anesthetics. Many divisions of anesthesia exist: inhalation anesthesia, regional anesthesia, intravenous anesthesia, precision anesthesia, pain management clinics, and so forth.

With regard to inhalation anesthetics, today nitrous oxide still enjoys applications in dentistry and surgery, although prolonged used of nitrous oxide is known to cause cancer. The use of ether is no longer in practice due to its irritancy and flammability. Chloroform is no longer used as an anesthetic at all because it occasionally causes asphyxia. The three old inhalation anesthetics have been largely replaced by nonflammable solvents such as halothane and fluoroethane.

How do nitrous oxide, ether, and chloroform exert their anesthetic effects? The mechanism of action is still not completely known despite their tremendous utilities. One popular theory is that anesthetics inhibit the ion channels, which are important physiological and biological "units." For instance, the poison curare kills people instantaneously via the blockage of sodium channels.

Anesthesia has become extremely reliable today. Anesthesia-related deaths have dramatically declined from 1 in 10,000 cases a generation ago to 1 in 250,000 now. For healthy patients, it is 1 in 400,000, a remarkable safety record. Nonetheless, hidden factors such as heart disease could increase the risk of a heart attack. Sometimes allergic reactions to other medications occur under anesthesia. Although the chance of this is quite small, it does occur. On January 17, 2004, Olivia Goldsmith, author of the novel *The First Wives Club,* died of complications during anesthesia before plastic surgery for a simple "tuck" to tighten facial skin.

CHAPTER 8

Anti-Inflammatory Drugs

Discovery consists of seeing what everybody else has seen and thinking what nobody else has thought.

—Albert Szent-Györgi (1893–1986)

Inflammation and immunity, like all other normal reactions of the body, are meant to preserve or restore health. They can nonetheless cause a range of uncomfortable symptoms. Inflammation is such a complicated process that one would have a hard time reaching a consensus on its definition. Historically, inflammation was one of the earliest recognized and defined diseases. Two thousand years ago, Roman physician and encyclopedist Aulus Cornelius Celsus (25 B.C.–50 A.D., not to be confused with Celsius, the unit for temperature) described the four cardinal signs of inflammation: *calor* (warmth), *dolor* (pain), *tumor* (swelling), and *rubor* (redness).[1] The fifth element of inflammation, *functio laesi* (loss of function or movement), was noted later. Classic inflammatory diseases include rheumatoid arthritis and Crohn's disease, an inflammatory bowel disease. However, evidence is mounting that inflammation is implicated in many diseases that are not normally considered inflammatory. For instance, when arterial plaques become inflamed they can burst open, prompting a myriad of heart diseases. Inflammatory bowel conditions greatly increase the risk of colon tumors. Even diabetes has been associated with a number of inflammatory compounds.

It was hard to define what inflammation was, but finding a remedy was even more challenging. Aspirin, available in 1880, represented possibly the first really effective treatment for inflammation, whereas cortisone and other corticosteroids were not available for the treatment of rheumatoid arthritis until the early 1950s.

Cortisone

Louis Pasteur stated, "Dans les champs de l'observation, le hazard ne favorise que les esprit préparés" [In the field of experimentation, chance favors the prepared mind].[2] Like numerous cases in drug discovery, Philip S. Hench's discovery of cortisone for the treatment of rheumatoid arthritis illustrates Pasteur's point.

Rheumatoid arthritis (RA) is a chronic inflammatory disease characterized by pain, swelling, and subsequent destruction of joints. Until the late 1940s, there was no viable treatment, and, understandably, pessimism prevailed in medical society about its prognosis. Even William Osler, one of the greatest physicians, said "When an arthritic patient walks in the front door, I want to run out the back door!"[3] The situation did not change much until Hench discovered a "miracle drug" in 1949.

Philip S. Hench (1896–1965), a rheumatologist, joined the Mayo Clinic in Rochester, Minnesota, in 1921.[4,5] A severe cleft deformity made his speech loud and difficult to understand. But Hench overcame his disability and became an excellent orator. He was appointed as a clinician to the permanent staff at the Mayo Clinic in 1926 and soon became head of the section on rheumatic disease. For the ensuing two decades, he followed clinical trials studying the spontaneous remission of rheumatoid arthritis. In 1929 Hench observed an excellent remission in one of his intractable rheumatoid arthritic patients who developed a concurrent attack of jaundice. The fortuitous observations made Hench suspect that jaundice might be capable of suppressing rheumatoid inflammation and, therefore, that there must exist an "anti-rheumatic substance X." In the early 1930s, Hench noticed that women with rheumatoid arthritis saw temporary respite of their symptoms during pregnancy, but the pains returned after childbirth. This observation once again led him to consider the possibility that the same "anti-rheumatic substance X" could control the activities of rheumatoid arthritis. "It would," he wrote, "be gratifying if one were able to repeat Nature's miracle."[5] In an experiment that would be appalling to today's bioethics experts, jaundice was artificially introduced to gauge its efficacy in combating rheumatoid arthritis, without much success.

Also in 1929 Hench's colleague at the Mayo Clinic, biochemist Edward C. Kendall (1886–1972), decided to transfer his research from thyroid hormones to hormones from the adrenal cortex (a small organ attached to the top of the kidney), in part due to Hench's influence. Collaborating with the Parke, Davis and Company, Kendall isolated "substance E" from bovine adrenal glands in 1936. By 1941, Hench and Kendall concluded that Hench's "anti-rheumatic substance X" and Kendall's "substance E" were the same molecule—cortisone. Hench named it *cortisone* to signify the adrenal cortex steroid hormone, although Kendall initially called it *corsone*.[6–8]

During World War II, cortisone became a top priority in the Allies' war research efforts. During the Battle of Britain, rumors from the Polish Underground indicated that fighter pilots of the Luftwaffe were able to fly at unusually high altitudes without oxygen deprivation because they were being treated with cortisone allegedly made available by German chemists (see chapter 4). Lewis H. Sarett at Merck first synthesized cortisone from desoxycholic acid, also known as bile acid, a steroid isolated from cattle bile. Without an assistant, Sarett single-handedly made 18 milligrams of cortisone, although his synthesis was 36 steps long! Max Tishler, the head of Merck Process Chemistry, supplied some of the intermediates required for Sarett's efforts. Tishler and his colleagues at Merck Process Development further optimized Sarett's original synthesis and prepared 100 g of cortisone.[9] Hench thought very highly of their achievements, stating that Merck was "writing a brilliant chapter in the history of pharmaceutical manufacturing, accomplishing the impossible."[10] Within 2 years, cortisone became available to every researcher and physician in the United States. Both Sarett and Tishler went on to become members of the National Academy of Sciences, a rare honor for industrial chemists.

On September 21, 1948, Hench received 5 g of cortisone from Merck. A dosage of 50 mg, then 100 mg, of cortisone was administered to desperately ill 29-year-old Mrs. Gardner, who looked 50 and had not been able to get out of bed without assistance for the previous 5 years because of rheumatoid arthritis. After 3 days of injections, Mrs. Gardner miraculously recovered. She even went to downtown Rochester and had a 3-hour shopping spree! In April 1949, at the annual meeting of the American Rheumatism Association (ARA), Hench presented his seminal work on cortisone in treating rheumatoid arthritis, illustrated by a before-and-after movie of a severe rheumatoid patient attempting to climb stairs and failing and then skipping up them the day after receiving an injection of cortisone.

The film was followed by a standing ovation from the entire audience. The dramatic effect of cortisone on patients incapacitated by rheumatoid arthritis was reported in the *New York Times* on April 20, 1949. The triumph was touted as one of the greatest achievements in medical history. Hench and Kendall went on to win the Nobel Prize in Physiology or Medicine in 1950, along with Swiss chemist Tadeus Reichstein. Reichstein (1886–1972), a Polish immigrant working at the Federal Institute of Technology (ETH) in Zürich, also isolated cortisone in 1935. Reichstein's other important contribution was his total synthesis of vitamin C in 1933. Reichstein's synthesis is so practical that it is still used today in commercial production.

Philip Hench was a charismatic, enthusiastic, and generous character. He had a wide range of interests outside medicine. The discovery of cortisone and the Nobel Prize brought him much fame. When his good friend John Glyn had a baby daughter, he enthusiastically suggested the name of Cortisona. With much skillful diplomacy, Glyn declined, and instead named the girl Philipa, the female version of Hench's first name.[11]

In 1950, cortisone supply from bovine adrenal glands was extremely limited. One ton of cattle adrenal glands gave only 1 kilogram of dry yield, which, in turn, afforded only 25 grams of cortisone. Merck's 36-step synthesis was obviously not amenable to commercialization. Therefore, a short and commercially viable synthesis of cortisone became greatly important. The problem attracted the cream of the crop in organic chemistry. Louis Fieser, Robert B. Woodward, Robert Robinson, E. R. H. Jones (of the Jones reagent fame), and Derek Barton all joined the foray. In 1951, Carl Djerassi at Syntex accomplished a 14-step total synthesis of cortisone from diosgenin, an ingredient isolated from wild Mexican yams. Although it was an excellent scientific contribution similar to Sarett's synthesis, Djerassi himself admitted that their synthesis did not "ever contribute to the treatment of even one

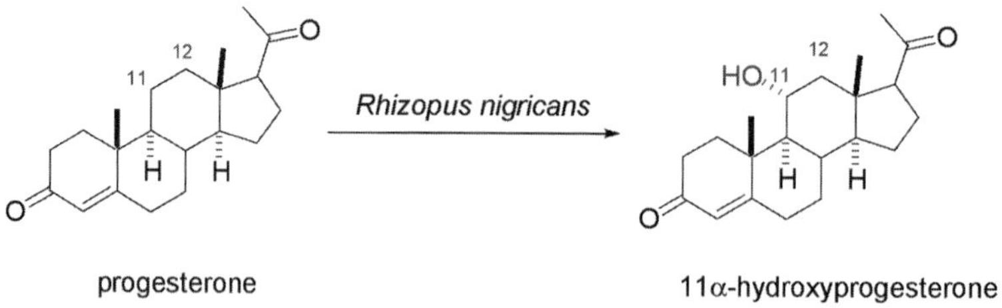

Chemical structures, progesterone and 11α-hydroxyprogesterone

arthritic patient."[12] Within months of Djerassi's publication, the Upjohn Company in Kalamazoo, Michigan, stunned the world with their publication in the *Journal of the American Chemical Society,* in which they reported the total synthesis of cortisone in 11 steps.[13] The key operation was exposure of the female sex hormone, progesterone, to a microorganism, *Rhizopus nigricans,* which in one single operation inserted, in high yield, an oxygen atom into position 11, which was required for the synthesis of cortisone.

After Hench's revelation of the miraculous effects of cortisone on rheumatoid arthritis patients in April 1949, Upjohn, like most drug firms at the time, began aggressively pursuing the synthesis of cortisone. Among six different approaches, the microbiological method received the most skepticism from many organic chemists. Chemist Durey H. Peterson and microbiologist H. C. Murray believed in that strategy. An agar plate placed in a windowsill of the oldest and dirtiest laboratory at the Upjohn Company by Murray yielded a *Rhizopus nigricans* fungus, which was then incubated with progesterone. Within 5 months, the introduction of oxygen at carbon-11 by the microorganism was achieved. The structure of the 11α-hydroxylprogesterone was then unambiguously established by X-ray diffraction studies and comparison with authentic samples.[14,15]

Upjohn's highly innovative approach using microbiological fermentation technology in steroid chemistry enabled a commercially viable process to functionalize the 11-position. Using 11α-hydroxylprogesterone, Upjohn chemists, led by John A. Hogg, accomplished a practical total synthesis of cortisone in 11 steps, a process that was commercialized.[14] Because the Upjohn synthesis utilized progesterone as the starting material and because Syntex was the only supplier of progesterone, the cortisone enterprise made Syntex very rich. Syntex prepared progesterone from inexpensive wild Mexican yams using Marker's process (see chapter 4).

As the understanding of the functions of cortisone increased exponentially, it was soon realized that cortisone itself is a prodrug that is reduced in vivo to cortisol (hydrocortisone). Cortisol, the actual active principle, is vital to the body's defense against inflammation. Because Addison's disease is due to adrenal cortex deficiency and is characterized by the failure of the adrenal glands and the inability to produce cortisol, cortisol became the drug of choice for Addison's disease. President John F. Kennedy (1917–1963) suffered from Addison's disease. During his presidential campaign in 1963, he was given large daily doses of hydrocortisone. The mental side effects of hydrocortisone overdose include elevated anxiety, panic

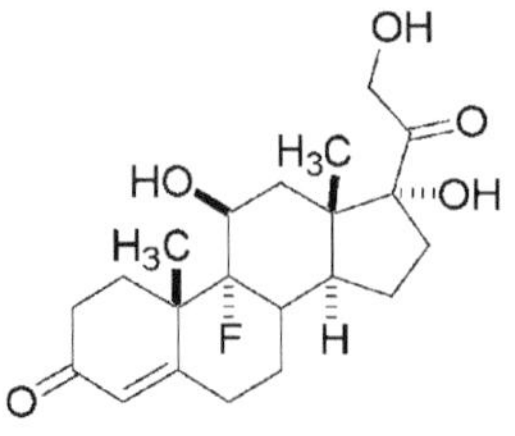

9-fluorocortisol

Chemical structure, 9-fluorocortisol

attacks, insomnia, and an increased libido (which might help explain his numerous extramarital exploits).

With the miraculous success of cortisone, it is not surprising that many incremental improvements quickly ensued. Prednisone was one, and fludrocortisone (9-fluorocortisol) was another. Fludrocortisone, 10 times more potent than cortisone, was discovered by Josef Fried at the Squibb Institute for Medical Research in the early 1950s, the golden age of steroid chemistry. When Fried and his associate Elizabeth Sabo synthesized 9-fluorocortisol in 1953, nobody at Squibb really believed that it would be of any interest. Not only was there no fluorine-containing drug on the market at the time, but fluoroacetate was also a highly toxic enzyme inhibitor. But history proved the critics wrong. 9-Fluorocortisol was not only much more potent than cortisone, but it also possessed appreciable mineralocorticoid activity similar to that of cortisone. 9-Fluorocortisol added a powerful weapon to a physician's arsenal in combating rheumatoid arthritis.

Unfortunately, the initial enthusiasm for corticosteroids was dampened by their severe side effects following chronic administration—notably osteoporosis, immune suppression, ulcerogenicity, adrenal suppression, and development of steroid dependence. Prednisone, a steroid in the cortisone family, saves lives but takes an awful toll in side effects, the most severe being weakening of the bones within months. Prednisone's other side effects can include diabetes, glaucoma, cataracts, blurred vision, high blood pressure, upset stomach, vomiting, headache, and dizziness, as well as insomnia, depression, anxiety, acne, increased hair growth, bruising, skin rash, and so forth. As a consequence, for the treatment of inflammatory diseases, corticosteroids have now been largely replaced by nonsteroidal anti-inflammatory drugs (NSAIDs).

Nonsteroidal Anti-Inflammatory Drugs

By contemporary definition, nonsteroidal anti-inflammatory drugs (NSAIDs) are drugs that exert their anti-inflammatory effects via blockage of prostaglandin synthesis. Aspirin, ibuprofen, and naproxen are NSAIDs, but acetaminophen is not.

Aspirin

Salicylic acid, an ingredient in the bark of willow tree, is aspirin's precursor. Although aspirin was marketed in 1899, salicylic acid was used medicinally as early as 1500 B.C. The Ebers papyrus referenced willow's medicinal properties in general and its treatment of rheumatism in pregnant women in particular. In Greece, the father of medicine, Hippocrates, recommended using the bark of the willow tree as an analgesic. In 1753, the Reverend Edward Stone from England experimented with the extraordinarily bitter willow bark in treating ague (fever from malaria) and intermitting disorders with satisfactory results. Five years later, he communicated his experience to the president of the Royal Society.[16]

In due time, the bitter-tasting salicin, the precursor of salicylic acid, was isolated from willow tree (*salix alba*) bark as yellow crystals. When salicin is ingested, it is hydrolyzed into glucose and salicylic alcohol, which are then oxidized to salicylic acid in the stomach.

In 1853, French chemist Charles Gerhardt at Montpellier University treated sodium salicylate with acetyl chloride and produced the first synthetic aspirin, although the reaction was messy and the product not pure. In 1859, German chemist Hermann Kolbe of Marburg University developed a reaction to synthesize salicylic acid by simply treating phenol with carbon dioxide under high pressure. The reaction is known as the Kolbe reaction in textbooks of organic chemistry. One of his students, Friedrich von Heyden, founded a company to profit from the process, with much financial success. In 1869, German chemist Karl Johann Kraut also synthesized aspirin in a similar manner.

Bayer and Company, a small dye company founded in 1856 by Friedrich Bayer, commercialized aspirin. At the end of the nineteenth century, Bayer established its own drug research laboratories, partially because the dye industry was no longer as profitable. The Bayer drug outfit was divided into a pharmaceutical section led by Arthur Eichengrün and a pharmacological section led by Heindrich Dreser. Felix Hoffmann was a chemist in Eichengrün's group. His father suffered from rheumatism, but the sodium salicylate he took upset his stomach so much that he had to discontinue its use. In order to find an analgesic that was easy on the stomach, Hoffmann played with salicylic acid analogs in the laboratory. On October 10, 1897, he prepared aspirin using an improved route in a purer form. Hoffmann gave some of his aspirin to his father, and it worked well in relieving the arthritis pain without severe gastrointestinal (GI) side effects. Further testing also confirmed aspirin's superior attributes. However, Dreser, as well as the head of the company, Carl Duisberg, were both more enamored of Heroin, another Hoffmann invention. They were both against the development and marketing of aspirin because they believed all salicylic acids had detrimental effects on the heart (now we know that aspirin has beneficial effects on the heart). Exasperated by the stone wall, Eichengrün tested aspirin on himself and did not suffer any heart malady. He smuggled some samples of aspirin to doctors, and the response was so positive that the company management had no choice but to market it. After aspirin's phenomenal success, Dreser published a paper detailing the clinical results of aspirin. Conspicuously, Dreser failed to even acknowledge any contributions by Eichengrün and Hoffmann. To Dreser's credit, he did discover the fact that aspirin is a prodrug of salicylic acid—aspirin is broken down by the body into salicylic acid, which is the real drug that exerts the beneficial effects. Obviously, aspirin, as a prodrug for salicylic acid, has less damaging effects on the stomach than salicylic acid.[17]

Because Eichengrün was Jewish, the Third Reich did its best, with much success, to minimize his contributions. The truth was not known until Walter Sneader, a Scottish medical historian, started digging into historical records in the 1990s.

Where did the name *aspirin* come from?

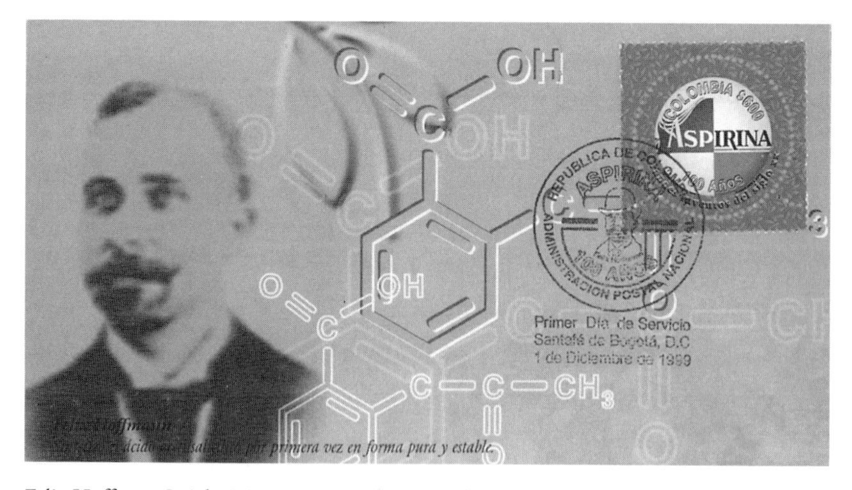

Felix Hoffman © Administracion Postal Nacional, Columbia

Because salicylic acid was initially also isolated from the meadowsweet plant in the *Spiraea* genus, its chemical name is *acetylspirsäure* in German. In those days, most drugs' names ended with "-in"; thus the name *aspirin* was derived.

In the mass production of aspirin, the reaction vessels were generally made of silver metal because acetic acid was a by-product of the process and neither steel nor iron could withstand its corrosiveness. Either gold or platinum could also have been chosen, but obviously they would be too costly. Today, novel synthetic polymer materials have long displaced precious metals as the material of choice.[18]

World War I ended on November 11, 1918, after 5 years of bloody conflict. The Allies, including France, the United States, England, and Russia, defeated Germany. In the Treaty of Versailles, signed in 1919, the Allies imposed on Germany harsh terms. One of these forced Bayer to relinquish the trademarks of Aspirin and Heroin as part of the reparations. The U.S. government also confiscated German companies in the United States, including Merck and Bayer, and auctioned them off to domestic bidders.

Interestingly, although aspirin was listed as the most commercially successful drug in the *Guinness Book of Records* in 1950, its mode of action was not known until 1971, when John R. Vane discovered that aspirin works by preventing the synthesis of prostaglandins. Vane's discovery is another marvelous example of serendipity.

Vane (1924–2004) began his career at the Royal College of Surgeons in London in the 1950s. By the early 1970s, he had developed a bioassay technique that was regularly used in his laboratory to investigate the generation and fate of agents such as catecholamines, bradykinin, angiotensin, and prostaglandins. Two seemingly unrelated events served as two pieces of a jigsaw puzzle that Vane was able to put together with his genius. One was his discovery of the rabbit aorta contracting substance; the other was Bergström's discovery of prostaglandins.

In 1969, in continuation of his bioassay research, Vane looked for novel biological agents. To that end, he exposed guinea pig lung to egg albumin (egg white). Because guinea pigs are violently allergic to egg white, the guinea pig lung was sent into immediate shock (anaphylaxis) on immersion in egg white. Vane then treated the solution passing over the lungs with tissues that included rat stomach, rat colon, rat stomach strop, chicken rectum, guinea pig trachea, guinea pig ileum, and rabbit aorta, which is the main blood channel from the heart. He discovered that the rabbit aorta contracted convulsively. Vane suggested that a chemical was responsible for the effect and named it the *rabbit aorta contracting substance* (RCS). At the time, Vane had a graduate student, Priscilla Piper, who had worked with Harry Collier at Parke-Davis Pharmaceuticals on the mode of action of aspirin. Piper suggested that they inject aspirin into the guinea pig lung at the beginning of the experiment. Surprisingly, they did not observe the twitching of rabbit aorta during this experiment. Clearly, aspirin stopped the generation of RCS. Unfortunately, Vane was not able to isolate and identify RCS, so he published a paper in *Nature*[19] without identifying RCS and let the matter rest, although it was always on his mind.

In the 1960s, K. Sune Bergström and his student Bengt I. Samuelsson at the Karolinska Institute in Stockholm, Sweden, isolated prostaglandins, a group of hormone-like chemicals, in minute quantities. The name *prostaglandin* signified prostate, because prostaglandins were initially isolated from human semen and thus were thought to be associated with the prostate. After having determined their structures, Bergström and Samuelsson discovered that prostaglandins are the downstream products of a biological cycle, the arachidonic acid cascade. This cascade transforms arachidonic acid to prostaglandins with the aid of an enzyme called *cyclooxygenase*. The arachidonic acid cascade plays an important role in the inflammation process. The discovery of the arachidonic acid cascade, as well as novel prostaglandins, generated a flurry of research, but nobody connected it with aspirin.

On Saturday, April 4, 1971, Vane was at home writing a review on the actions of prostaglandins. A flash of enlightenment struck him, and he asked himself: could RCS and prostaglandin be the same compound? The following Monday, he gathered together his students and excitedly told them, "I think I know how aspirin works. It is working by inhibiting the enzyme that makes prostaglandins, and maybe RCS is a prostaglandin too."[16] He then proceeded to announce, "I am going to do an experiment." Although he rarely performed hands-on laboratory work, he still insisted on doing the crucial experiment himself. As soon as one knew what one was looking for, it would be easier to find it. Vane personally isolated the cyclooxygenase enzyme using a method described by Bengt Samuelsson and carried out one experiment with aspirin and another without. The cyclooxygenase enzyme indeed transformed arachidonic acid to prostaglandins. Indomethacin, aspirin, and other salicylic acids all inhibited prostaglandin synthesis in a dose-related fashion. The results were a resounding confirmation of his proposal—RCS was indeed a prostaglandin. On June 23, 1971, Vane published their monumental discovery in the journal *Nature*,[20] which sent shock waves across the world. Seventy years after its commercialization, the mode of action of aspirin was finally unveiled. Soon it was found that other NSAIDs, such as ibuprofen and naproxen, also exert their anti-inflammatory effects by blocking cyclooxygenase (COX). In 1975, Bergström and Samuelsson confirmed that RCS was actually thromoxane A_2, a member of the prostaglandin family.

In 1982, the Nobel Prize for Physiology or Medicine was awarded to Vane, Bergström, and Samuelsson for their discoveries concerning prostaglandins and related biologically active substances.

When aspirin was first marketed, physicians feared that it might cause heart problems, as salicylic acid did. Trying its best to dispel the misconception, Bayer often labeled its aspirin bottle with the statement "aspirin does not affect the heart." Ironically, few would have dreamed that aspirin would one day become a preventive drug for cardiovascular diseases. The genesis of this discovery came not from a professor in a high-power research institute but from a practicing physician, Lawrence Craven.

Craven was a family doctor in Glendale, California. In the 1940s, he noticed that patients who took large doses of Aspergum (a chewing gum containing aspirin) sticks for pain after surgery all had difficulty in stopping bleeding. Craven was intrigued by the phenomenon and wondered whether aspirin possessed anticlotting properties. Because many of his affluent middle-aged patients were overweight, he recommended that they take 325–650 mg of aspirin as a preventative anticoagulant. Miraculously, among 1,465 healthy male participants who took aspirin, none suffered coronary occlusion or coronary insufficiency. Craven published his results in an obscure journal called the *Mississippi Valley Medical Journal* in 1953. In a footnote, Craven indicated that this article had won the *third*-place prize in the 1952 Mississippi Valley Medical Society Essay Contest. Unfortunately, not only did the journal not have a large readership, but Craven's astute clinical insight also was often discredited as not being scientifically rigorous enough. However, many scientists began to explore the antiplatelet effect of aspirin, and massive clinical trials were carried out to gauge the statistical significance. By the early 1980s, scientific data overwhelmingly confirmed that aspirin does prevent strokes, myocardial infarction (heart attack), and ischaemia brain stroke (IBS). In 1985, Margaret Keckler, the U.S. Secretary of Health and Human Services, announced that an aspirin a day helped prevent a second heart attack. The use of baby aspirin (81 mg) in preventing heart attacks has become very popular. Puzzlingly, a 10-year study involving 40,000 women, which came out in July 2005, seemed to suggest that women respond less favorably than men to aspirin's cardioprotective effect. More alarming is the fact that large doses of aspirin cause a significant occurrence of GI tract bleeding, and every year thousands of patients die because of it. The wonder drug is obviously not perfect.

Another popular analgesic, acetaminophen (e.g., Tylenol), is not an NSAID. It is biologically active as an analgesic possibly through inhibition of cyclooxygenase-3 (COX-3), a subtype of the cyclooxygenase enzyme that was not identified until just a few years ago. Unfortunately, a metabolite of acetaminophen is very toxic to kidneys and liver. Therefore, both acetanilide and acetaminophen are toxic in large doses. In fact, acetaminophen is one of the prevalent means of committing suicide in the United Kingdom. The victim simply swallows a copious amount of acetaminophen and lets the toxic metabolite exert its effects on the liver. As a consequence, acetaminophen carries a warning on the label that one should not exceed recommended doses. On the other hand, an overdose of aspirin is not as lethal.

In the summer of 1982, three people died from cyanide laced into capsules of Tylenol. The event became the impetus for putting seals on bottles of over-the-counter (OTC) drugs.

Ibuprofen

After World War II, thanks to aspirin's tremendous commercial success, drug firms began to look for super-aspirins—anti-inflammatory drugs with better potency but fewer side effects. Four companies made significant contributions to the field at the time: Geigy in Switzerland, Parke-Davis and Merck in the United States, and Boots in the United Kingdom. At that time, animal models were becoming more and more important, because the old-fashioned way in which scientists made and ingested compounds themselves was no longer viable. Not only was using human guinea pigs out of fashion, but also compounds were being systematically made in greater numbers.

Charles Winter at Merck, Sharp, and Dohme (he later moved to Parke-Davis) developed the cotton string granuloma test as a model of inflammatory pain. Using this model, Merck screened about 350 indole compounds and identified indomethacin as a potent anti-inflammatory drug. Indomethacin was initially synthesized by medicinal chemist T. Y. Shen (who later became vice president of inflammation research at Merck) as a plant growth regulator. It was also found to be particularly active in another model of inflammatory pain, carrageenan-induced rat paw edema. Indomethacin was introduced in 1964 and is still regarded as a gold standard that combines both anti-inflammatory and analgesic activity. In all, Merck synthesized more than 500 salicylate compounds that led eventually to diflunisal (Dolobid), a 5-fluorophenyl salicylate, in 1971.

In 1949, pharmacologist Gerhard Wilhelmi at Geigy in Basel developed a novel animal model for inflammation. His method measured edema elicited in the depilated skin of guinea pigs following irradiation with ultraviolet (UV) light. Wilhelmi published his method in 1950. Geigy identified phenylbutazone as a particularly active anti-inflammatory drug, using Wilhelmi's UV erythema model. The commercial success of phenylbutazone, in turn, catalyzed the merger between Geigy and Ciba. Ultimately, Ciba Geigy merged with Sandoz to form today's Novartis.

At the end of 1958, a paper published by Steve Winder of Parke-Davis reported that he had developed a UV erythema technique for screening anti-inflammatory drugs. Winder's model was similar to, and likely inspired by, Wilhelmi's model. Using this model, Parke-Davis successfully identified a group of fenamates, which also became a commercial success.

Coincidentally, pharmacologist Stewart S. Adams at Boots in the United Kingdom was taking a similar approach.[21] After obtaining his Ph.D. in 1952 at the University of Leeds, Adams began working for the Boots Company in Nottingham, England, founded by chemist Jesse Boots in 1921. Inspired by Wilhelmi's elegant and pragmatic model, Adams spent the next 3 years perfecting his version of the UV erythema model using shaved guinea pigs. Up to that point, he was working as a lone biologist with a couple of assistants. By 1956 he had gathered enough information to request chemical support, and management assigned John Nicholson to him. The duo would have a 20-year-long collaboration that brought many drugs and much fortune to Boots, transforming a small drug firm into an internationally renowned powerhouse in the field of anti-inflammatory drugs. At first they identified phenoxylalkanoic acids, originally made as selective herbicides in the Agricultural Division. Among 600 phenoxylalkanoic acids, Adams, using his own model, found that BTS8402 was 10 times more potent than aspirin. Unfortunately, it was less potent as an analgesic when tested using the technique in which pressure was applied to a rat foot with edema. In 1960, Nicholson synthesized *tert*-butylphenylacetic acid, which was proven to be effective in rheumatoid arthritis. Sadly, it caused rash in some patients and had to be abandoned. A very similar drug, isobutylphenylacetic acid (ibufenac) did not cause a rash but did cause liver toxicity in a small number of patients after long-term use. Ibufenac was withdrawn in the United Kingdom but was in use in Japan for quite some time because it did not cause liver damage in the Japanese, a striking example of ethnic differences in adverse drug reactions. In the 1960s, Boots sent more than 38 tons of ibufenac to Japan. Finally, isobutylphenylpropionic acid (ibuprofen) was found to possess the best safety profile, although it was not the most potent.[22] Boots marketed ibuprofen in the United Kingdom in 1969 with the trade name Brufen, and Upjohn marketed it in the United States in 1974 with the trade name Motrin. The active ingredient of Wyeth's Advil and Bristol-Myers Squibb's Nuprin is also ibuprofen. It is still widely used today and is regarded as the gold standard for over-the-counter analgesics. More impressively, millions of patients have taken ibuprofen in doses up to 1,600 milligrams without negative gastric effect. It has certainly lived up to its aspiration to be a super-aspirin.

The strikingly different safety profiles for ibufenac and ibuprofen may serve as a salient example of how much impact can be made on pharmacological effects by a small perturbation to a molecule. Ibuprofen (isobutylphenylpropionic acid) has only one carbon more than ibufenac (isobutylphenylacetic acid). Simply adding a methyl group to ibufenac provides ibuprofen, which is devoid of hepatotoxicity.

In the same class, naproxen is also a propionic acid with a similar pharmacology to that of ibuprofen. But naproxen is twice as potent as ibuprofen and has a much longer half-life in the body, enabling a once-daily regimen. Naproxen was discovered under the leadership of John Fried, who moved from Merck to Syntex to lead their medicinal chemistry program

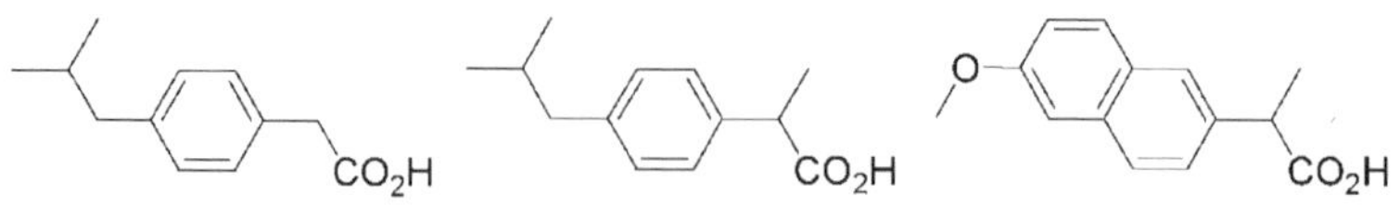

CO_2H CO_2H CO_2H

ibufenac ibuprofen naproxen

Chemical structures, ibufenic, ibuprophen, naproxen

and eventually rose to be the president of the Syntex research division. Naproxen, introduced in 1976, is sold in the United States under the trade names Naprosyn and Aleve. At the height of naproxen's popularity, its annual sales exceeded $1 billion, and Syntex became the only pharmaceutical company that flourished in the second half of the twentieth century until it was ultimately acquired by Hoffmann-La Roche in the late 1990s for over $6 billion. The patent for naproxen expired in December 1993, and the FDA approved naproxen sodium as an over-the-counter drug in 1994. Many newer drugs, such as the COX-2 selective inhibitors, have been tested against naproxen, whose fate is closely associated with them as well.

In all, about 30 NSAIDs have entered the market during the past 40 years, but none of them are devoid of gastrotoxic effects.[23] In the United States alone, between 10,000 and 20,000 deaths are caused by NSAID-associated toxicity, such as perforations and bleeding of the stomach. An NSAID that is devoid of the gastrotoxic effect is still desired.

Celebrex

In the early 1970s, John Vane and his colleagues demonstrated that NSAIDs work by blocking the cyclooxygenase enzyme. By the mid-1980s, evidence began to emerge that two subtypes of cyclooxygenase exist. Slowly but surely, the two isoforms of cyclooxygenase were deciphered thanks to the elegant work of several groups of scientists, including Philip Needleman of Washington University at St. Louis and later G. D. Searle, Daniel L. Simmons of Brigham Young University, Harvey R. Herschman of UCLA, and Donald Young of the University of Rochester.[24,25] Between the two subtypes of cyclooxygenases, one is inducible and the other is constitutive. The inducible enzyme associated with the inflammatory process was named cyclooxygenase II (COX-2), whereas the constitutive enzyme was called COX-1. COX-2 is localized mainly in inflammatory cells and tissues and becomes up-regulated during the acute inflammatory response; COX-1 is mainly responsible for normal physiological processes such as protecting the gastric mucosa and maintaining dilation of blood vessels. In general terms, COX-1 is the "good" enzyme and COX-2 is the "bad" one. Therefore, it was envisioned that a COX-2 selective inhibitor would be beneficial in both inhibiting prostaglandin production and reducing adverse gastrointestinal and hematological side effects.

Using this novel approach, under the leadership of Philip Needleman, the first COX-2 selective inhibitor, celecoxib, was synthesized on October 4, 1993, in the Skokie, Illinois, laboratories of G. D. Searle. Searle began comarketing it with Pfizer as Celebrex in June 1999. Six months later Merck received approval from the FDA to market their version of a COX-2 selective inhibitor, rofecoxib (Vioxx). Both Celebrex and Vioxx quickly became blockbuster drugs for treatment of osteoarthritis (OA) and rheumatoid arthritis (RA). In 2003, Bextra was comarketed with Pfizer by Pharmacia, the name of the newly formed company that arose from a merger between Pharmacia, Upjohn, and Monsanto (the parent company of Searle).

In 1992, the University of Rochester had filed a patent on Donald Young's invention covering the gene in humans that is responsible for producing COX-2.[25] The patent also had blanket coverage of all COX-2 selective inhibitors that were discovered through approaches that involved selective blockage of COX-2. The U.S. Patent and Trademark Office granted the patent in April 2000 with patent number U.S. 6,048,850. The University of Rochester sued G. D. Searle and its marketing partner, Pfizer, charging that Celebrex was an infringement of their patent. In 2003, U.S. District Judge David G. Larimer in Rochester ruled that the patent was invalid.[26] In July 2004, a federal patent appeal court in Washington, D.C., upheld the 2003 district court ruling by Judge Larimer, ending the first case of high-profile legal wrangling around the COX-2 selective inhibitors. Judge Larimer's argument was worth noting in appreciating patent laws: to qualify for a patent, inventions must be useful, nonobvious, and novel. The Rochester patent did not contain the requisite sufficient written requirement: U.S. 6,048,850 contained no description or claims of compounds that selectively block COX-2. The inventors did not have possession of a substance; therefore, the invention was merely theoretical. Judge Larimer finally concluded that "what the '850 does *not* do . . . is to provide the necessary link between these two steps: actually finding a compound that works."[26] Sadly, the *Rochester v. Searle* lawsuit has by now shrunk in significance compared with other lawsuits currently going on today involving Vioxx.

Indeed, the fate of Vioxx represents one of the most dramatic medical sagas of our time. Merck's VIGOR (Vioxx Gastrointestinal Outcome Research) clinical trials, published in 2000, found a fivefold increase in myocardial infarction (heart attack) among patients treated with Vioxx over those treated with naproxen. There could be three possibilities, according to John Vane:[27] (1) that it was a chance effect; (2) that naproxen protects the cardiovascular system; or (3) that Vioxx has a deleterious cardiovascular effect. In September 2004, APPROVe (Adenomatous Polyp Prevention On Vioxx) showed that Vioxx increased the risk of myocardial infarction and stroke. Merck voluntarily withdrew Vioxx from the market on September 30. The stock market reacted violently, wiping out $30 billion of Merck stock in a matter of a few months.

On February 18, 2005, a 32-member advisory panel for the FDA that evaluated the three COX-2 selective inhibitors on the market recommended keeping all of them on the market. The vote was 31–1 for Celebrex, 17–13 for Bextra, and 17–15 for Vioxx. In April 2005, acting against the vote of the advisory panel, the FDA asked Pfizer to voluntarily withdraw Bextra. Thus far, Celebrex is the only COX-2 selective inhibitor remaining on the market.

After all is said and done, COX-2 selective inhibitors are just like any other drug. One has to evaluate the risk versus the benefit of a particular drug when taking or prescribing it. For instance, a COX-2 selective inhibitor is a good choice for treating pain for a patient with gastrointestinal problems but is probably not a wise choice for another patient with cardiovascular problems. It is not surprising, then, that John Vane proposed a "Back to an Aspirin a day" after COX-2 selective inhibitors failed to meet their initial promises.[27]

Antiasthmatics

Asthma is a chronic lung inflammatory condition that afflicts about 15 million Americans, nearly a third of whom are children. The disease can cause episodes of wheezing, coughing, and breathing difficulty. It is characterized by bronchial hyperresponsiveness and reversible airway obstruction. Bronchial mucosal inflammation is present in all patients. The primary goal of asthma management is to maintain control of the disease process by reducing symptoms and improving lung function.

A related disease is chronic obstructive pulmonary disease (COPD), which is often caused and exacerbated by smoking. COPD affects over 5% of the adult population and is worse in Europe and in third-world countries in which smoking is not actively discouraged. It is one of the few conditions for which the mortality and morbidity are still increasing. Similar to asthma, COPD is also a chronic inflammatory condition of the airway. Therefore, asthma drugs often also work for COPD.

Flonase

One of the popular methods of pharmacological control of asthma is inhaled glucocorticoids, a subgroup of corticosteroids. Earlier in this chapter I mentioned that corticosteroids became miracle drugs for rheumatoid arthritis, an inflammatory condition of joints. Because corticosteroids are anti-inflammatory, it was logical that they should be effective in relieving asthma, an inflammatory condition of the lungs.

Initially, glucocorticoids were administered orally. The mechanism of action of these agents is not well understood. The clinical efficacy of these agents is probably the result of their inhibitory effect on leukocyte recruitment into the airways. However, as one would expect, therapeutic doses of oral glucocorticoids are associated with a wide range of adverse effects, such as Cushing's syndrome, altered lipid and bone metabolism, bone erosions, and vascular effects. The introduction of inhaled preparations made this class of drugs the most suitable for the treatment of asthmatic patients. They alleviate the major symptoms of the disease by reducing airway reactivity while restoring the integrity of the airways. The first inhaled glucocorticoid, beclomethasone dipropionate (BDP), revolutionized asthma therapy when it was found that topical delivery to the lung resulted in reduced systemic side effects (adrenal suppression, osteoporosis, and growth inhibition) typically seen with oral steroid treatments.

Unfortunately, glucocorticoids such as beclomethasone dipropionate have significant bioavailability, and when one considers the surface area of the tracheobronchial mucosa, significant plasma levels and systemic side effects occur at therapeutic doses. The onset of osteoporosis and reduced bone growth in children is by far the most serious adverse event.

The solution to the high bioavailability of these agents was the development of fluticasone propionate, which GlaxoSmithKline sold under the trade names of Flonase and Flovent, a prodrug that results in much lower systemic bioavailability. The evolution of this drug stemmed from observations with the steroid 17-carboxylates that showed that these esters were active topically when esterified, whereas the parent acids were inactive.

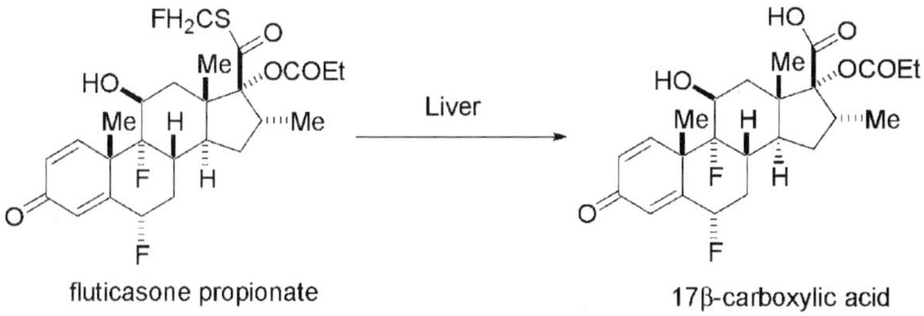

Liver

fluticasone propionate

17β-carboxylic acid

Chemical structures, fluticasone propionate and 17β-carboxylic acid

Thus researchers realized that enzymatic hydrolysis of the ester would lead to systemic deactivation. Structure-activity relationship (SAR) studies led to a series of carbothioates, which were very active in vivo when topically applied to rodents but were inactive after oral administration. In the end, GlaxoSmithKline scientists arrived at Flonase, which is hundreds-fold more active than BDP and thousands-fold more active than cortisol, the active form of cortisone. Moreover, Flonase was designed to be pulmonary selective. As a consequence, Flonase has only a 1% oral bioavailability, whereas cortisol has an 80% oral bioavailability.

It was shown that fluticasone propionate underwent first-pass metabolism in the liver to the corresponding inactive 17β-carboxylic acid. This observation was confirmed by experiments that showed that it was rapidly metabolized by mouse, rat, or dog liver homogenates.

Right now corticosteroids are considered the mainstay of asthma therapy.

Serevent

Another asthma drug, salmeterol xinafoate, works through a different mode of action from that of Flonase. GlaxoSmithKline has marketed salmeterol xinafoate with the trade name Serevent since 1990. A highly lipophilic drug, Serevent rapidly crosses the bronchial epithelium and is retained in the lung surface. It has a slow onset with prolonged duration of action and is virtually resistant to washout.

Serevent is a beta-adrenoceptor agonist. It works by dilating the lung's bronchial tubes, which become constricted and make it difficult for asthmatics to breathe. Believe it or not, beta-adrenoceptor agonists, one of the oldest classes used in medicine, have been known for more than 5,000 years to relieve bronchoconstriction by mimicking the effect of adrenaline.

As early as 3000 B.C., the Chinese used sympathomimetic agents to relieve breathing difficulties. The active principle, an alkaloid now identified as ephedrine, was originally extracted from the plant *Ephedra equisetina* and known as Ma Huang. However, Ma Huang was not introduced into Western medicine until 1924. In 1948, Raymond P. Ahlquist at the Medical College of Georgia speculated that there were two types of adrenergic receptors (adrenoceptors for short), which he termed *alpha-adrenoceptor* and *beta-adrenoceptor* (see chapter 3). Whereas beta-blockers are excellent drugs for lowering blood pressure, beta-adrenergic agonists are the most prescribed class of drugs for the treatment of asthma. Beta-adrenergic agonists are preferred both for the rapid relief of symptoms and for the level of bronchodilation achieved in patients with bronchial asthma. They have now become standard bronchodilators in emergency rooms and in day-to-day use as reliever medicine to help the patient breath.

These drugs produce their effects through stimulation of specific β2-adrenergic receptors located in the plasma membrane, resulting in alterations in adenylyl cyclase and elevations in intracellular AMP. Long-acting β2-adrenergic agonists, such as salmeterol xinafoate (Serevent), are very lipophilic and have a high affinity for the receptor by a different mechanism. However, these treatments also suffer from a variety of side effects. The widespread distribution of β2-adrenergic receptors results in a number of undesired responses when these agents are absorbed into the systemic circulation. Tremor is the most common side effect and results from stimulation of the β2-receptor in skeletal muscle. The most serious side effects are cardiac in nature (increased heart rate, tachyarrhythmias) that result from stimulation of the β2-receptor in the heart. Most of these side effects disappear with long-term use and do not have any long-term health consequences.

GlaxoSmithKline combined two asthma drugs with different mechanisms and arrived at Advir, which is composed of Serevent (a beta-adrenergic agonist) and Flonase (a glucocorticoid). In 2004, Advir became very successful in controlling asthma and brought in $4.5 billion in annual sales for GlaxoSmithKline. It was the company's best-selling drug. In addition, formoterol (Foradil), marketed by Novartis and Schering-Plough, is another β-adrenergic agonist for asthma. In August 2003, the FDA added a black-box warning on Serevent and Advir, stating that they possess "small but significant increased risk of life-threatening asthma attacks or asthma-related deaths seen in patients taking salmeterol in a recently completed U.S. study." However, in July 2005, a panel of experts for the FDA evaluated the risk versus benefit of these drugs and recommended that Foradil, Serevent, and Advir be kept on the market.

Singulair

The newest therapy available for the treatment of asthma arises from the recognition of the role of the leukotrienes in the initiation and propagation of airway inflammation. Merck's montelukast sodium (trademark Singulair) is an antagonist of leukotriene receptors.

As early as 1938, Australian physiologist Charles H. Kellaway isolated a slow-reacting substance anaphylaxis (SRS-A) after he sensitized guinea pigs with cobra venom. SRS-A was the active principle that led to a slow, prolonged contractile response of the animal's intestinal muscles. Because SRS-As are formed in only trace amounts and are intrinsically unstable, their identities remained unknown for four decades. By 1979, Bengt I. Samuelsson's group at the Karolinska Institute in Stockholm, Sweden, identified one of the essential components of SRS-A (leukotriene E; LTE), consisting of the amino acid cysteine linked to a 20-carbon fatty acid. The realization shed light on the structure of SRS-A, but even at that point there were many possibilities, including approximately 128 possible stereochemical structures. In 1979, Elias J. Corey at Harvard University and coworkers reported a stereospecific synthesis of all leukotrienes and, in collaboration with Samuelsson, demonstrated functional identity between synthetic and the natural leukotriene C_4 (LTC_4). Subsequently, groups led by Samuelsson, Corey, Robert A. Lewis, and K. Frank Austen of Harvard showed that biologically generated SRS-A was composed not only of LTC_4 but also of leukotriene D_4(LTD_4) and leukotriene E_4(LTE_4), which are formed from leukotriene C. Corey's group also completed a stereochemically specific synthesis of all intermediates of the leukotriene pathway, thereby definitively establishing the structures. At that time, synthetic leukotrienes were in such high demand that they were valued at about $1,000 per milligram. Because Corey's group possessed the world's supply, they provided leukotrienes to researchers around the world for their leukotriene investigations.

Samuelsson and his mentor Bergström, along with John Vane, won the Nobel Prize for Physiology or Medicine in 1982 for their prostaglandin work, and Corey won the Nobel Prize for Chemistry in 1990 for the development of the theory and methodology of organic synthesis.

The leukotrienes exert their effects through G-protein-coupled receptors (GPCRs) that regulate a signal transduction pathway that ultimately causes calcium release from the cells. There are two classes of leukotriene receptors, BLT_1 receptors and cysteinyl leukotriene (CysLT) receptors 1 and 2. It is these latter receptors that mediate the actions of the cysteinyl leukotrienes in asthma. One of the first drugs that came out of the leukotriene research was Merck's montelukast sodium (Singulair), an LTD_4 antagonist for treatment of asthma.

Back in 1981, Merck Frosst in Montreal, Canada, hired Anthony W. Ford-Hutchinson from Kings College Hospital Medical School, London, to serve as director of pharmacology. Under Ford-Hutchinson's leadership, Merck Frosst established biological assays and animal models for modulation of leukotriene receptors in search of a treatment of asthma. In order to find leukotriene receptor antagonists, they hand-screened tens of thousands of compounds from Merck's compound library at a time when high-through-put (HTS) screening was not yet available. They then decided to use quinolein as their lead compound for their studies on structure-activity relationship (SAR). They arrived at MK-571, which is a thousandfold more potent than quinolein. The clinical trials in 1989 demonstrated that leukotriene receptor antagonists were effective for treating asthma, thereby confirming the pivotal role of leukotrienes in respiratory disease clinically. Unfortunately, MK-571 caused a large increase of liver weight in mice. Merck Frosst scientists discovered that only one enantiomer of MK-571 had the liver side effect but that the other, surprisingly, did not. In April 1991 they finally produced montelukast sodium, which possessed desirable attributes such as high intrinsic potency, good oral bioavailability, and long duration of action for a once-daily regimen for asthma. The "mont" in its name is a tribute to the place in which it was discovered, Montreal. In 1998, the FDA approved montelukast sodium for marketing, and Merck sold it under the trademark Singulair, which had annual sales of $2.62 billion in 2004. In the course of 19 years of quest for Singulair, Merck Frosst published more than 800 papers on related topics. The three medicinal chemists who discovered Singulair were Robert N. Young (current vice president of medicinal chemistry), Robert J. Zamboni (past vice president of medicinal chemistry), and Marc Labelle. They were named the 2003 heroes of chemistry by the American Chemical Society.

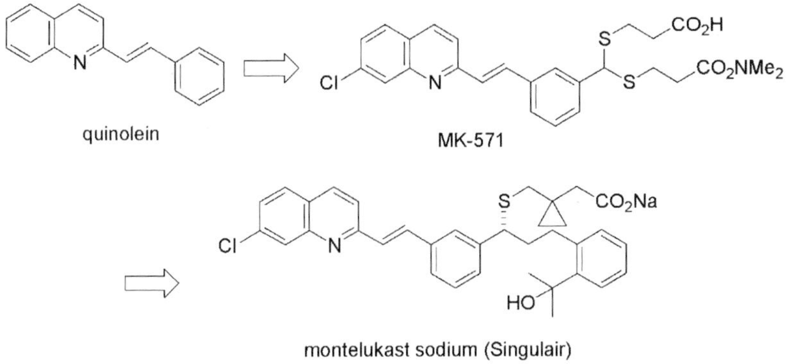

quinolein

MK-571

montelukast sodium (Singulair)

Chemical structures, Singulair

Anthony Ford-Hutchinson was promoted to executive vice president for worldwide basic research at Merck.

Two additional asthma drugs work via the same mode of action as that of Singulair. They are AstraZeneca's zafirlukast (Accolate) and Ono Pharmaceutical's pranlukast (Onon). These three $CysLT_1$-selective antagonists have become another important class of drugs for managing asthma, as well as allergic rhinitis, another prevalent inflammatory disease.

Biologics

The past two decades have witnessed phenomenal advances in biotechnology, which has yielded many revolutionary biologic drugs. In the anti-inflammatory arena, biologics have found widespread applications in the treatment of rheumatoid arthritis, Crohn's disease, and psoriasis. The three most prominent biologics are tumor necrosis factor (TNF) inhibitors, including etanercept (Enbrel), infliximab (Remicade), and adalmumab (Humira).[28] All of these three drugs work through binding to the TNF-α protein, a soluble cytokine receptor. TNF is a protein that consists of 157 amino acids with a molecular weight of 26 kDa. In 1973, Lloyd J. Old and his colleagues at the Memorial Sloan-Kettering Cancer Center in New York isolated a protein that is produced by the body in the course of bacterial infection and that kills tumors in mice. The name *tumor necrosis factor* signifies its tumor-killing properties.[29] By 1984, the amino acid sequence of TNF was determined. Meanwhile, a few biotechnology companies succeeded in cloning the gene that encodes TNF. TNF was first

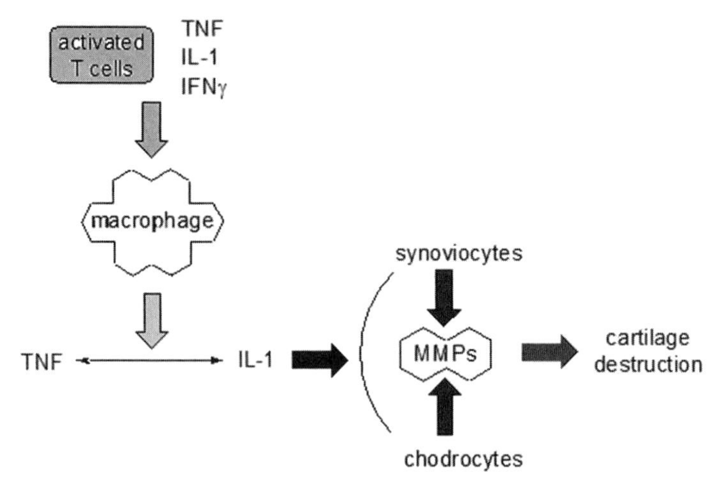

Role of TNF in rheumatoid arthritis

identified because of its anticancer activity; now the factor is recognized to be one of a family of proteins that orchestrate the body's remarkable response to injury and infection.

To understand the mode of action of TNF inhibitors, it is necessary to appreciate the inflammatory cascade for rheumatoid arthritis. Inflammation takes place when an internal or external stimulus changes the level of cytokines. Cytokines are regulatory proteins that are secreted by white blood cells and several other cell types in the body.[30] They are important in modulating our bodies' immune and inflammatory responses. Since the 1950s, many cytokines have been identified, including interleukin 1 (IL-1), nerve growth factor (NGF; see chapter 1), interferon (IFN), and TNF. TNF plays a key role in the inflammatory cascade that leads to rheumatoid arthritis.

Marc Feldmann and Ravinder Maini discovered that anti-TNF therapy is an effective treatment for rheumatoid arthritis and other autoimmune diseases while they were working at the Kennedy Institute of Rheumatology at Imperial College School of Medicine in London. In 1984, when the cytokine field was relatively new, Feldmann (a basic immunologist) and Maini (a clinical rheumatologist) collaborated to investigate the hypothesis that cytokines were important in the pathogenesis of rheumatoid arthritis. They uncovered the cytokine cascade in which TNF stimulates the production of other inflammatory cytokines (such as IL-1, IL-6, and granulocyte-macrophage colony stimulating factor), which in turn stimulate TNF itself. Feldmann and Maini then tested their theory in experimental animals and showed that inhibition of TNF indeed blocked collagen-induced arthritis.[31]

As is often the case, blocking the TNF protein is easier said than done. TNF-α receptor is a huge protein, and blocking it is difficult using a small molecule. Therefore, most successful approaches in blocking the TNF receptor have been accomplished using large biologics. The first successful drug that came out of this approach was Immunex's etanercept (Enbrel), approved by the FDA in 1998 for the treatment of rheumatoid arthritis. Immunex was a biotechnology company in Seattle founded in 1981 by Steve Duzan, a hard-driving entrepreneur with little experience in drug discovery. But Duzan recruited an experienced immunologist and a Ph.D. in biotechnology, and the company was in business. The first product Immunex discovered was Leukine, which was not a commercial success, selling a meager $23 million a year. In 1993, American Cyanamid acquired the controlling shares of Immunex, merging it with its Lederle oncology unit. Enbrel, Immunex's second product, became a blockbuster, with annual sales of $1.9 billion in 2004. Enbrel changed the fate of Seattle's largest biotechnology company, which was acquired by Amgen for $16 billion. The deal was the biggest merger in biotech's history at that time.

How is Enbrel synthesized? Why was manufacturing large quantities an issue initially? To answer these questions, we look at the structure of the TNF receptors.

Two cell surface TNF receptors mediate the biological responses to TNF-α. They are type I, or p55 (TNF-RI), and type II, or p75 (TNF-RII). The names p55 and p75 were given simply because these two receptor proteins have molecular weights of 55 kilo-Dolton (kDa) and 75 kDa, respectively. Enbrel is a fully human anti-TNF inhibitor with 934 amino acids and a molecular weight of 150 kDa. It is a dimer fusion protein consisting of the extracellular domain of the human TNF-RII combined with the Fc portion of human IgG_1, and it is manufactured using recombinant DNA technology in a Chinese hamster ovary (CHO) mammalian cell line.

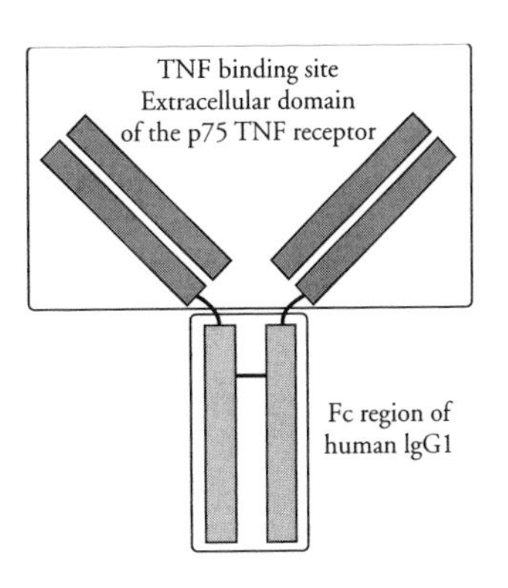

Etanercept (Enbrel), Fc = crystallizable fragment, which is the constant domain

Clinical trials for etanercept began in 1983 and finished in 1998. FDA approval for a small-molecule drug is processed via a new drug application (NDA). In the case of a biologic, its approval also involves a process called biologic license application (BLA), in addition to an NDA. Etanercept's BLA was granted in 1998, and Immunex sold it under the trade name Enbrel. The balance of safety, efficacy, and dosing made Enbrel a gem of a drug. Enbrel's sales were initially hampered by Immunex's manufacturing capacity, which had not been able to keep up with demand since the drug's introduction. Many rheumatoid arthritis patients had to wait 2 weeks or longer to get their Enbrel shots. After the merger, Amgen built an additional manufacturing facility, and Enbrel's supply problem was soon overcome.

On the heels of Enbrel came another TNF inhibitor, Centocor's Remicade, also approved by the FDA and available to rheumatoid arthritis patients in 1999. Interestingly, the popularity of Remicade traces back to Marc Feldmann and Ravinder Maini, the two scientists who advanced the idea that blocking the TNF protein would provide a means of treating rheumatoid arthritis. In the early 1990s, Feldmann and Maini obtained successful preclinical data to support their idea and proceeded to persuade Centocor, Inc., a biotechnology company near Philadelphia, to provide them with Remicade. Initially, Centocor's Remicade was developed for the indication of sepsis, a much smaller market. Feldmann and Maini convinced Centocor to try Remicade on rheumatoid arthritis. A remarkably successful initial clinical trial in 1992 was followed by the development of

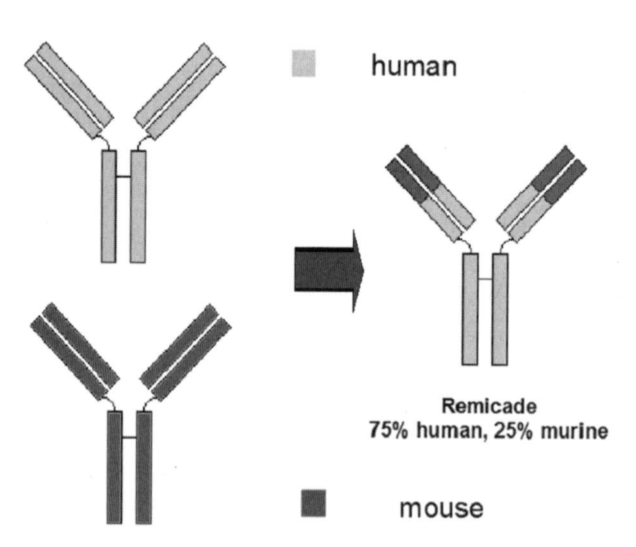

Remicade, a chimeric monoclonal antibody

a clinical program, leading in 1998 to the approval of the TNF blocker for the treatment of rheumatoid arthritis. Since its launch in 1999, Remicade has overtaken Enbrel as the market leader, with annual sales of $3 billion in 2004. Johnson and Johnson is Centocor's comarketing partner for Remicade.

Although it works through the same mechanism, Remicade is a chimeric monoclonal antibody, different from Enbrel. Chimeric antibody (chMAb), in this case, indicates that it is part human and part mouse. Mouse protein tends to cause the human immune system to respond with its own antibody to destroy the foreign one, not unlike the rejection response to organ transplant. Humanization of a mouse antibody can be achieved using a technique called site-directed mutagenesis by splicing a human gene into Chinese hamster ovary (CHO) cells, which are brewed in giant batches in bioreactors. The humanized monoclonal antibody was later named infliximab (Remicade), which was 75% human and 25% murine.[32,33]

The third TNF inhibitor is adalimumab (Humira). Abbott licensed it from Cambridge Antibody Technology and launched it in 2003. Because Humira can be given to patients via injection once a week, a convenient regimen, it has done well in the market, with annual sales of $852 million in 2004.

Unfortunately, the enthusiasm for anti-inflammatory biologics is somewhat tempered by concern over safety and cost.[34] The safety concerns arise from the mechanism itself. The safety of their long-term use needs to be

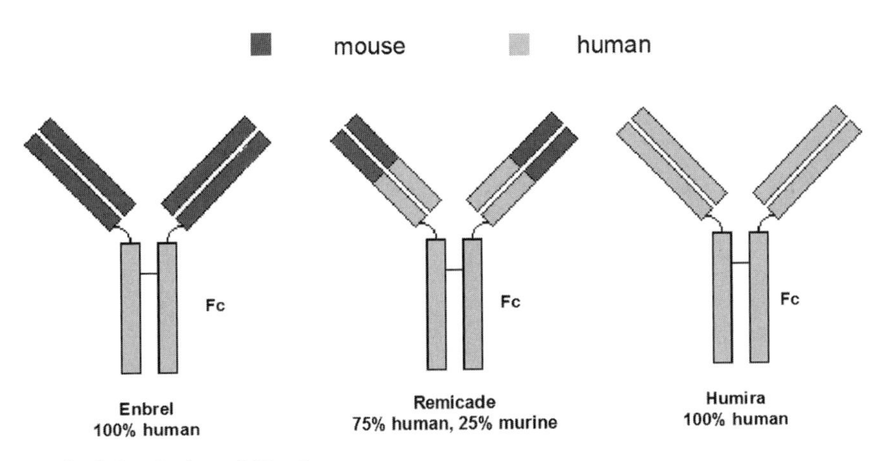

Enbrel, Remicade, and Humira

closely observed. Moreover, protein and antibody drugs are much more expensive to make than small-molecule drugs. Treatment with Enbrel, Remicade, or Humira costs $12,000 to $15,000 per year, whereas methotrexate, a small-molecule drug for rheumatoid arthritis, costs only about $900 a year. Therefore, a small-molecule TNF-α inhibitor would be more economical for patients. Nonetheless, the availability of Enbrel, Remicade, and Humira has been a blessing to thousands of patients with rheumatoid arthritis, Crohn's disease, psoriasis, and many other inflammatory diseases.

CHAPTER 9

Reflections

The Master-Apprentice Relationship

In this book, I have chronicled eight categories of medicines. As a testament to the changing times, the old European master-apprentice relationship between drug discoverers is a thing of the past. This truth is exemplified by the feud between Selman Waksman and Albert Schatz (see chapter 2). I have no doubt that Waksman sincerely believed that he was the one responsible for the discovery of streptomycin. After all, streptomycin was the fruit of decades of his endeavor with soil microbiology in general and actinomycetes in particular. Schatz happened to be at the right place at the right time. Waksman's conviction would have been completely acceptable if it had taken place just a century ago.

"Me-Too" Drugs

Since the new millennium, vilifying the pharmaceutical industry has become fashionable. One of the crimes that the pharmaceutical industry is accused of committing involves the so-called me-too drugs. Even Merck's former head of research Ed Scolnick once declared: "We at Merck do not do me-toos, if it is not innovative, we are not interested."[1] But history is replete with examples in which incremental improvements of a prototype yielded much better drugs.

The first ACE inhibitor was teprotide, a peptide with nine amino acids, inspired by a Brazilian snake venom extract. Peptides did not survive in the stomach juice, which broke them down into amino acids. Therefore, the prototype teprotide could be used only by IV injection. With brilliant insight, David Cushman and Miguel Ondetti at Squibb Pharmaceuticals designed and synthesized captopril. Captopril was the first *oral* ACE inhibitor, which contributed tremendously to the management of hypertension. In theory, captopril is indeed a "me-too" drug to teprotide, but most patients would certainly prefer to take an oral drug than to have injections for the same purpose. Because captopril has a short duration of action, it has to be taken more than once a day. It possessed a trio of shortcomings: bone marrow suppression (due to a decrease in circulating white blood cells), skin rash, and a loss of taste. Art Patchett and his chemists at Merck worked diligently and discovered another me-too to captopril, called enalaprilat. Not only was enalaprilat devoid of the undesirable side effects, but it was also long acting, enabling a once-daily regimen. It is true that society might not need hundreds more ACE inhibitors, but clumping all me-too drugs together and labeling them all as small increments and unnecessary is unfair.

Another case in point is the drug progression in the area of SSRI antidepressants. Initially, AB Astra had Zimeldine, the prototype SSRI, for the reuptake (removal) of the neurotransmitter serotonin. Unfortunately, a rare but serious side effect, Guillain-Barré syndrome, started to surface when Zimeldine was approved and administered in a large patient base. AB Astra pulled it out of the market in the early 1980s. In 1987, Eli Lilly launched Prozac, which worked via the same mechanism as Zimeldine. Prozac alone would be sufficient in the SSRI area, according to critics of me-too drugs. But in reality, people with inherent genetic differences respond differently to individual drugs. Not everybody's depression can be controlled by Prozac. Some may respond to either Zoloft, Paxil, or Wellbutrin more positively or may tolerate their unique side effects better.

Serendipity and Persistence

As E. J. Corey and Phil Baran observed in their forewords to this book, serendipity plays an important role in the drug discovery process. Indeed, serendipity is a recurring theme in many categories of drugs described in this book. Gone are the days of trial-and-error approaches, but serendipity has always been a staple in many discoveries. Modern drugs that have been discovered by serendipity include Cisplatin as an effective cancer chemotherapeutic, penicillin as an antibiotic, Viagra for erectile dysfunction, lithium for bipolar disorder, and many more. One could not help wondering how times have changed. For one thing, the era when a drug was discovered by one single individual is definitely a thing of the past. Nowadays, the pharmaceutical industry involves dozens of disciplines, years of research and development, and hundreds of millions of dollars for a single drug (one statistic revealed an average $1 billion for one drug). Although the majority of the drug discovery programs tackle diseases at the molecular level, luck always will be a factor in our drug discovery process. Louis Pasteur's words are still true: "In the field of experimentation, chance favors the prepared mind."

Persistence is another frequently recurring theme in the stories that I have chronicled herein. One example is Napoleone Ferrara's Avastin, a humanized monoclonal antibody against vascular endothelial cell growth factor (VEGF) receptor. Not only is the success a testimony to Ferrara's persistence, but its genesis could also be traced back to Judah Folkman's initial idea of killing cancer cells by cutting off the blood supply to the tumor. Both Valium and Cipro owe their genesis to individual scientists' persistence despite the doubts of their managers. Even today, persistence is still one of the most important attributes required for discovering novel medicines.

Innovation

Looking back at the history of drug discovery, innovation is a common theme. The fact that the pharmaceutical industry has grown to the scale it has today is largely the result of innovation. In the 1960s, James Black showed the world that one could target a specific molecule, an enzyme or a receptor, and find a selective molecule that modulates only the particular target without affecting other targets. The so-called rational drug design enabled the drug industry to tackle a myriad of molecular targets and to discover many drugs that have saved many lives. Almost half a century later, many low-hanging fruits using the rational drug design approach have been plucked. The drug industry, along with the society as a whole, will have to innovate even further in finding new targets and approaches in discovering more lifesaving medicines.

A case in point for the importance of innovation in drug discovery is the importance of biologics, which are also the "bread and butter" of the biotechnology companies. The monoclonal antibody technique created many protein kinase inhibitors such as Herceptin, Erbitux, and Avastin that have saved many cancer patients. Moreover, TNF-α inhibitors such as Enbrel, Remicade, and Humira give arthritis patients newfound freedom.

Innovation comes in different forms. When sulfa, penicillin, lithium, insulin, and paclitaxel (Taxol) were discovered through serendipity, these innovations opened up new fields of medicine. Another form of innovation comes from finding a novel use of a known medicine. Take Viagra as an example. It was initially developed as a hypertension drug. Although it failed for the initially intended therapeutic purpose, Pfizer scientists discovered the first efficacious drug for erectile dysfunction by thinking out of the box.

Another example of innovation is the HIV protease inhibitors. After discovery of the HIV protease, Roche, Abbott, and Merck came up with saquinavir, ritonavir, and indivir, respectively, with the help of crystal structures from X-ray, computer-aided drug design (CADD), and well-honed process chemistry. Now AIDS is a disease that can be controlled rather a death sentence.

Finally, a conspicuous, yet somewhat controversial, example of innovation is the COX-2 inhibitor Celebrex. Whereas nonsteroidal anti-inflammatory drugs (NSAIDs) such as aspirin and ibuprofen inhibit both COX-1 and COX-2, Celebrex hits COX-2 selectively. While treating inflammatory symptoms, Celebrex also has fewer of the gastrotoxic effects that are associated with COX-1. History will show the merits of COX-2 inhibitors with optimal selectivity profiles over time.

APPENDIX A

Chemical Structures

When both generic names and trade names are given, generic names precede trade names, which are in parentheses.

Chapter 1. Cancer Drugs

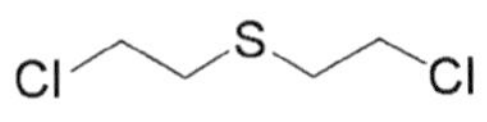

Mustard gas

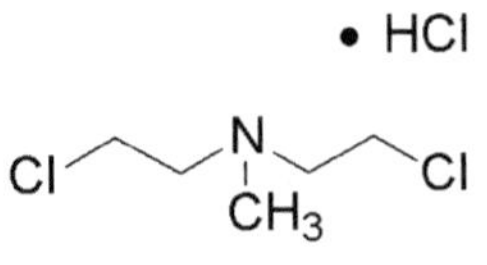

mechlorethamine
(Nitrogen mustard)

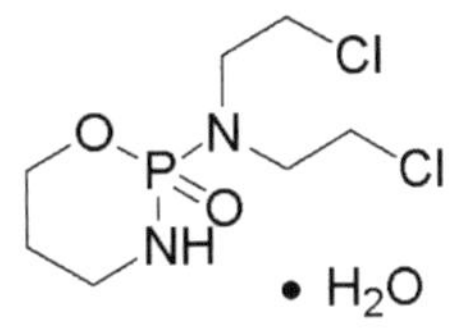

cyclophosphamide
(Cytoxan)

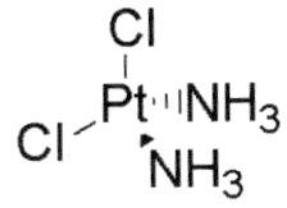

cisplatin (Briplatin)

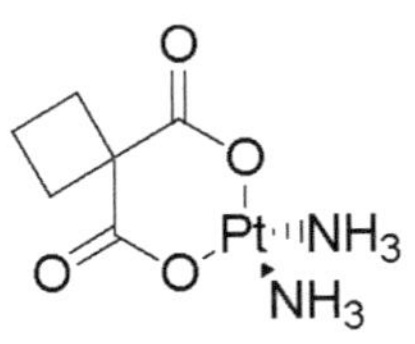

carboplatin (Paraplatin)

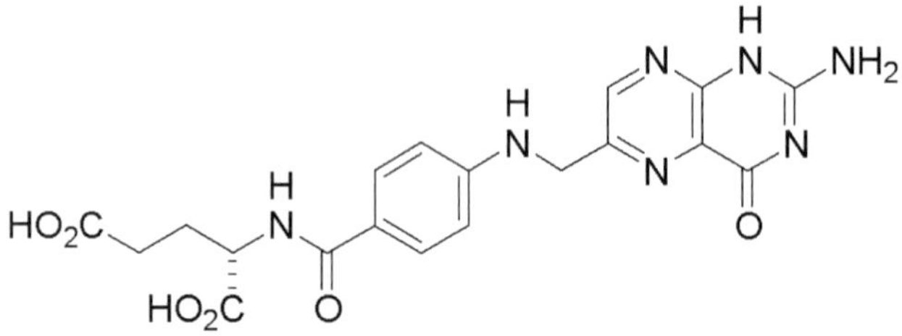

H
N N NH$_2$
H
N
N N
H
HO$_2$C N O
HO$_2$C O

folic acid

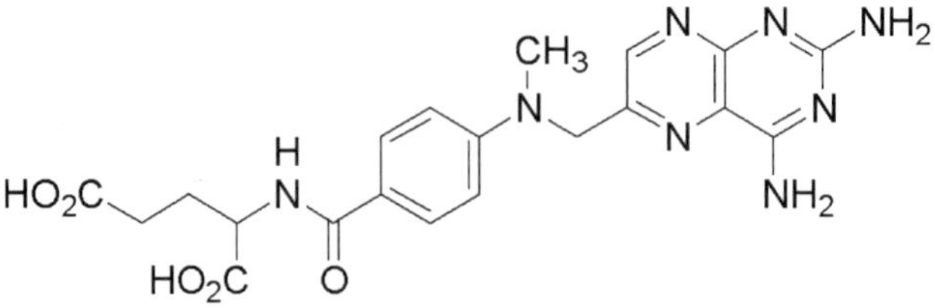

N N NH$_2$
CH$_3$
N N
H N
HO$_2$C N NH$_2$
HO$_2$C O

methotrexate (Emtrexate)

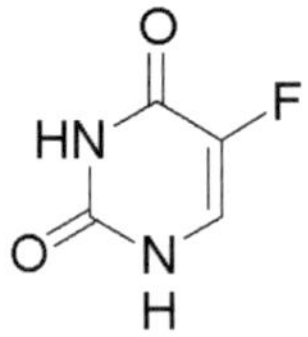

O
F
HN
O N
H

fluorouracil (5-FU)

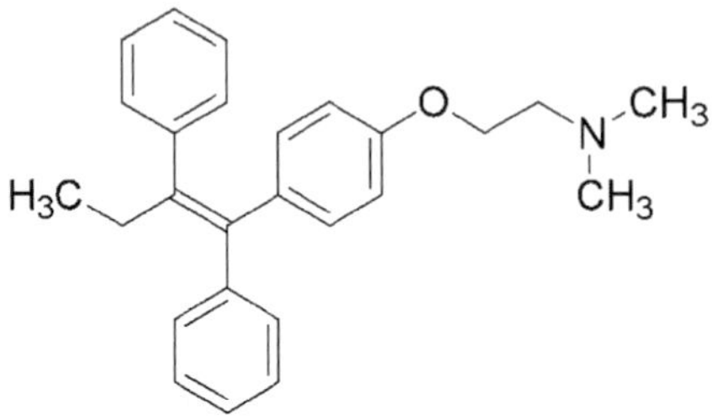

O CH$_3$
N
H$_3$C CH$_3$

tamoxifen (Nolvadex)

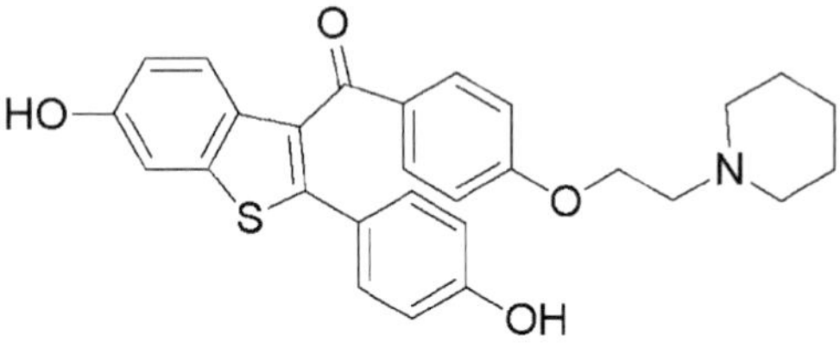

O
HO
N
S O
OH

raloxifene (Evista)

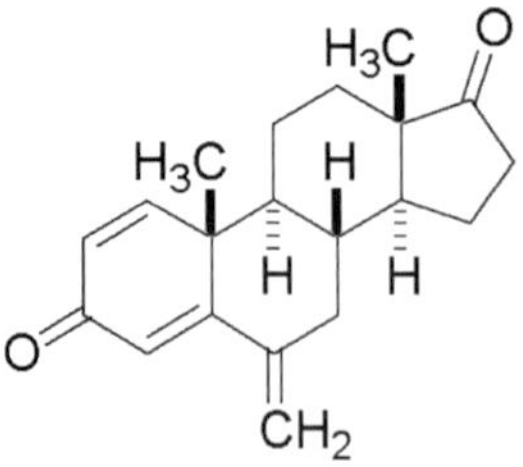

H$_3$C O
H$_3$C H
H H
O
CH$_2$

exemestane (Aromasin)

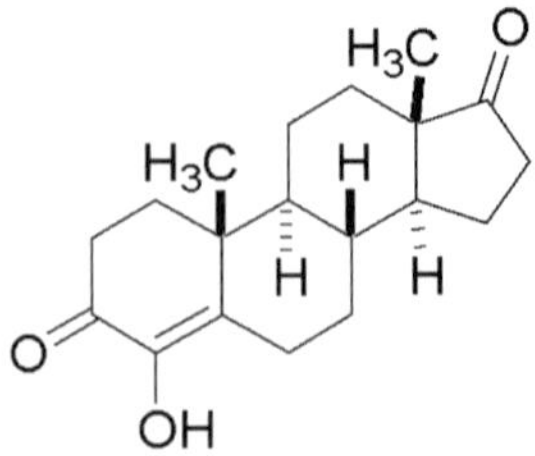

H$_3$C O
H$_3$C H
H H
O
OH

formestane (Lentaron)

H$_3$C O
H$_3$C
CH$_3$ H
H H
O

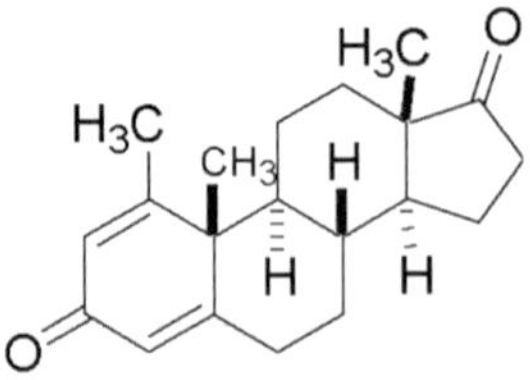

atamestane

aminoglutethimide

anastrozole (Arimidex)

letrozole (Femara)

prednisone

colchicine

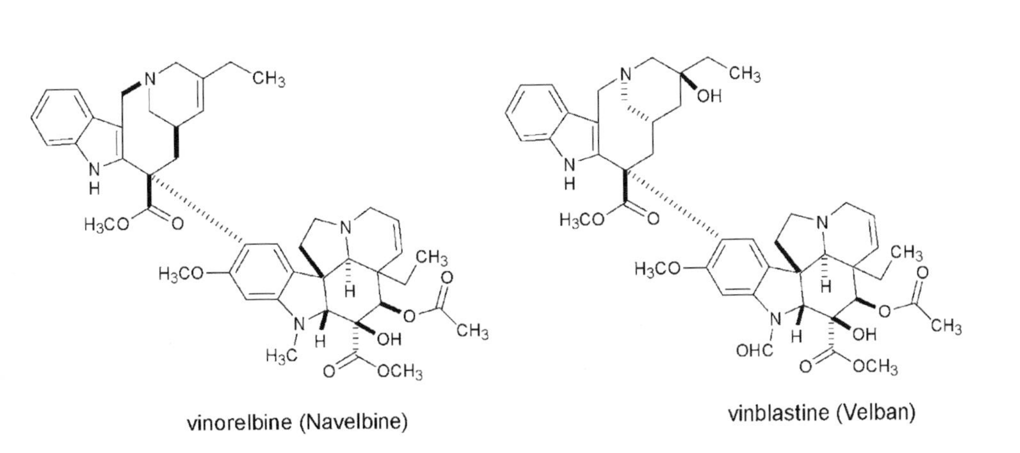

vinorelbine (Navelbine)

vinblastine (Velban)

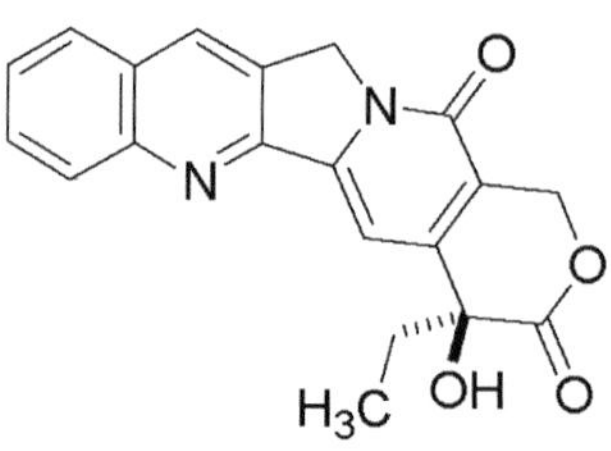

camptothecin

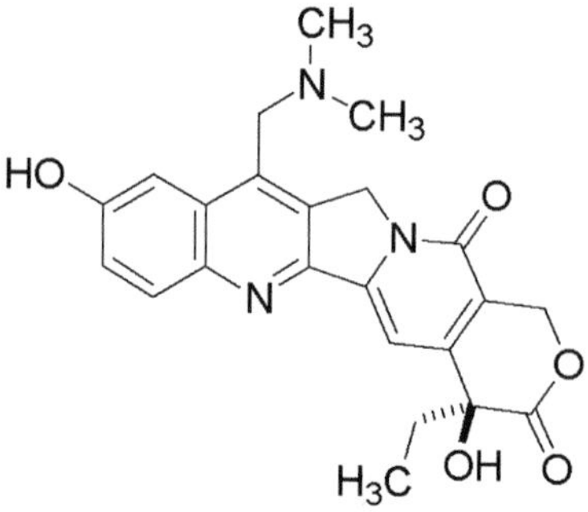

topotecan (Hycamtin)

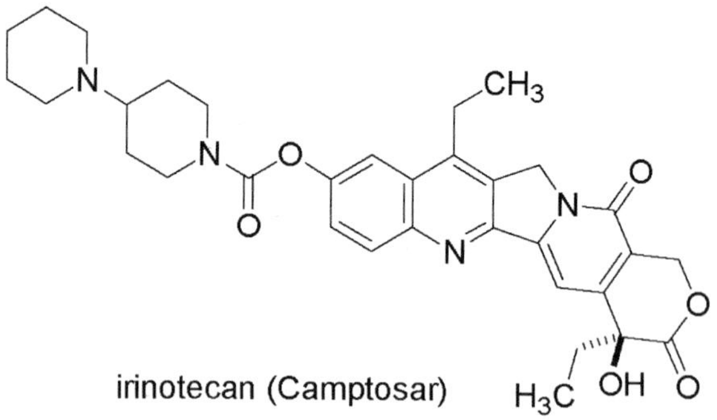

irinotecan (Camptosar)

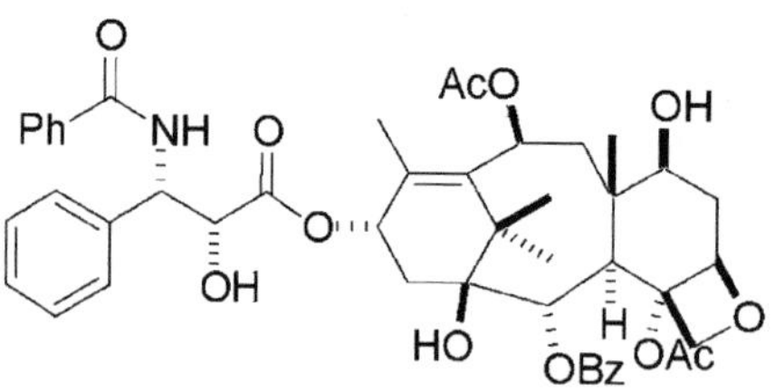

paclitaxel (Taxol)

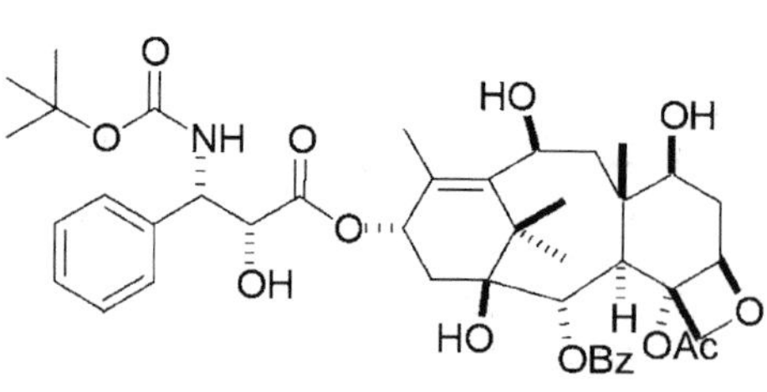

docetaxel (Taxotere)

thalidomide

gefitinib (Iressa)

imatinib mesylate (Gleevec)

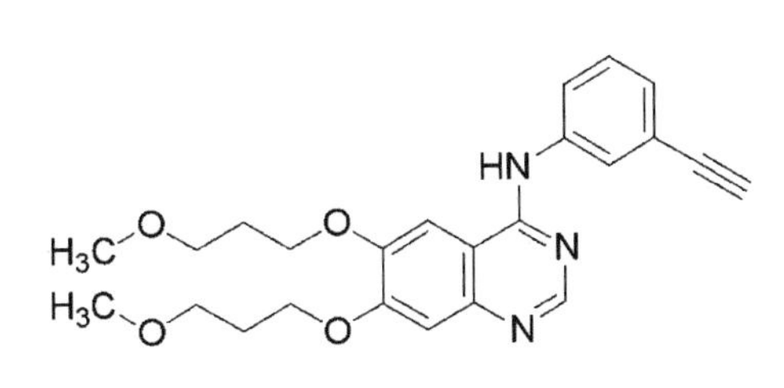

erlotinib (Tarceva)

Chapter 2. Drugs to Kill Germs

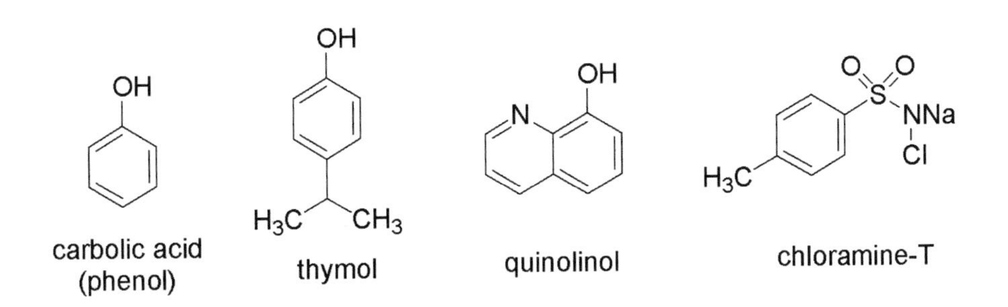

carbolic acid (phenol)

thymol

quinolinol

chloramine-T

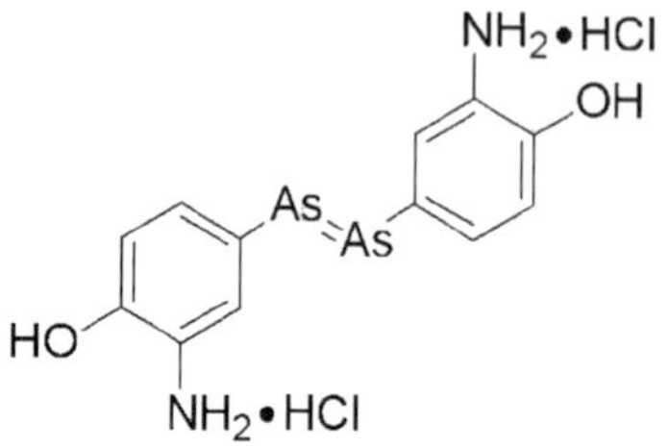

Ehrlich's 606
arsphenamine
(Salvarsan)

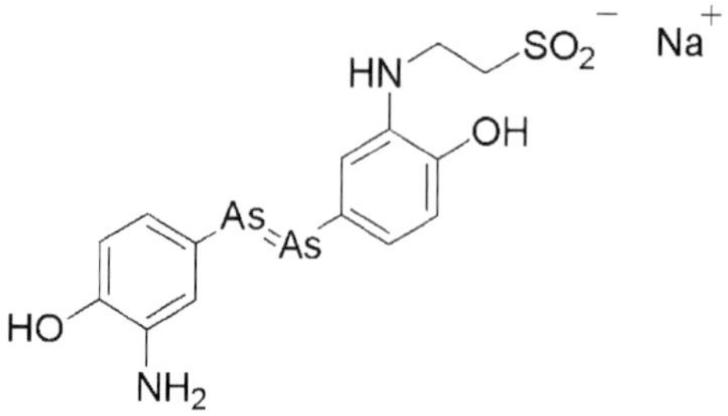

neoarsphenamine
(Neosalvarsan)

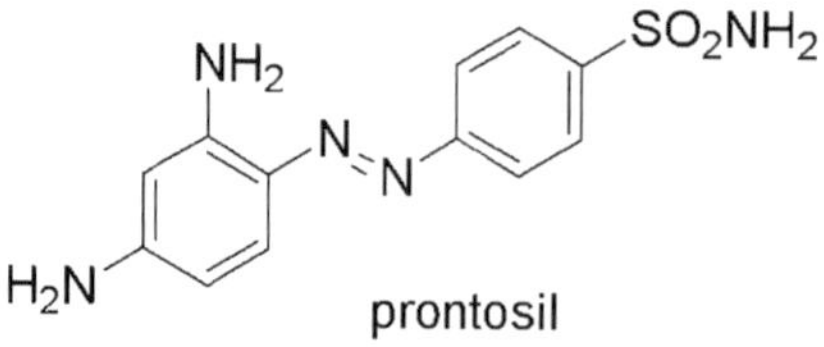

prontosil
(a pro-drug)

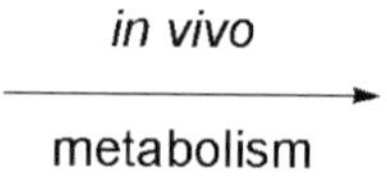

in vivo
metabolism

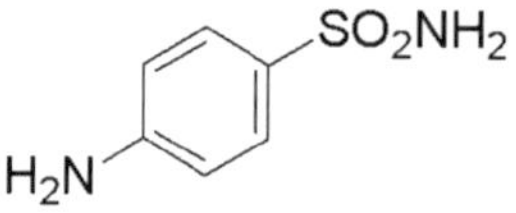

sulfanilamide
(the actual active drug)

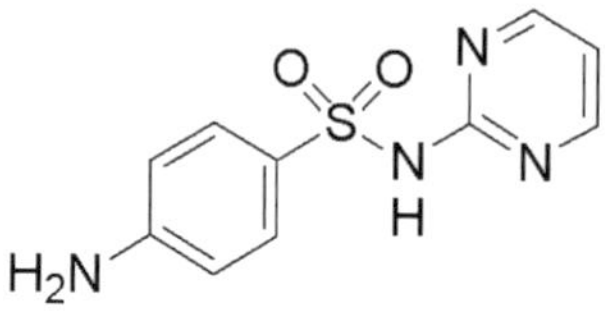

sulfadiazine

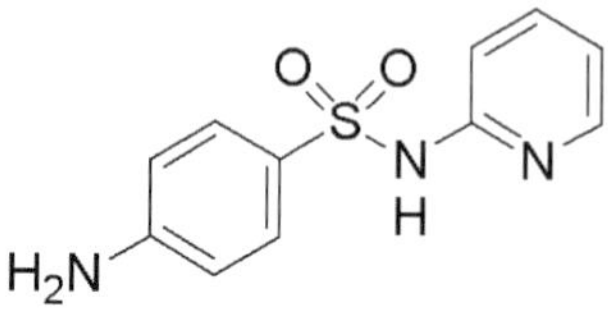

sulfapyridine

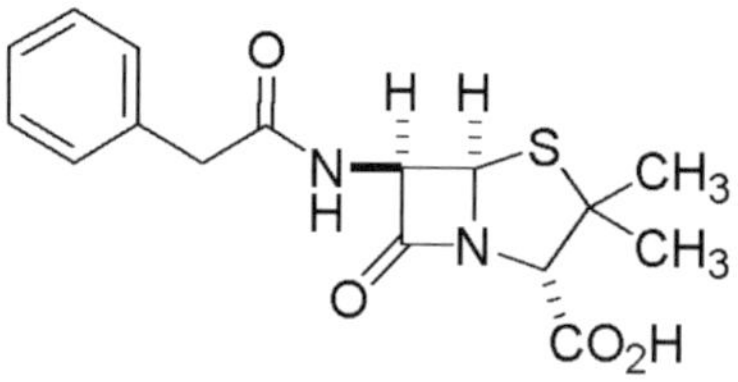

penicillin G

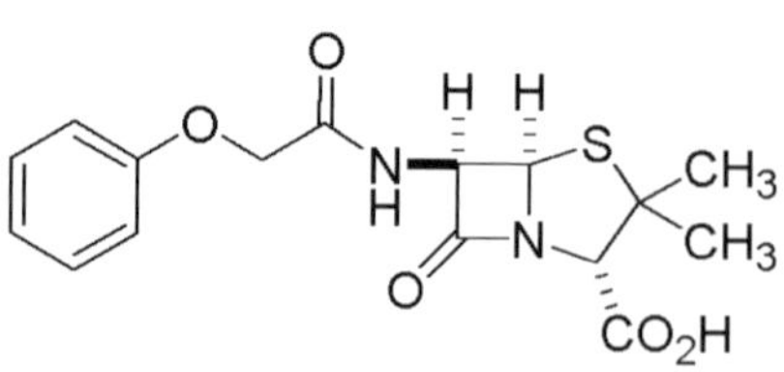

penicillin V

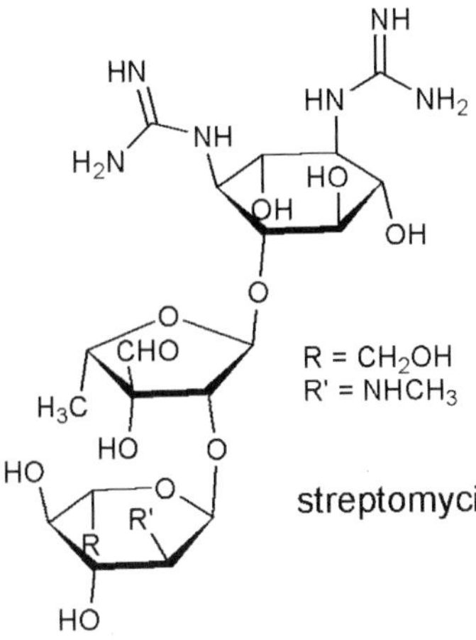

NH
HN
HN NH$_2$
H$_2$N NH
HO
OH
OH
O
O
CHO
R = CH_2OH
R' = $NHCH_3$
H_3C
HO O
HO
O
R'
R
HO

streptomycin

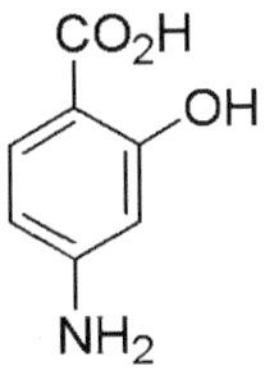

CO_2H
OH
NH_2

para-aminosalicyclic acid (PAS)

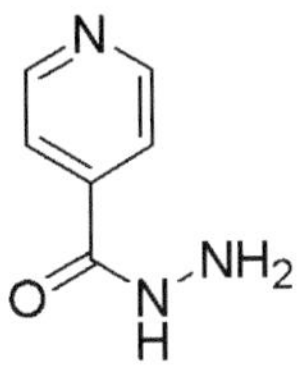

N
O N NH$_2$
H

isoniazid

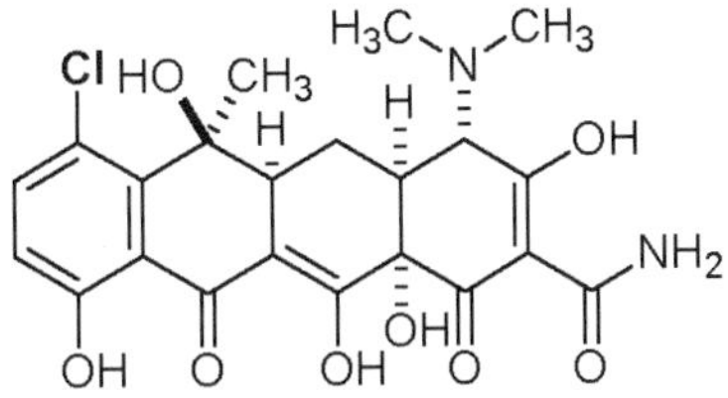

H_3C N CH_3
Cl HO CH_3
H H
OH
NH_2
OH
OH O OH O O

chlortetracycline

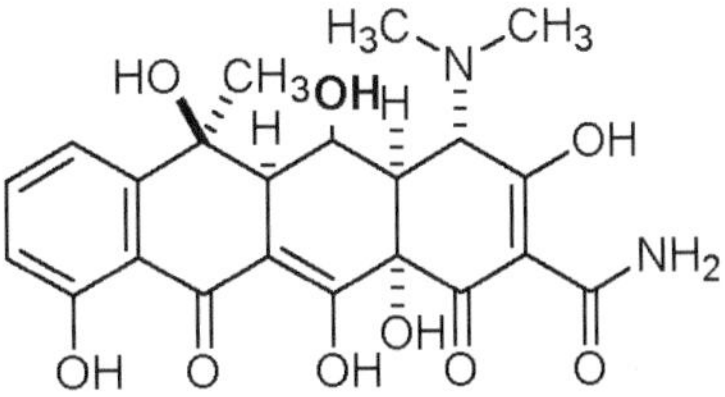

H_3C N CH_3
HO CH_3 OH H
H
OH
NH_2
OH
OH O OH O O

oxytetracycline

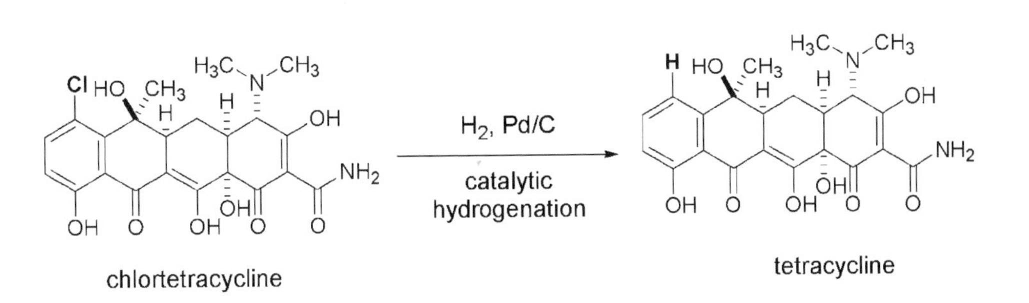

First generation quinolones

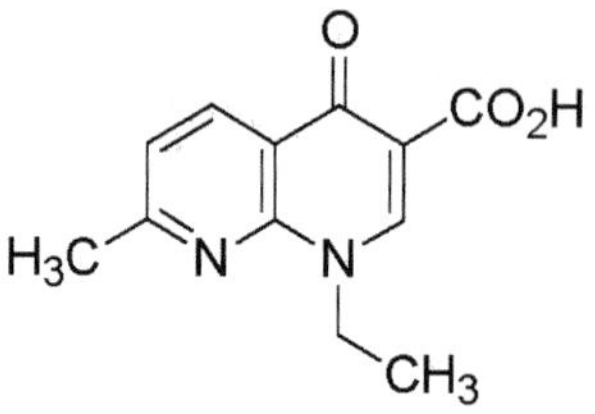

nalidixic acid

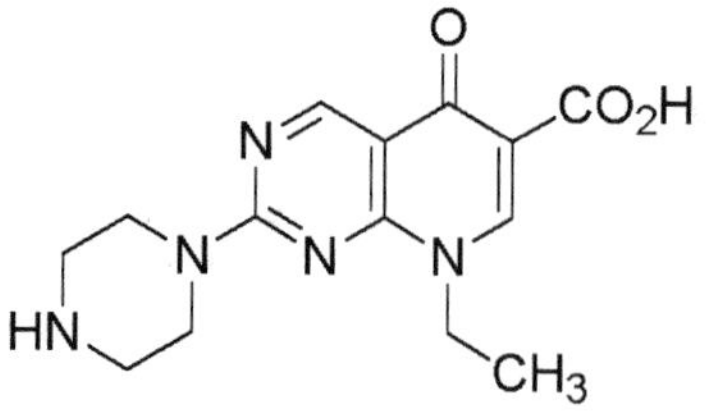

pipemidic acid

Second generation quinolones

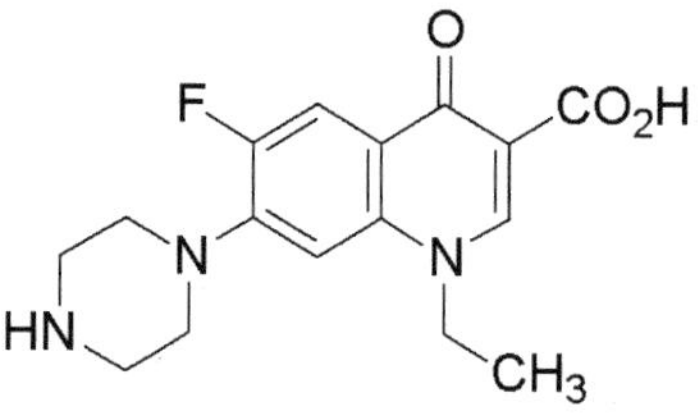

norfloxacin (Noflo)

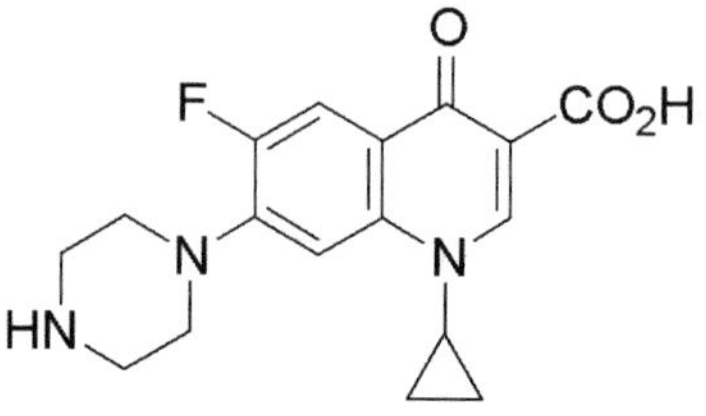

ciprofloxacin (Cipro)

Third generation quinolones

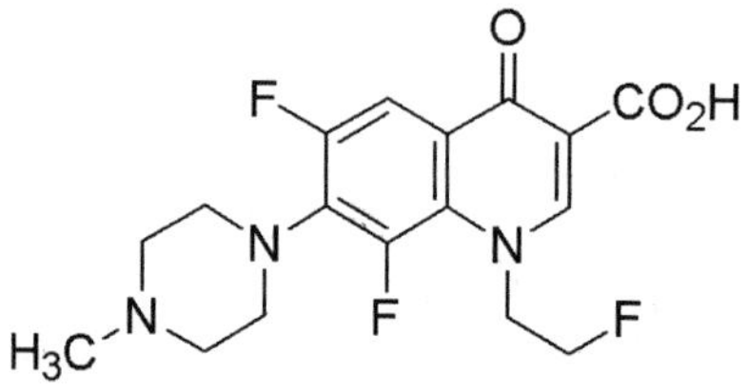

fleroxacin

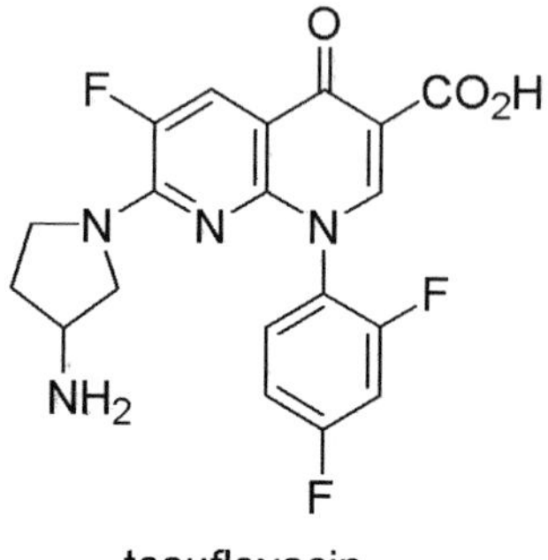

tosufloxacin

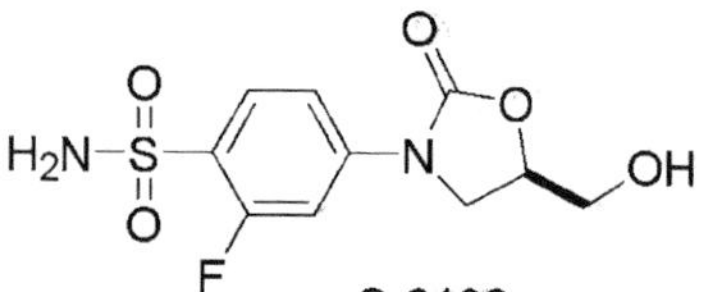

S-6132

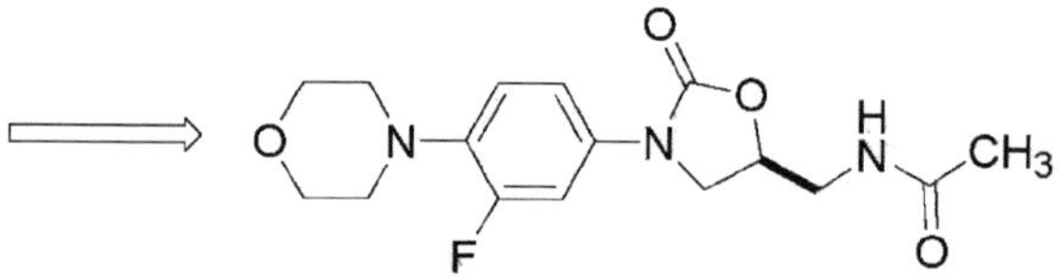

(linezolid) Zyvox

Chapter 3. Cardiovascular Drugs

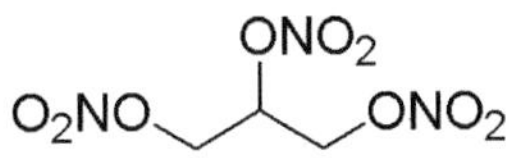

nitroglycerin

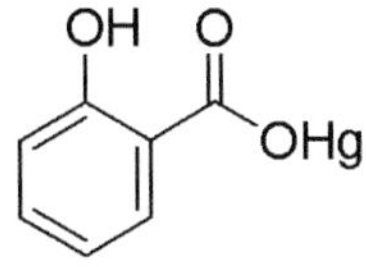

mercuric salicylate

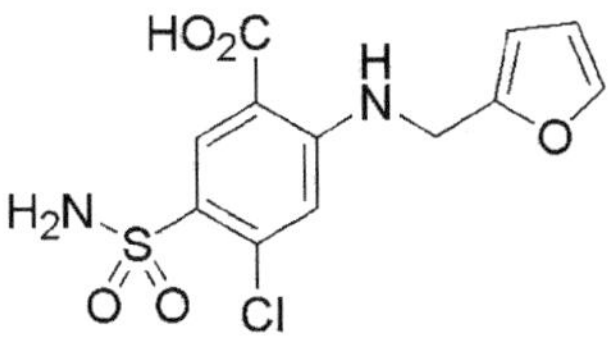

furosemide

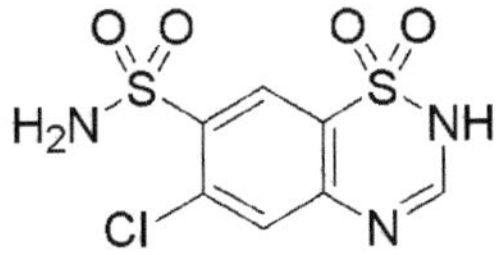

chlorothiazide

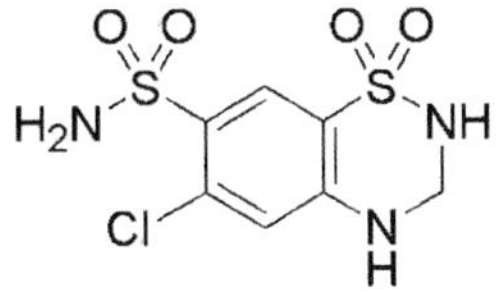

hydrochlorothiazide

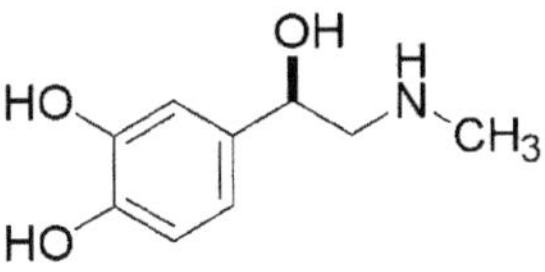

andrenaline

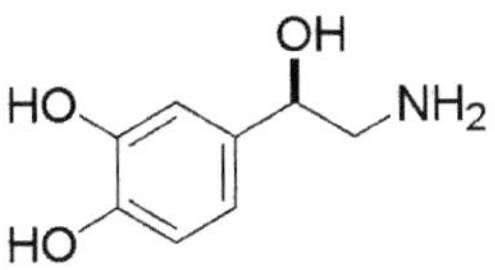

norandrenaline

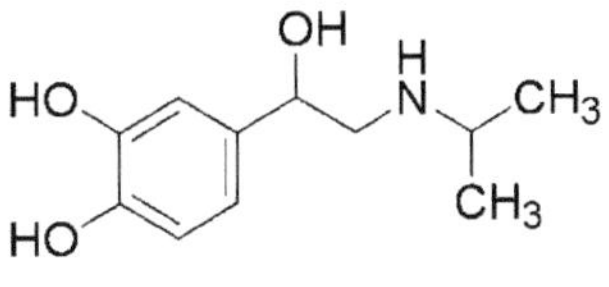

isoprenaline

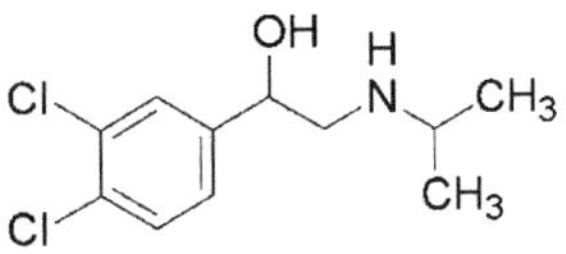

dichloroisoprenaline (DCI)

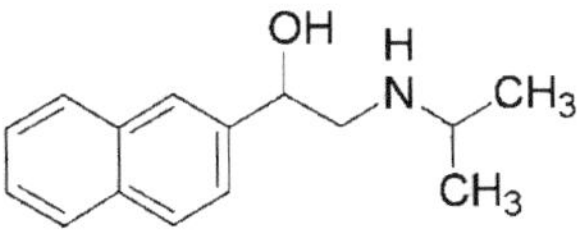

pronethalol

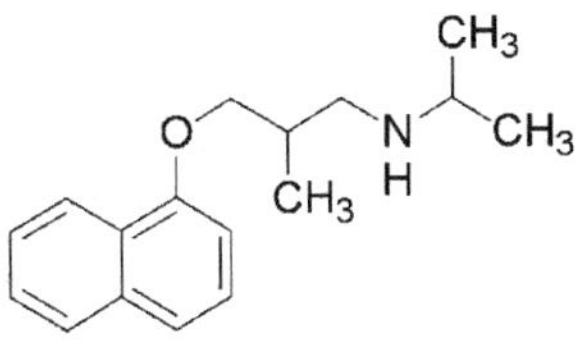

propranolol (Inderal)

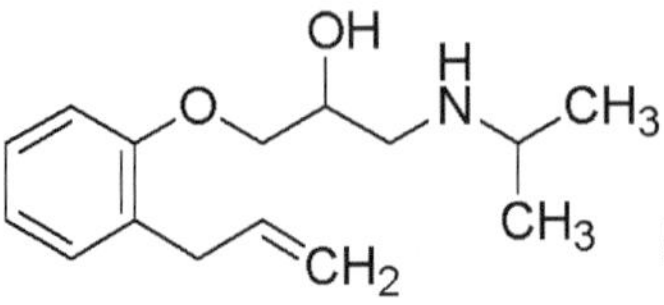

alprenolol

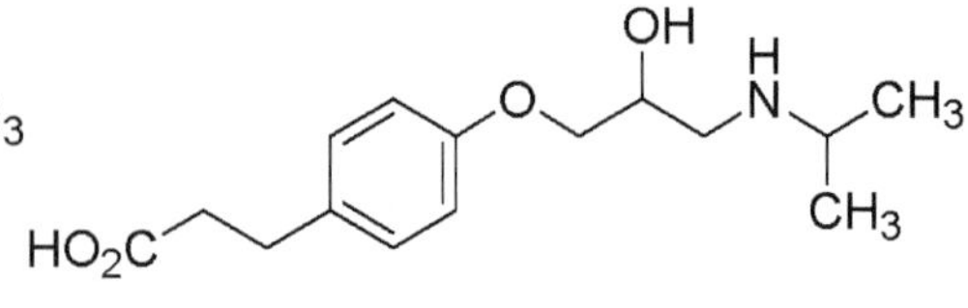

metoprolol

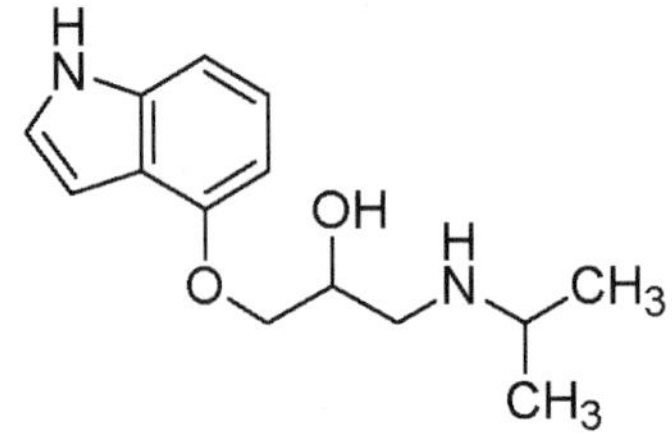

pindolol

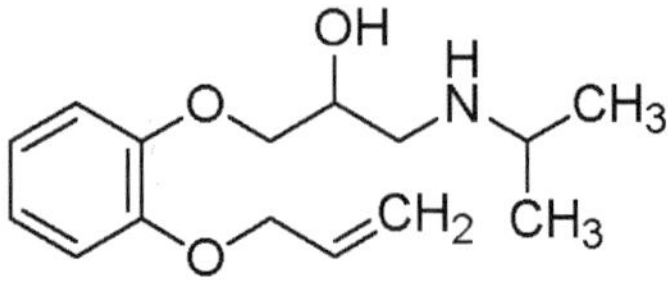

oxprenolol

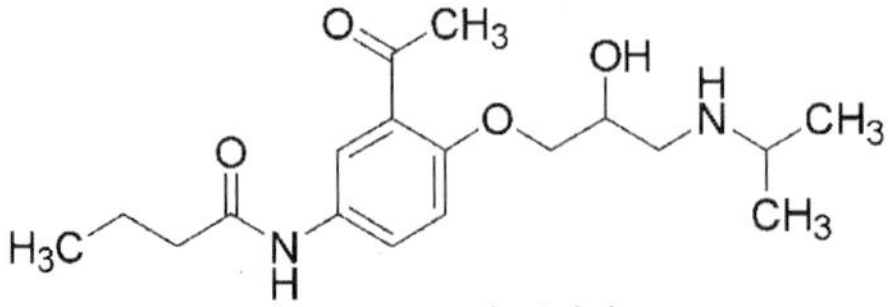

acebutolol

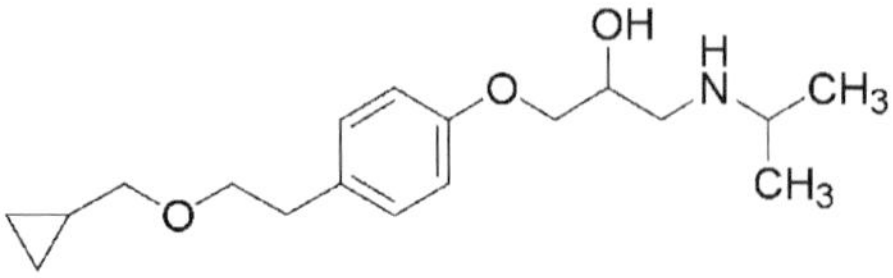

betaxolol

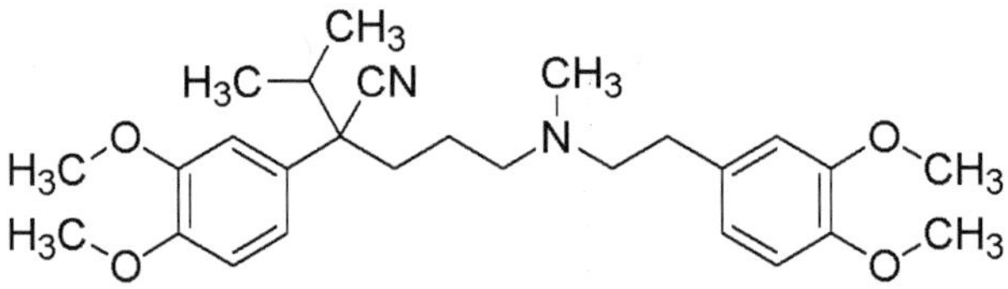

verapamil

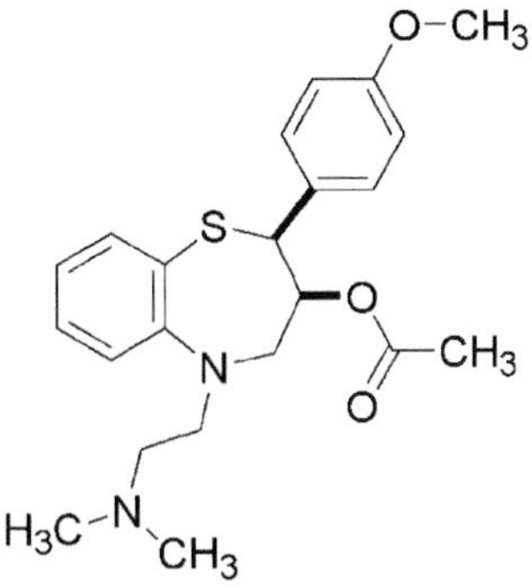

O–CH_3
S
O
N
CH_3
O
H_3C–N–CH_3

diltiazem

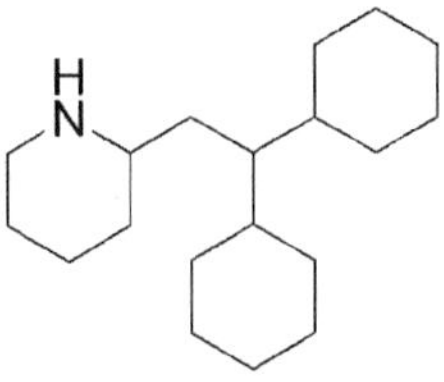

H
N

perhexiline

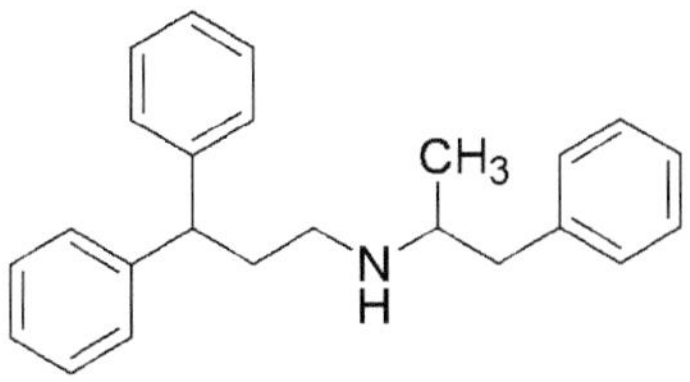

CH_3
N
H

prenylamine

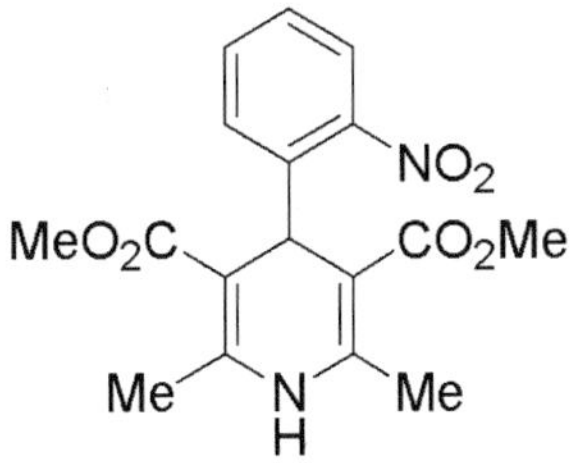

NO_2
MeO_2C
CO_2Me
Me
N
H
Me

nifedipine (Adalat)

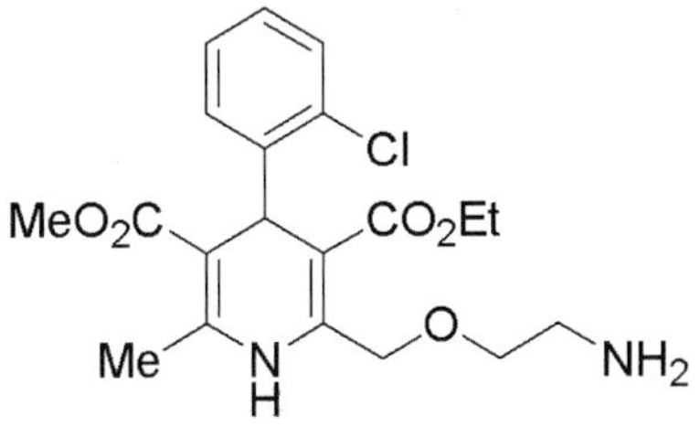

Cl
MeO_2C
CO_2Et
Me
N
H
O
NH_2

amlodipine (Norvasc)

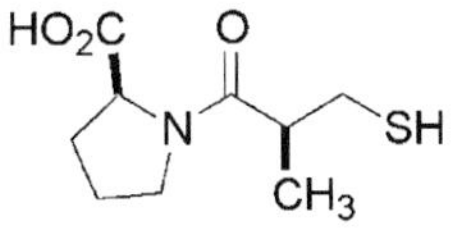

HO_2C
O
N
SH
CH_3

captopril (Capoten)

EtO
O
CH_3
N
H
O
N
O
OH

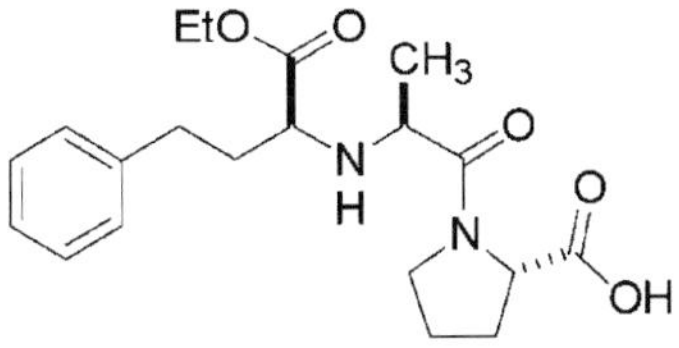

enalapril (Enalaprilat)

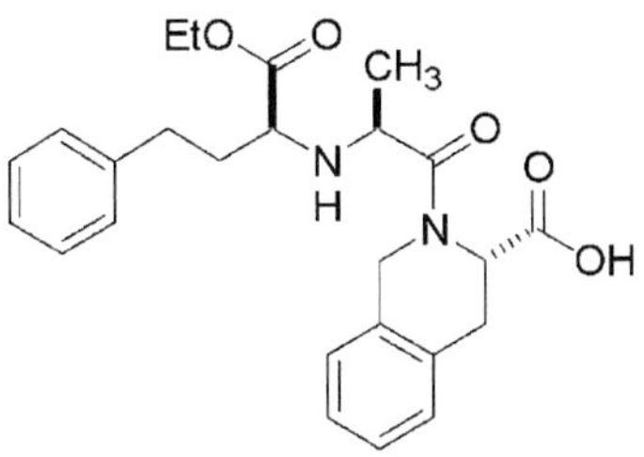

EtO
O
CH_3
N
H
O
O
N
OH

quinapril (Accupril)

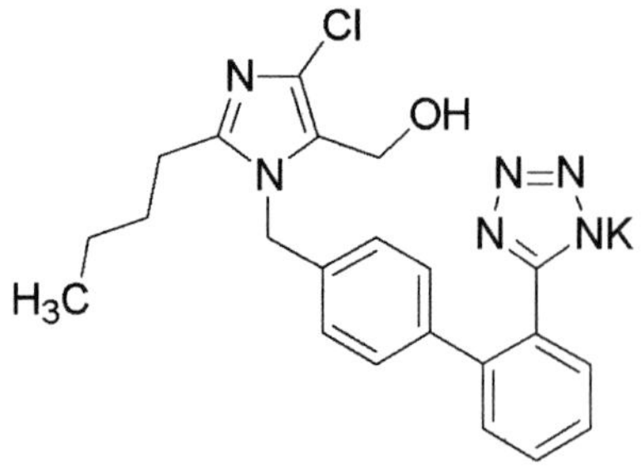

losartan (Cozarr)

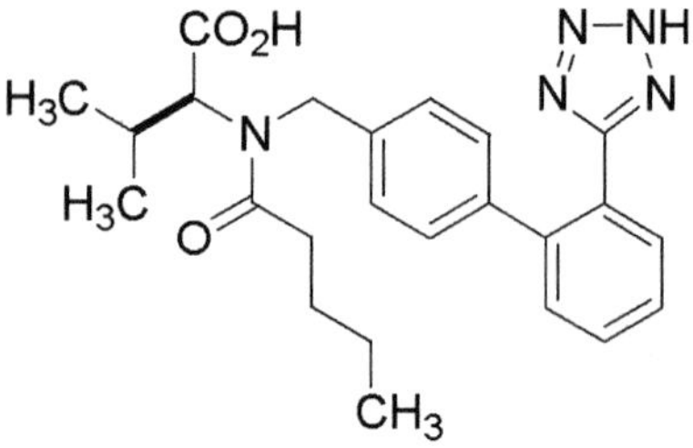

valsartan (Diovan)

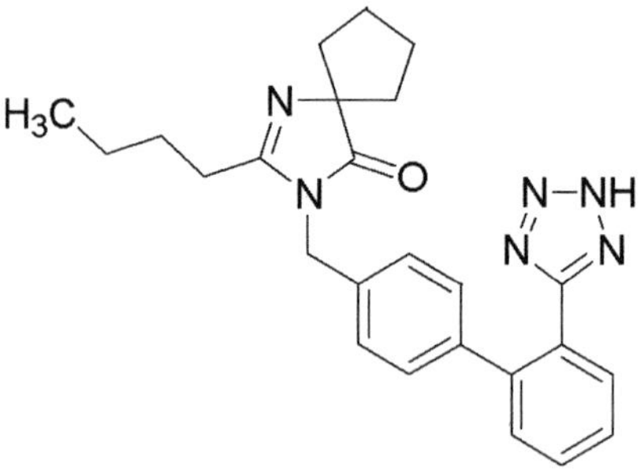

irbesartan (Avapro)

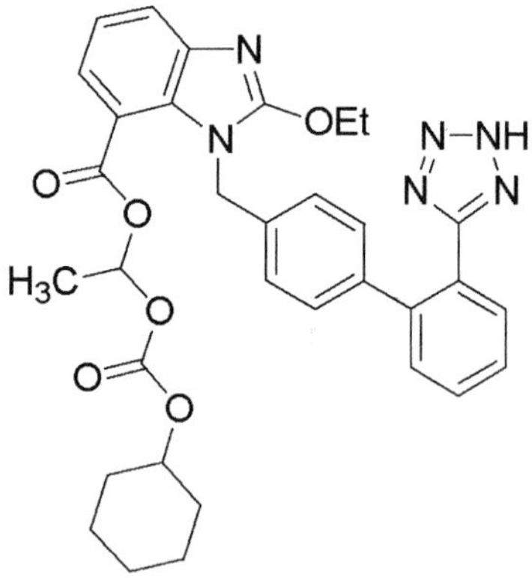

candesartan (Atacand)

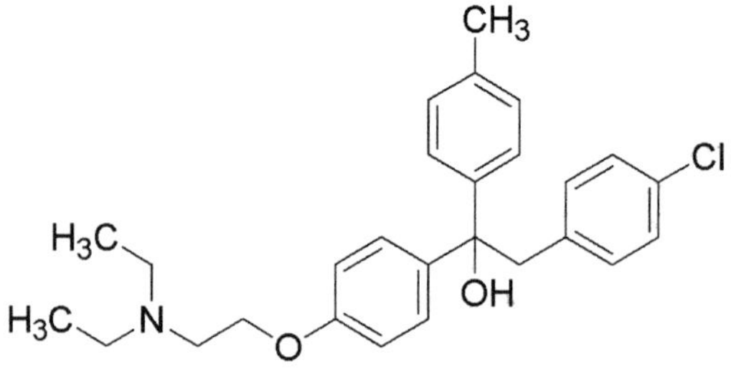

triparanol

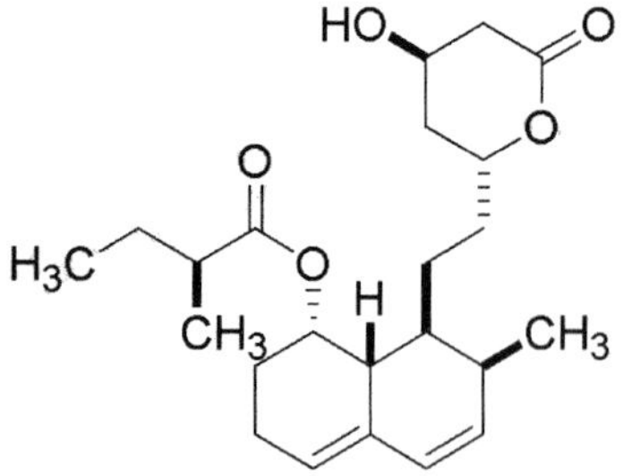

mevastatin = compactin

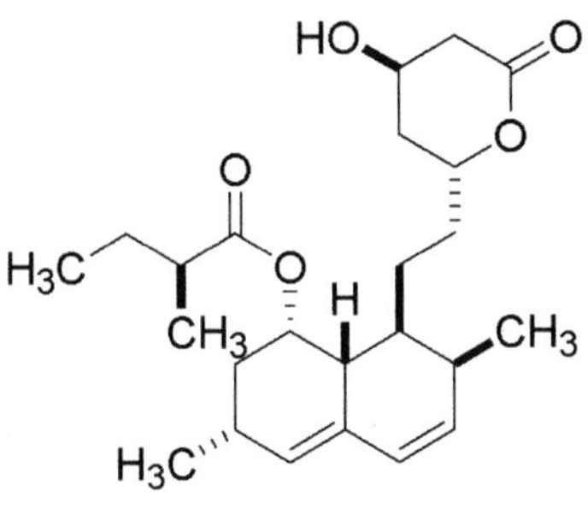

lovastatin (Mevacor)

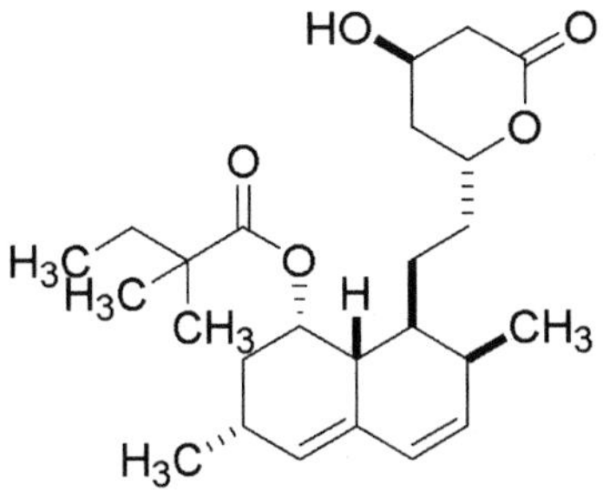

simvastatin (Zocor)

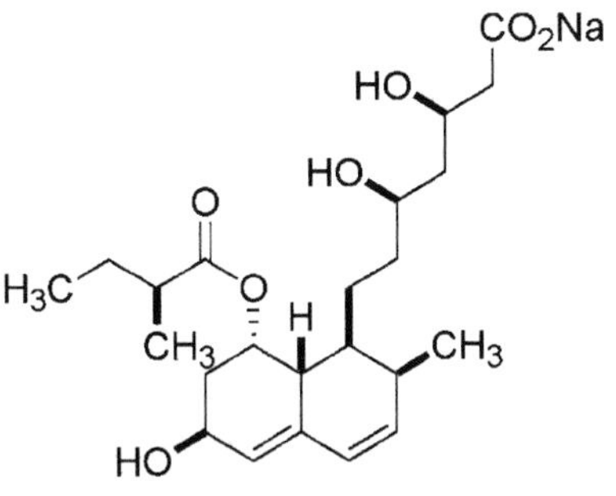

pravastatin (Pravacol)

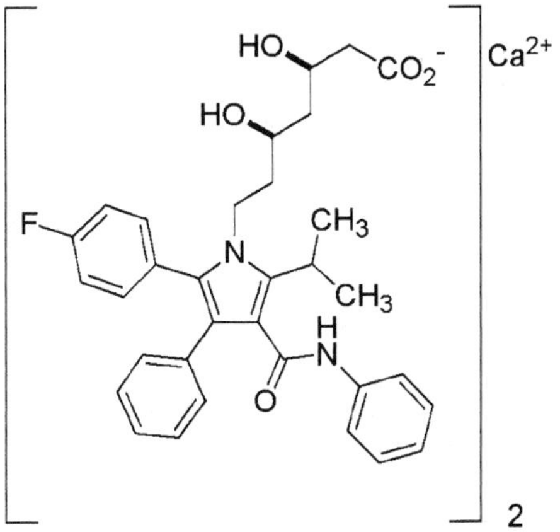

atorvastatin calcium (Lipitor)

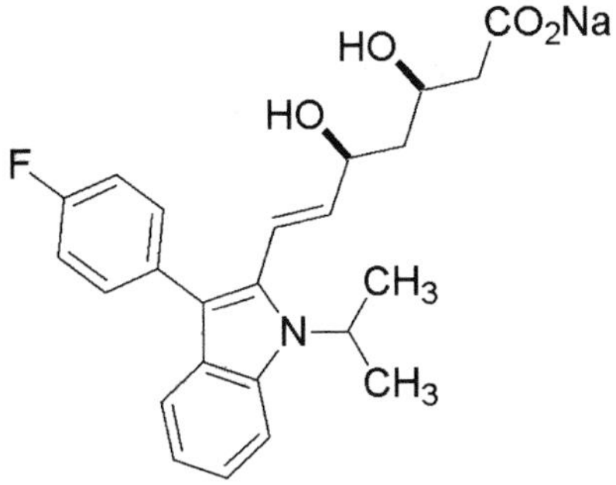

fluvastatin (Lescol)

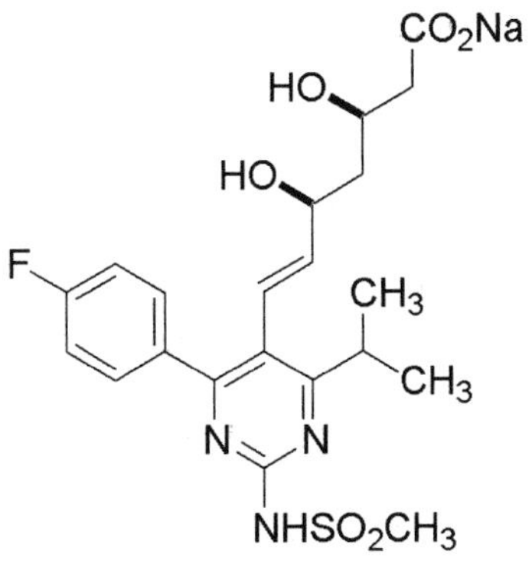

rosuvastatin (Crestor)

Chapter 4. Sex and Drugs

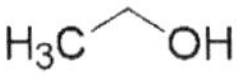

ethanol

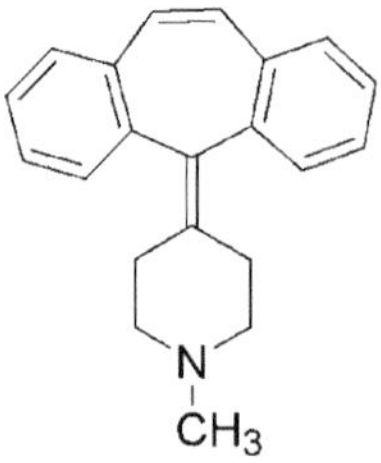

cyproheptadine

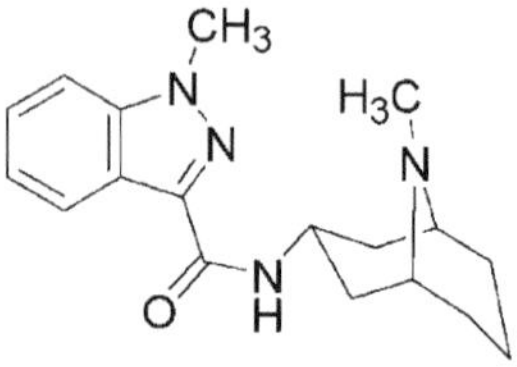

granisetron

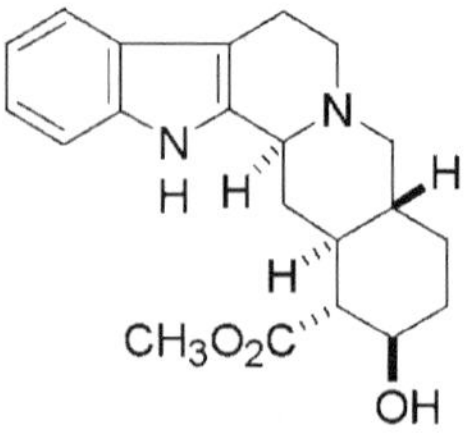

yohimbine

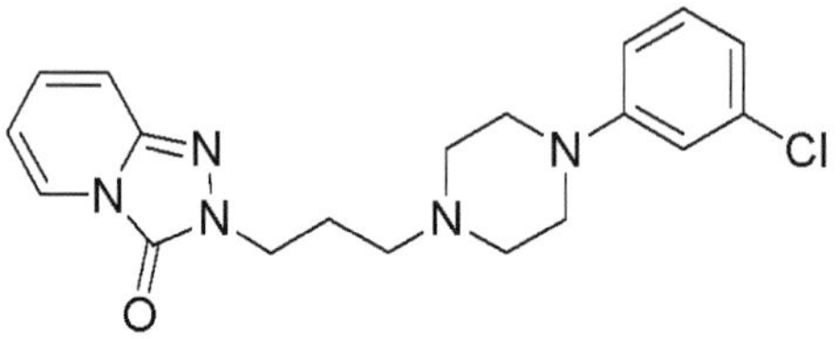

trazodone (Desyrel)

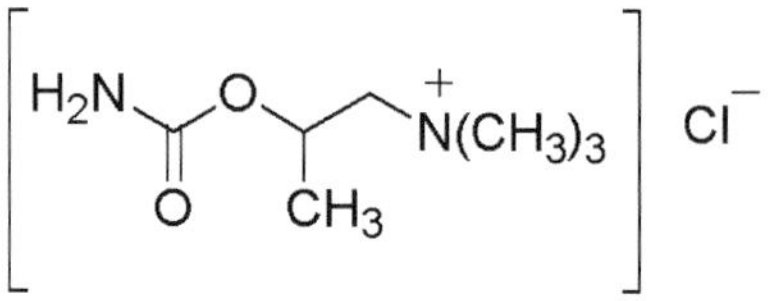

bethanechol chloride

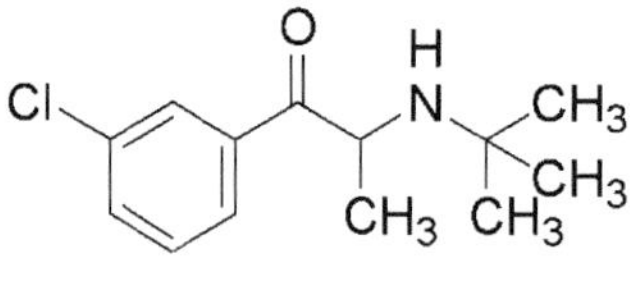

bupropion (Wellbutrin, Zyban)

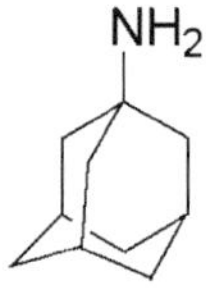

amantadine

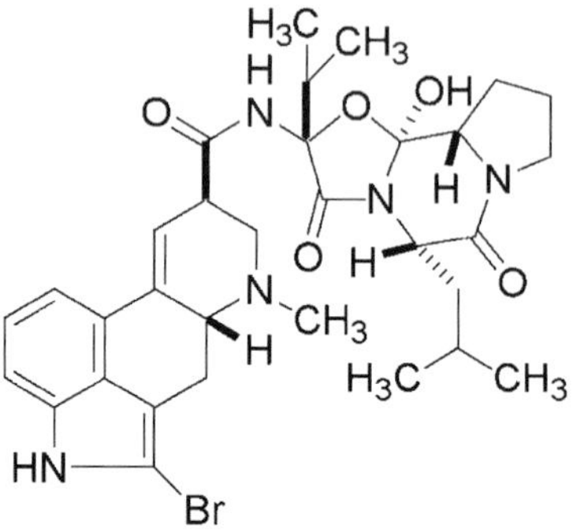

bromocryptine (Parlodel)

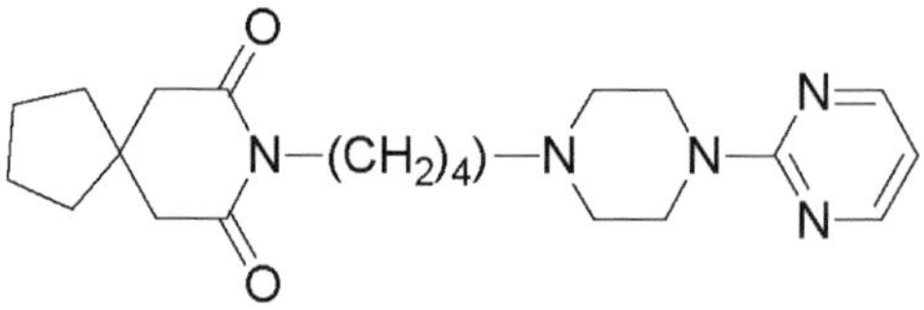

buspirone (Buspar)

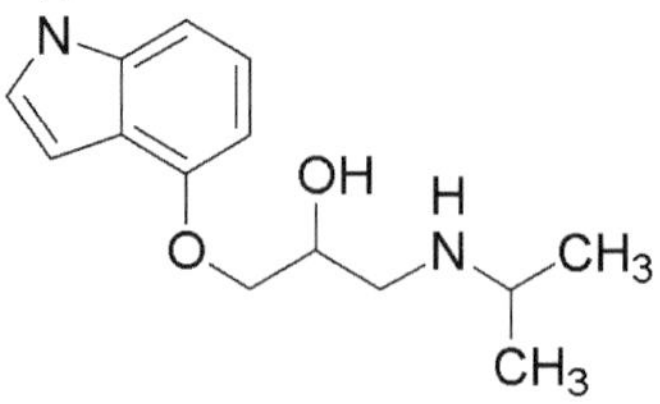

pindolol

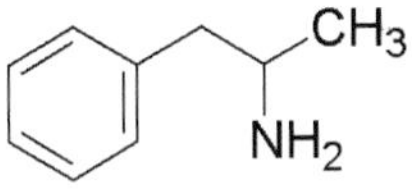

amphetamine

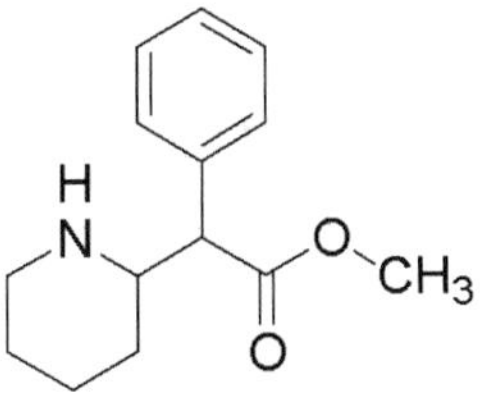

methylphenidate (Ritalin)

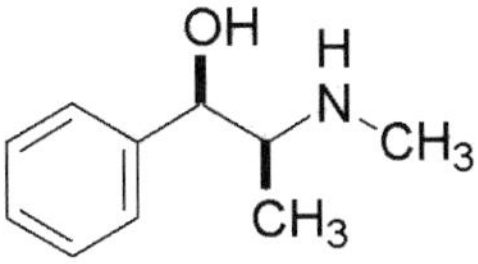

ephedrine

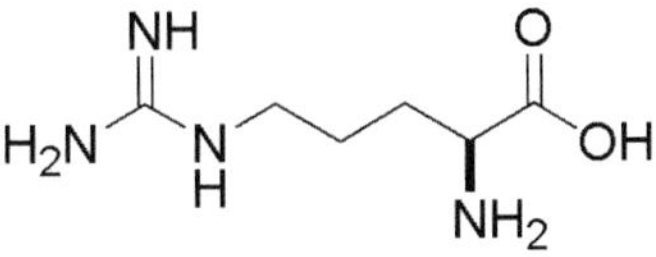

L-arginine

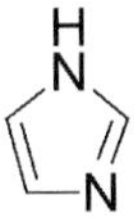

imidazole

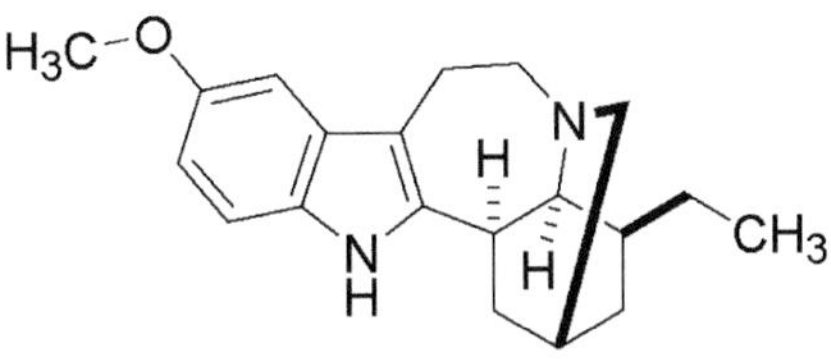

ibogaine

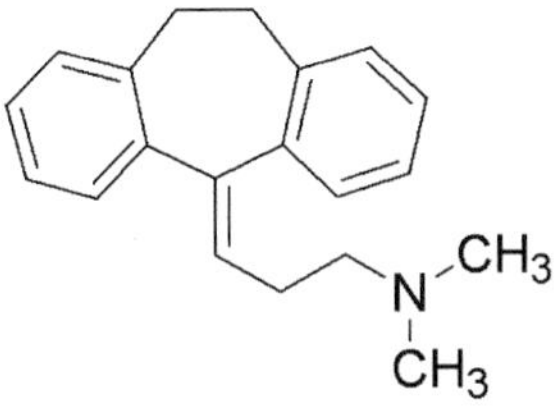

amitriptyline

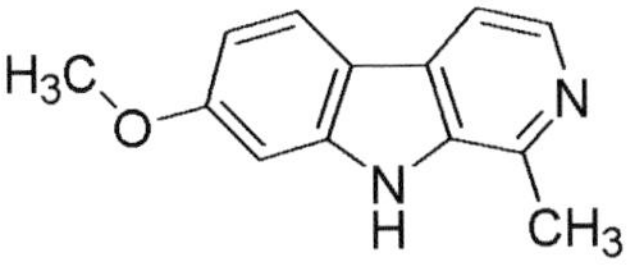

harmine

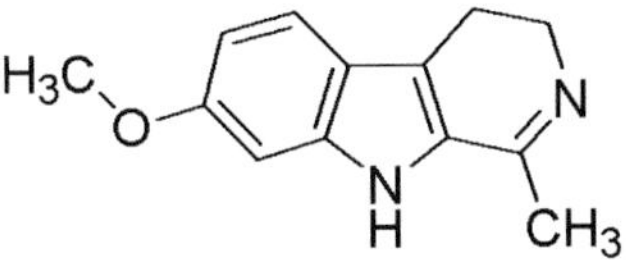

harmaline

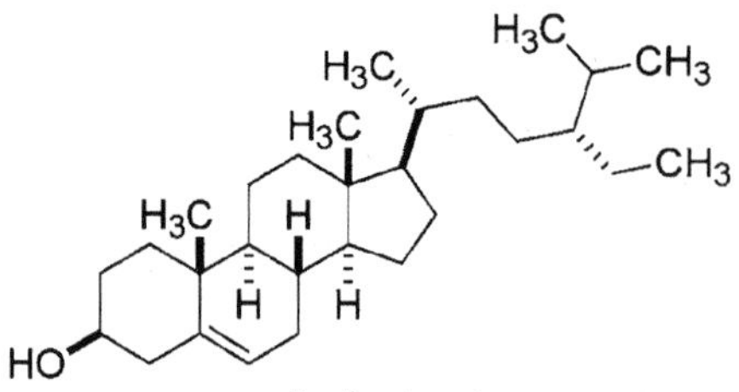

β-sitosterol

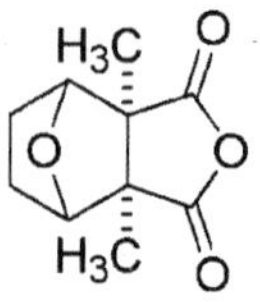

cantharidin

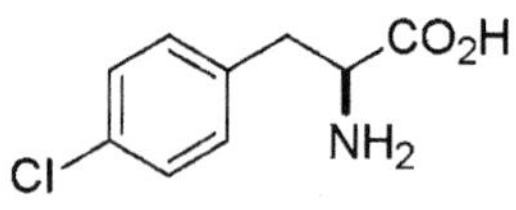

p-chlorophenylalanine

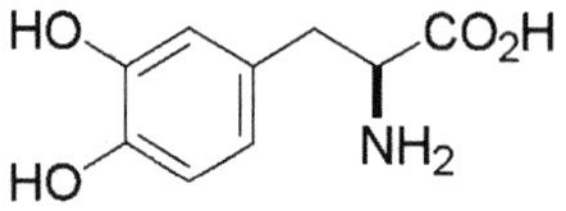

L-dopa

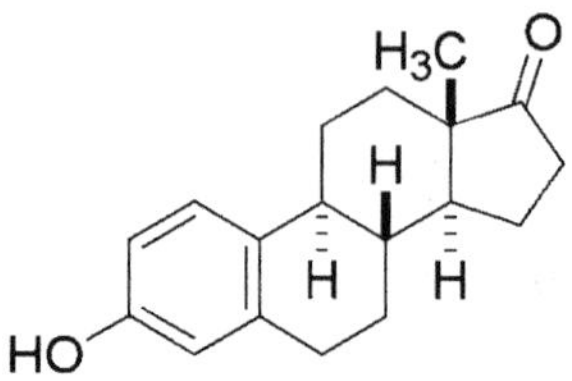

estrone

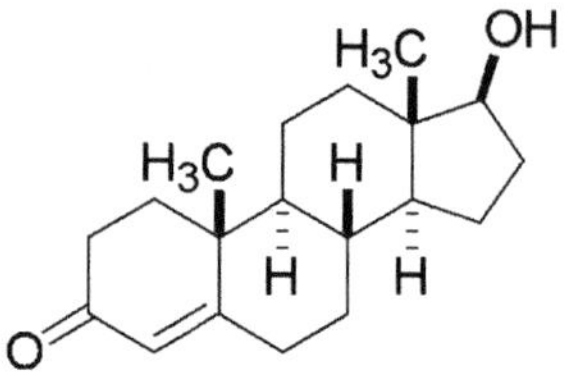

testosterone

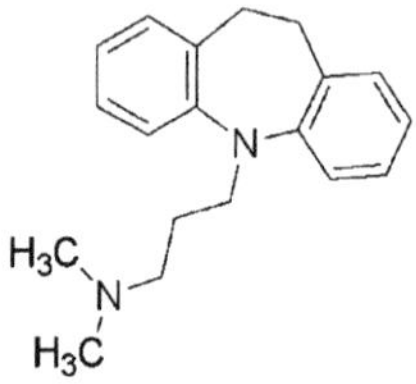

imipramine (Tofranil)

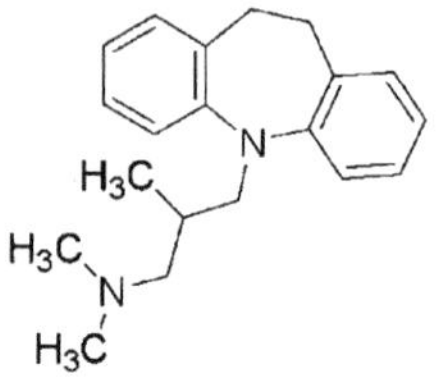

trimipramine

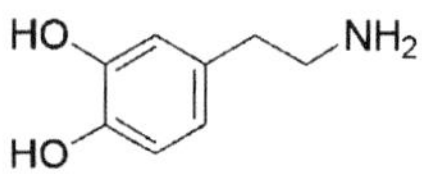

dopamine

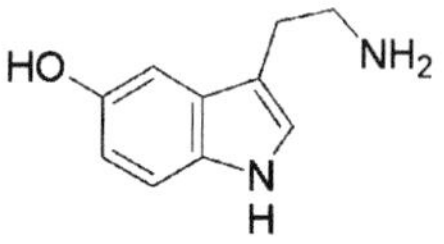

serotonin

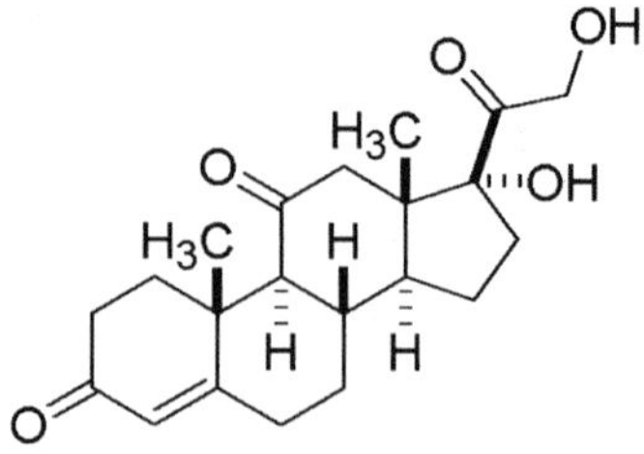

cortisone

The Djerassi route to norethindrone from estrone methyl ether

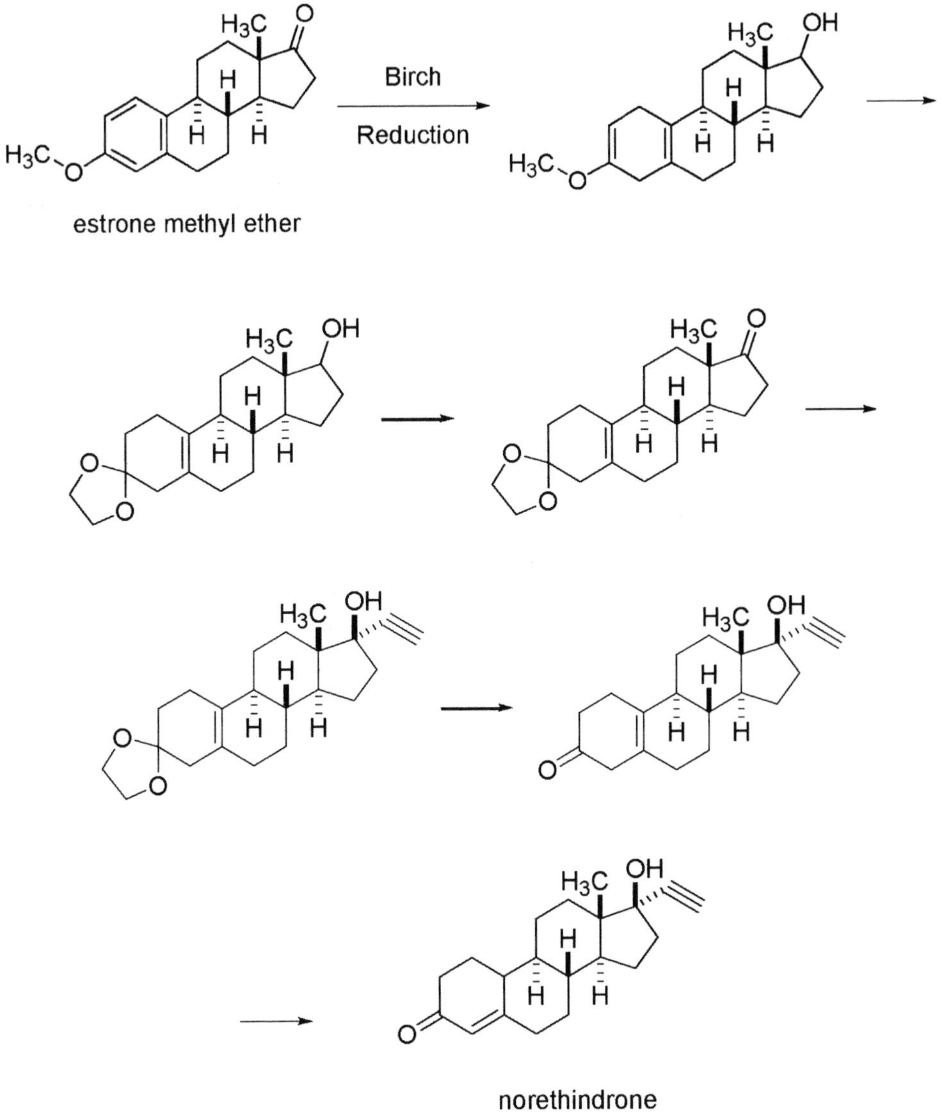

Birch
Reduction

estrone methyl ether

norethindrone

Oxidative degradation of cholesterol to estrogens

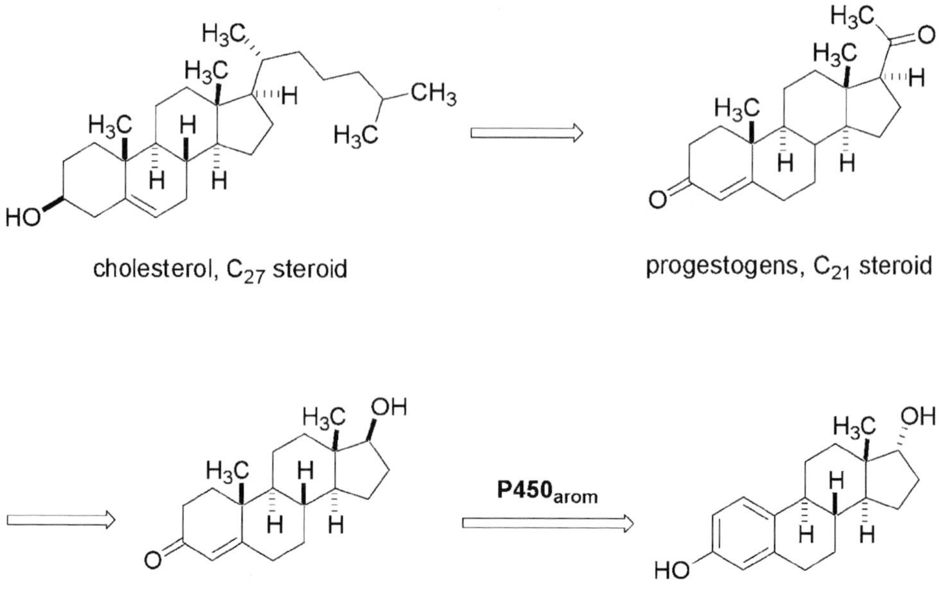

H_3C H_3C H CH_3 H_3C H H_3C HO H H

cholesterol, C_{27} steroid

H_3C O H_3C H H_3C O H H

progestogens, C_{21} steroid

H_3C OH H_3C H O H H

androgens, C_{19} steroid

P450$_{arom}$

H_3C OH H HO H H

estrogens, C_{18} steroid

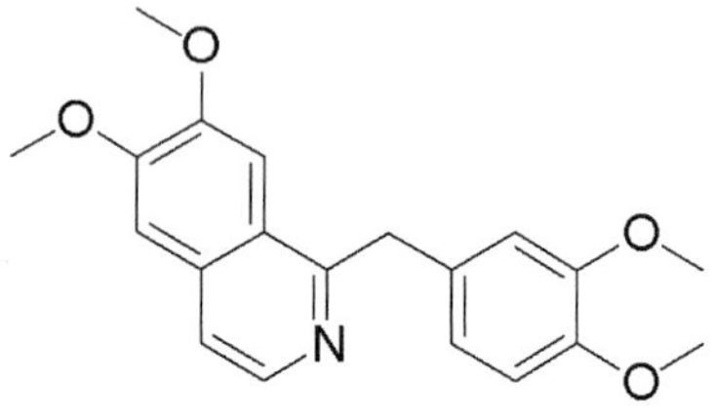

O O N O O

papaverine

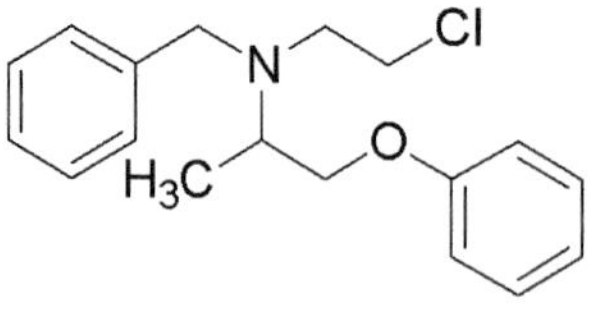

N Cl H_3C O

phenoxybenzamine

O CO_2H CH_3 HO OH

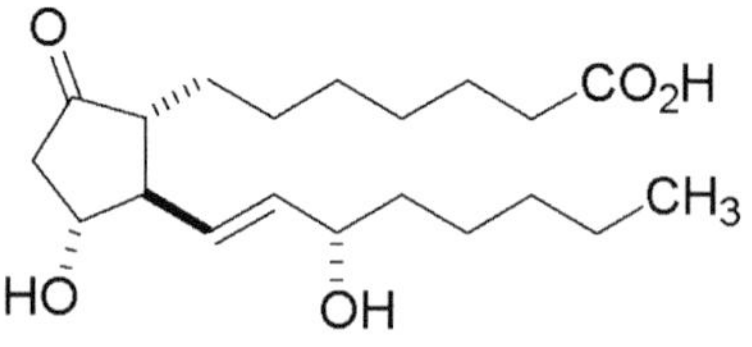

prostaglandin E_1 (alprostadil, Caverject)

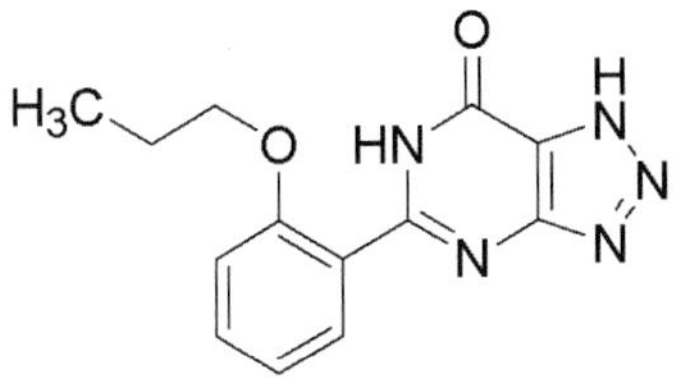

zaprinast

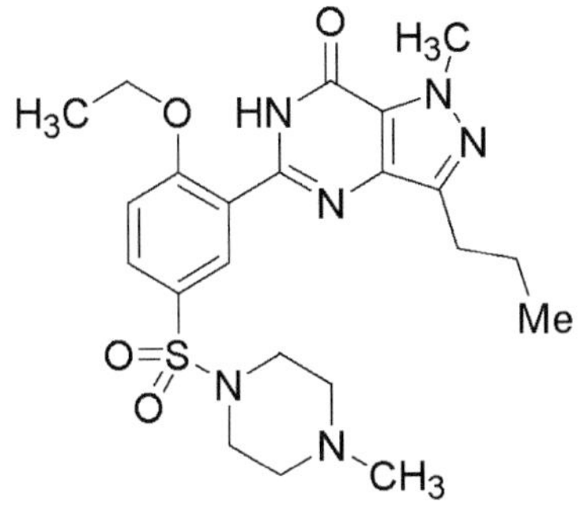

sildenafil citrate (Viagra)

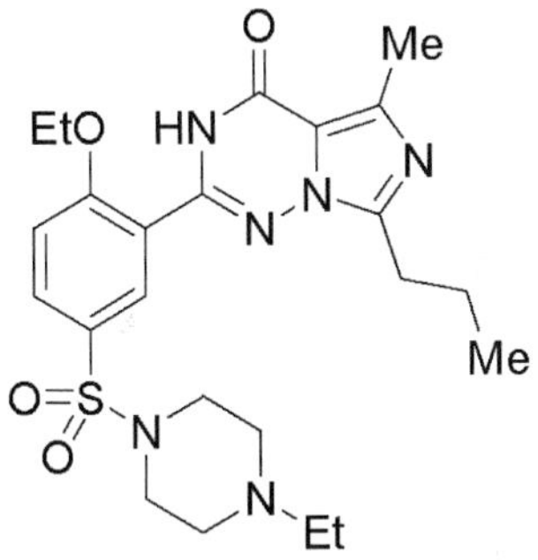

vardenafil hydrochloride (Levitra)

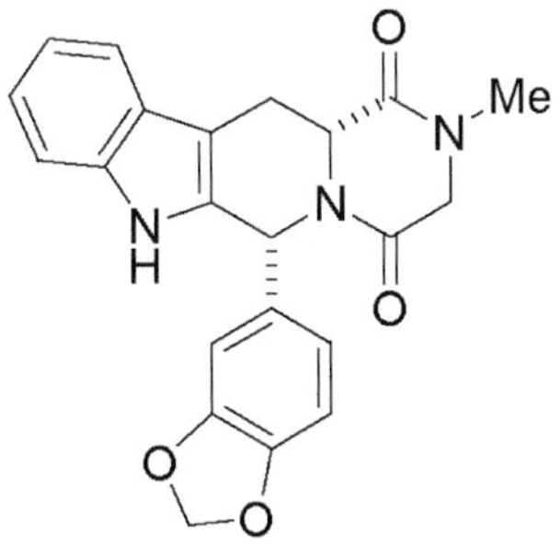

tadalafil (Cialis)

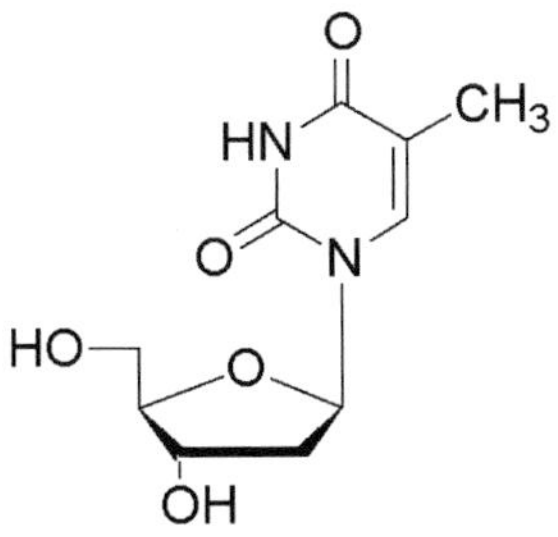

azidothymidine (AZT)

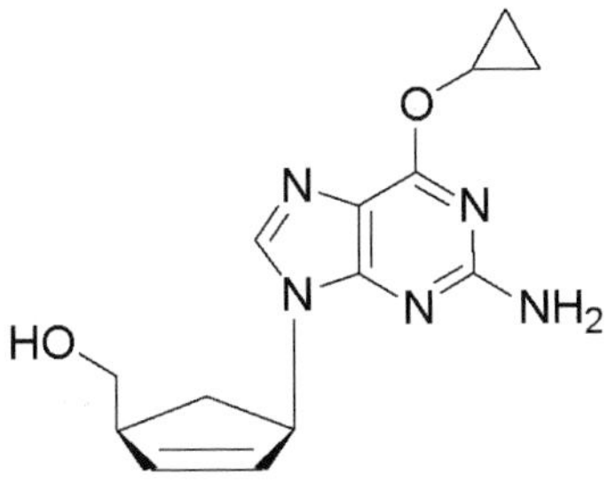

abacavir (Ziagen)

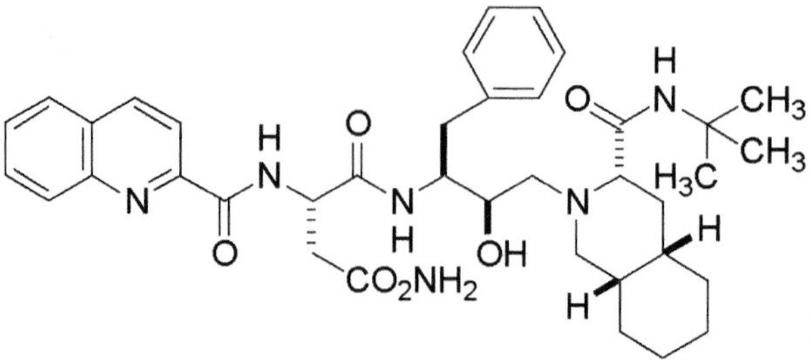

saquinavir (Invirase)

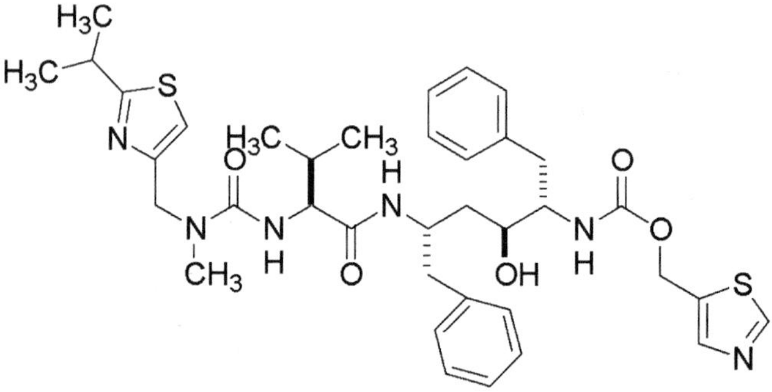

ritonavir (Norvir)

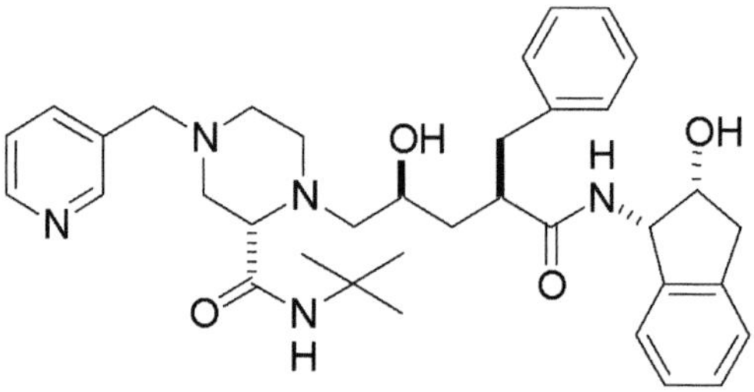

indinavir (Crixivan)

Chapter 5. Drugs of the Mind

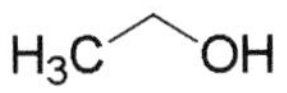

ethanol

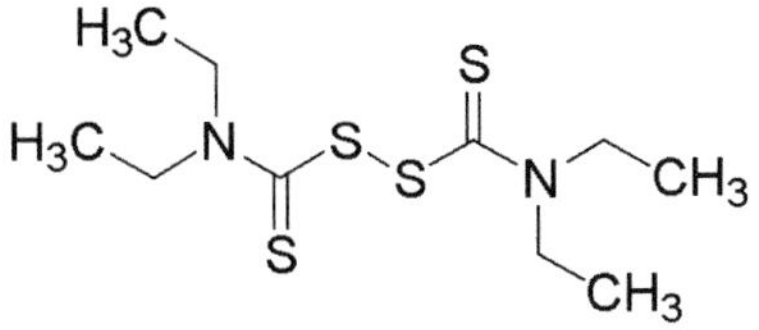

disulfiram (Antabuse)

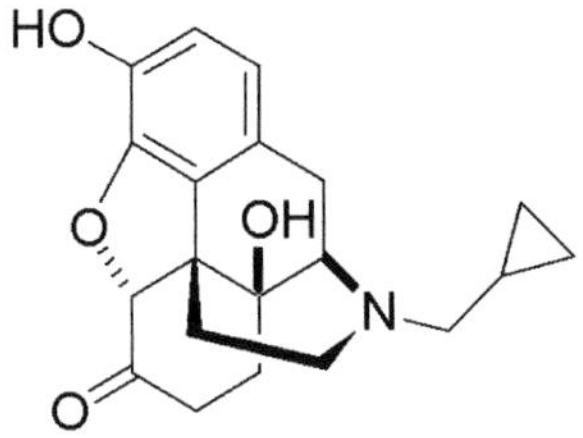

naltrexone

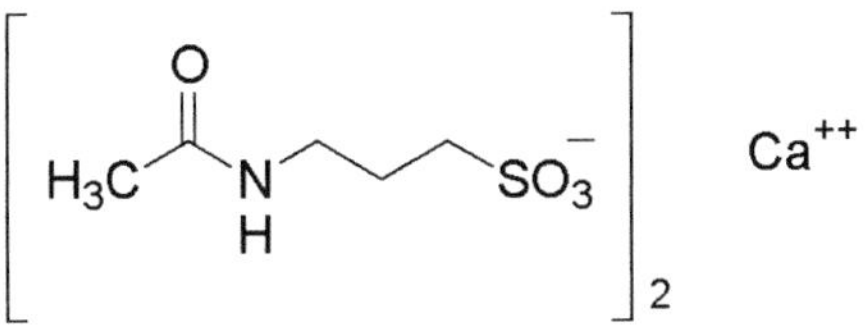

acamprostate calcium (Campral)

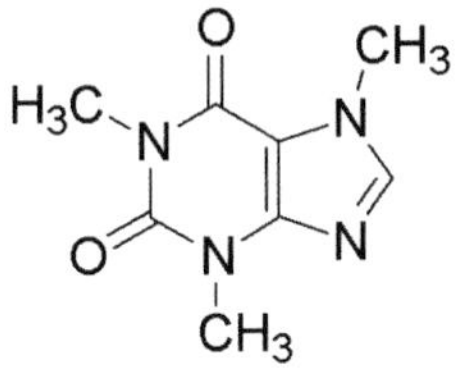

caffeine

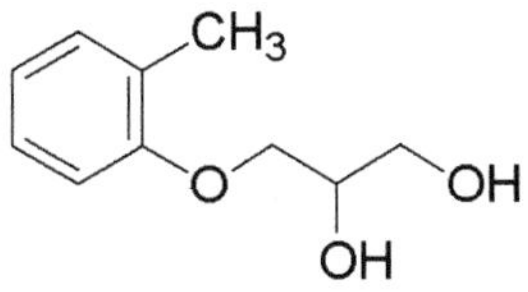

mephensin

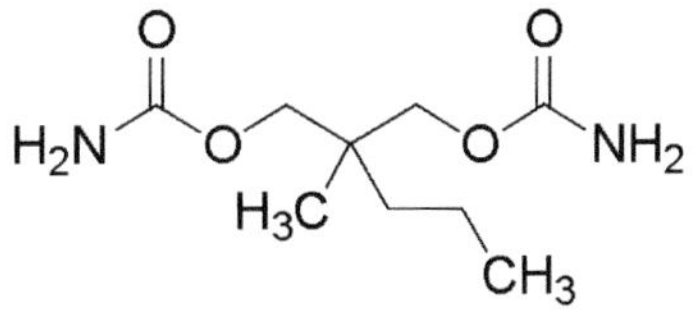

meprobamate (Miltown)

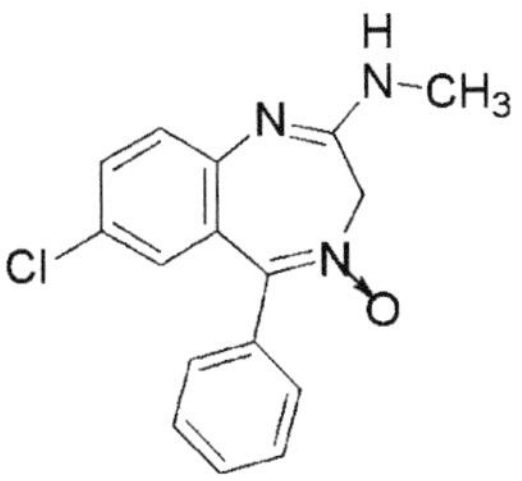

chlordiazepoxide (Librium)

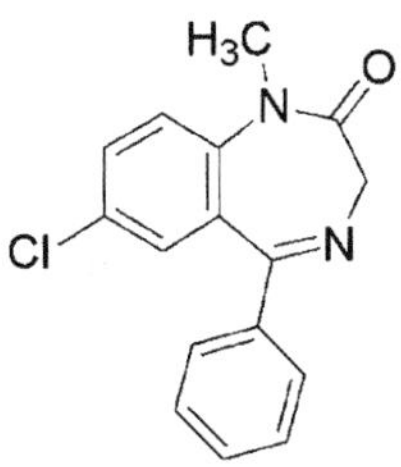

diazepam (Valium)

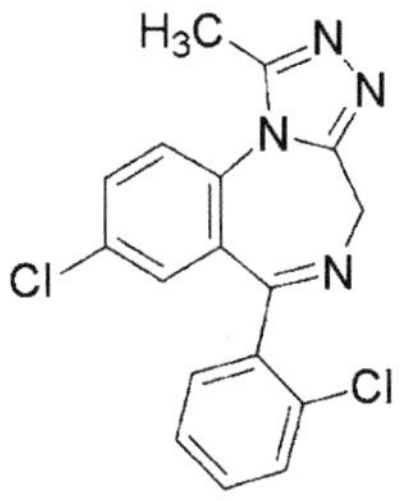

triazolam (Halcion)

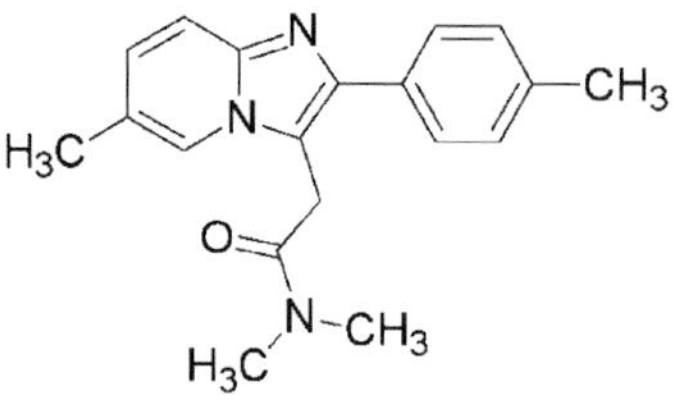

zolpidem (Ambien)

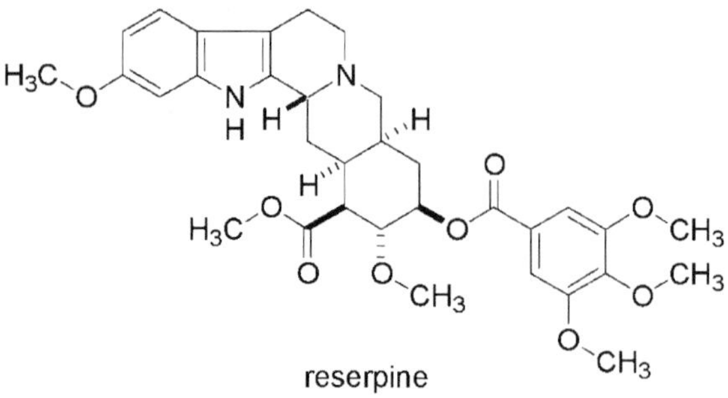

reserpine

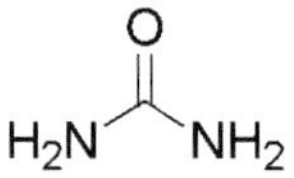

urea

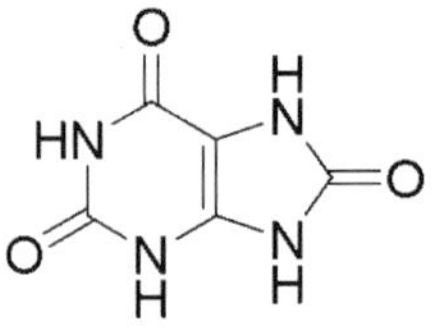

uric acid

Li_2CO_3

lithium carbonate

LiOAc

lithium acetate

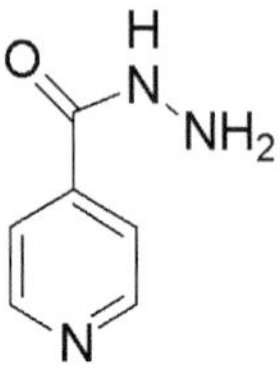

isoniazid

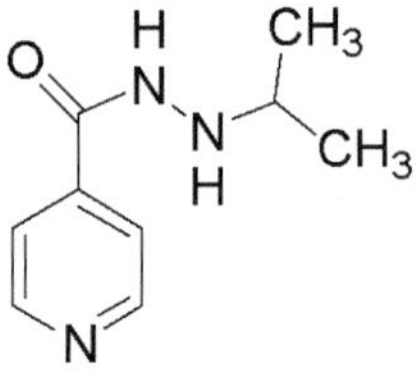

iproniazid (Marsilid)

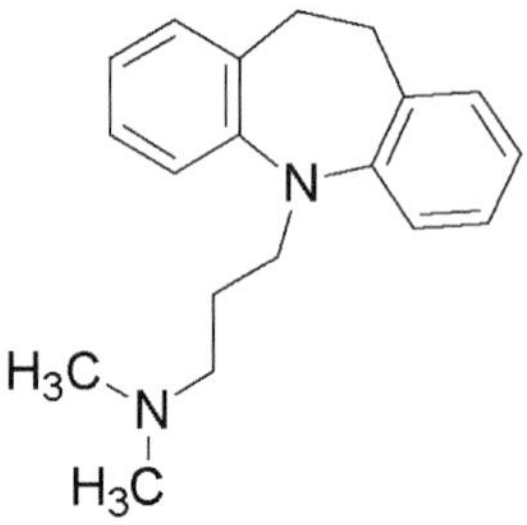

imipramine (Tofranil)

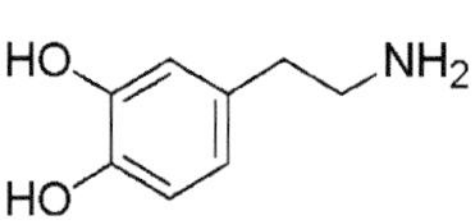

dopamine

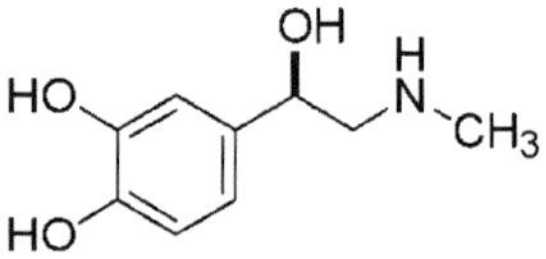

epinephrine (adrenaline)

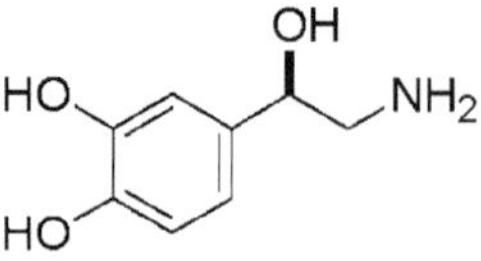

norepinephrine (noradrenaline)

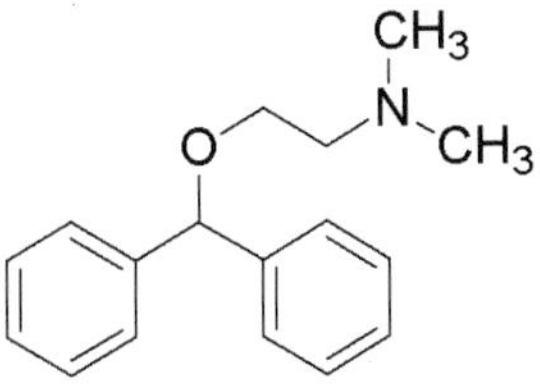

diphenhydramine
(Benadryl)

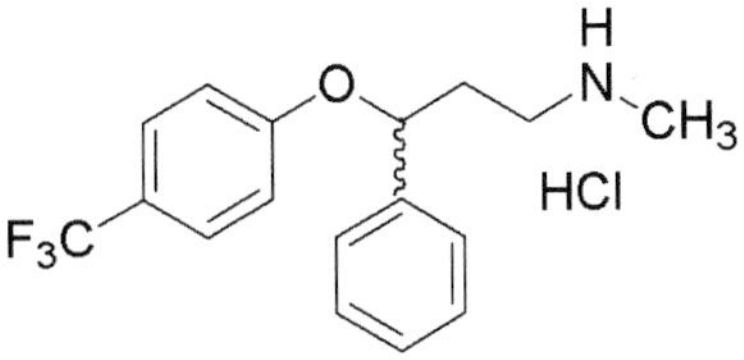

fluoxetine hydrochloride
(Prozac)

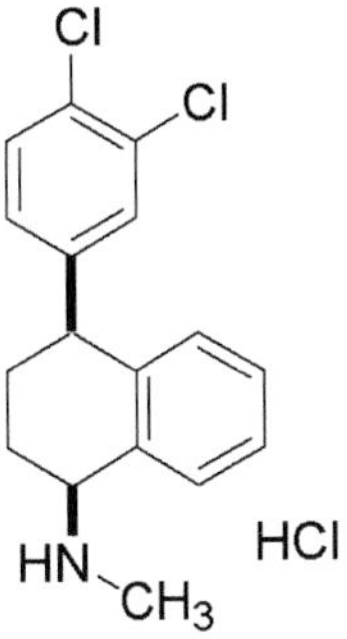

sertraline hydrochloride
(Zoloft)

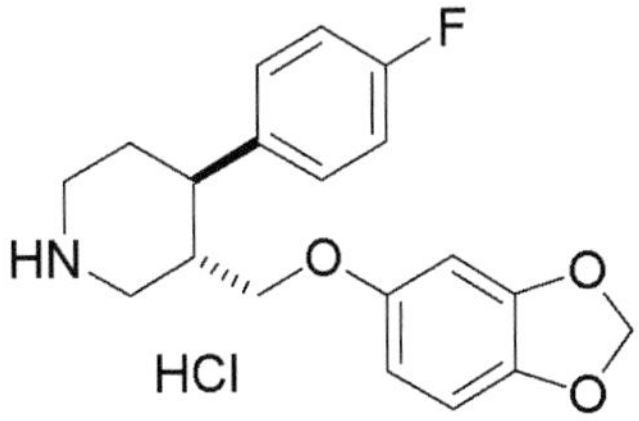

paroxetine hydrochloride
(Paxil)

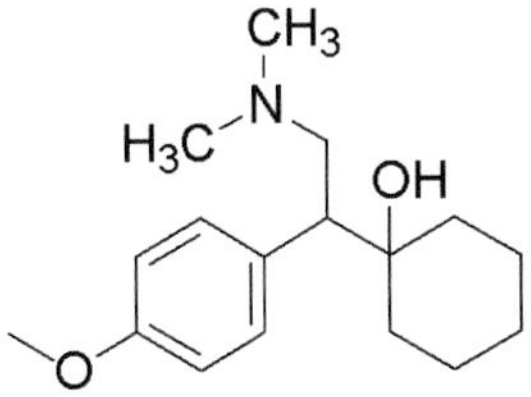

venlafaxine (Effexor)

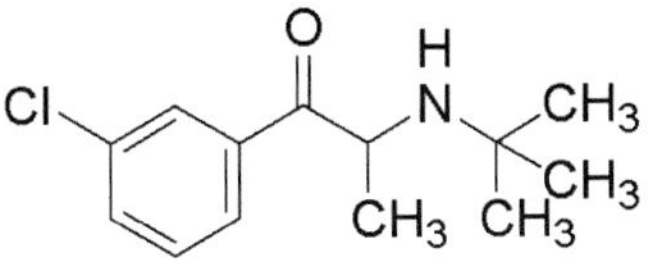

bupropion (Wellbutrin, Zyban)

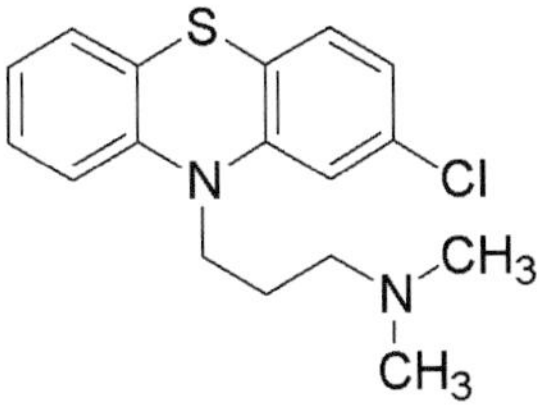

chlorpromazine (Thorazine)

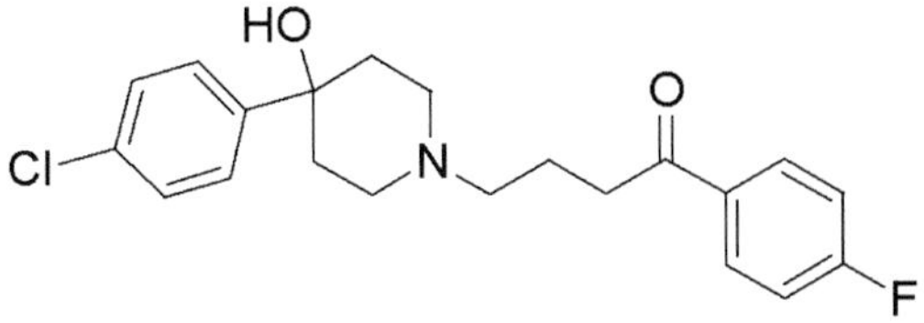

haloperidol (Haldol)

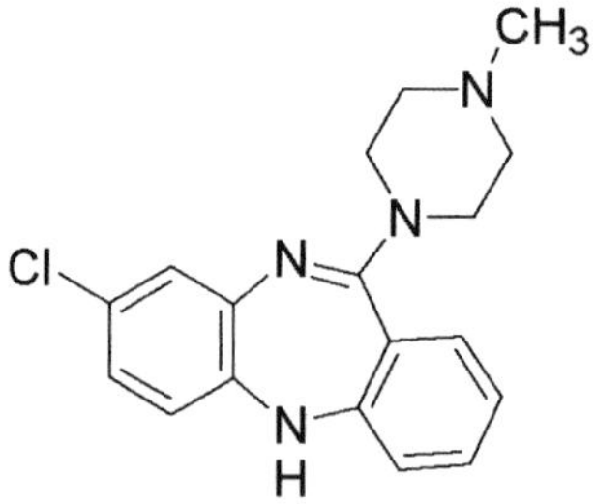

clozapine (Clozaril)

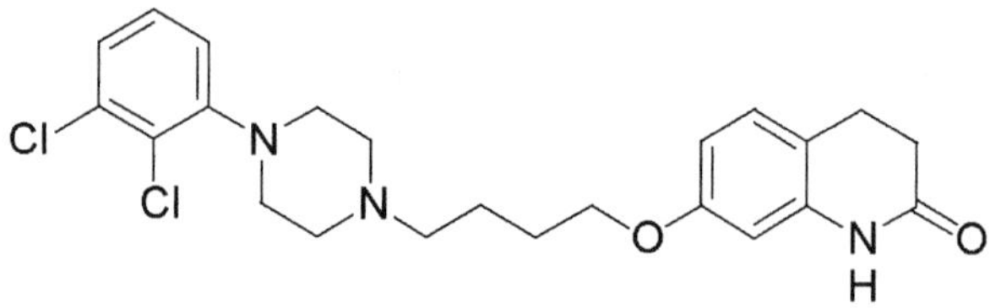

aripiprazole (Abilify)

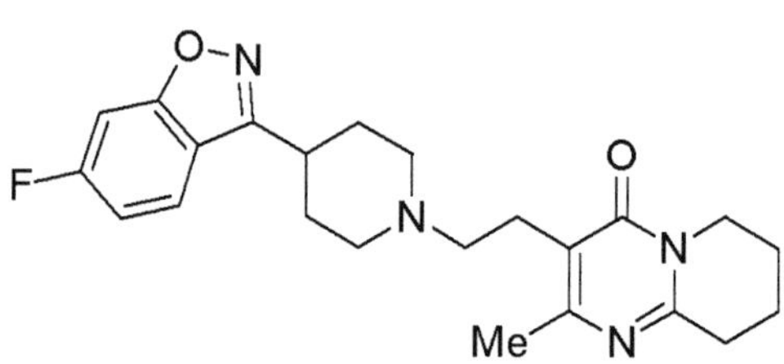

risperidone (Risperdal)

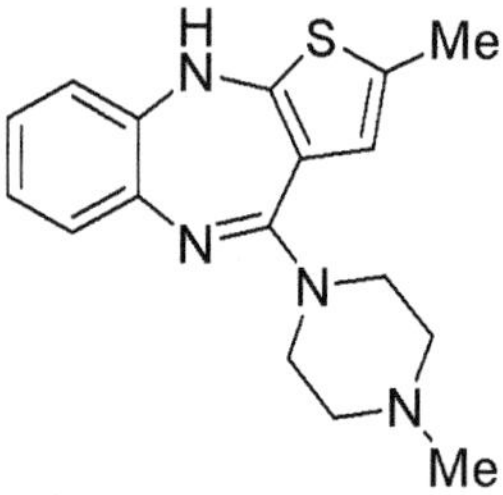

olanzapine (Zyprexa)

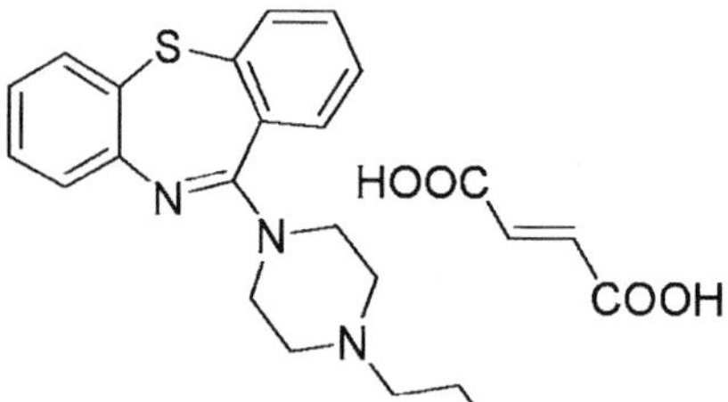

quetiapine fumarate (Seroquel)

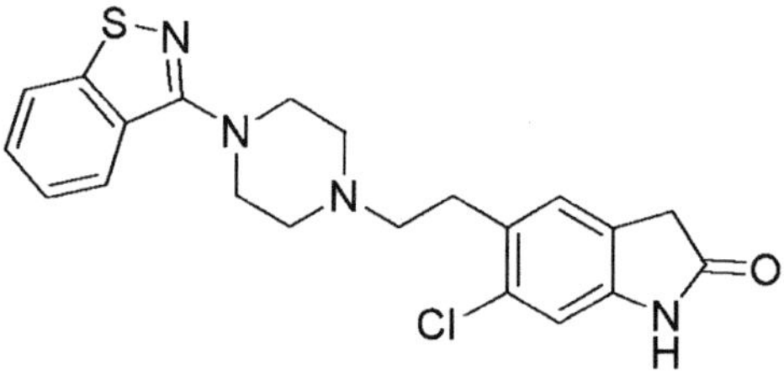

ziprasidone (Geodon)

lysergide (LSD)

amphetamine

methamphetamine

pseudoephedrine

reduction

methamphetamine

phenylephedrine

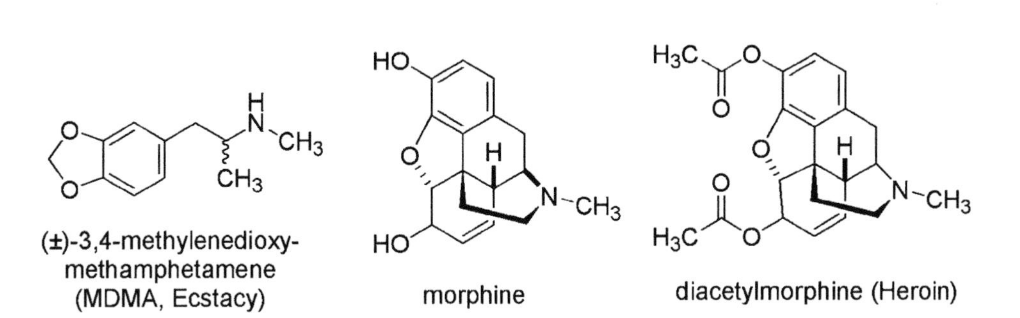

(±)-3,4-methylenedioxy-
methamphetamene
(MDMA, Ecstacy)

morphine

diacetylmorphine (Heroin)

Chapter 6. Diabetes Drugs

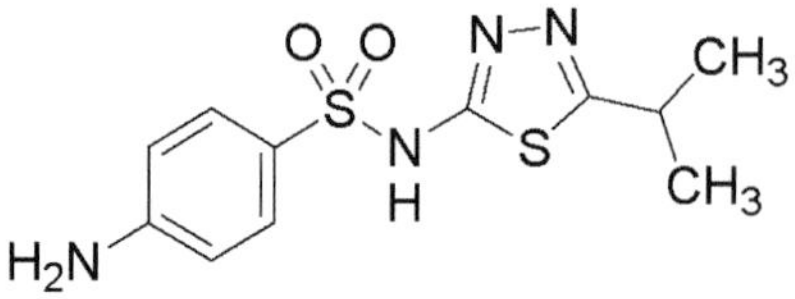

isopropylthiadiazole (IPTD)

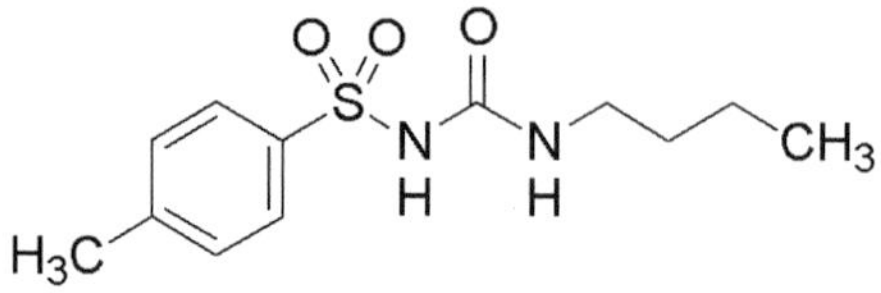

tolbutamide

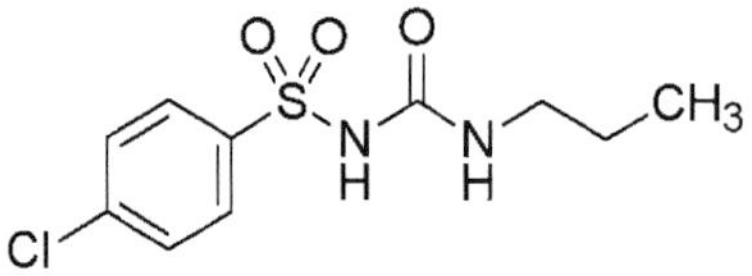

chlorpropamide

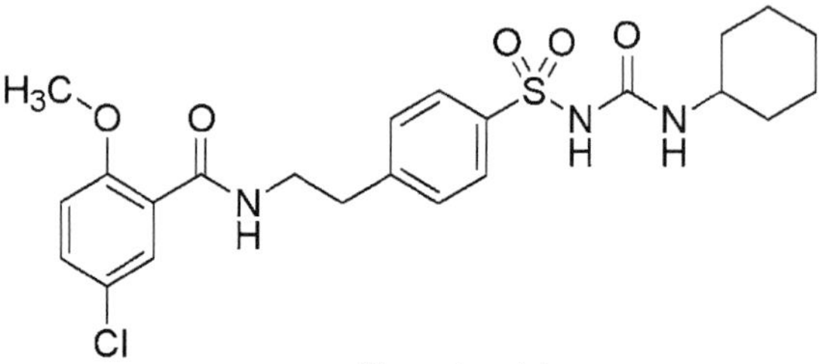

glibenclamide

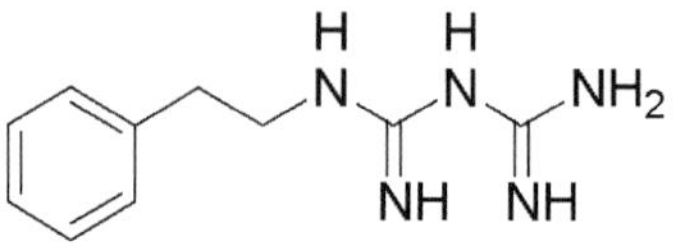

phenformin

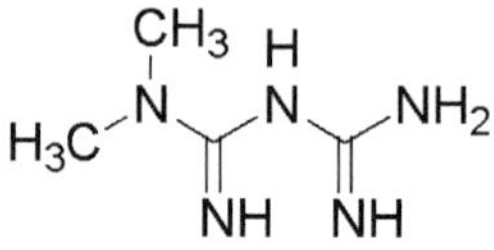

metformin (Glucophage)

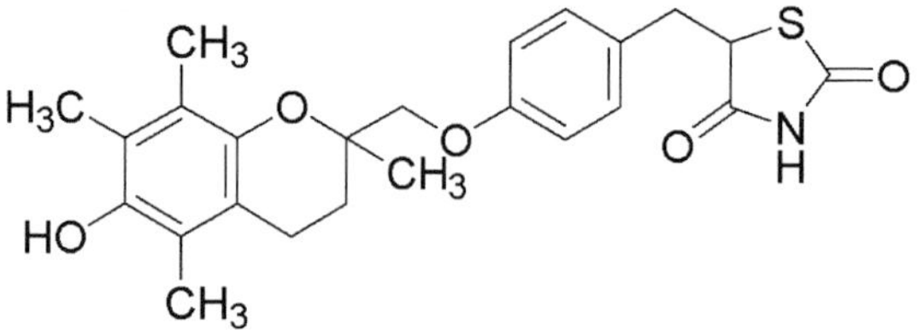

troglitazone (Rezulin)

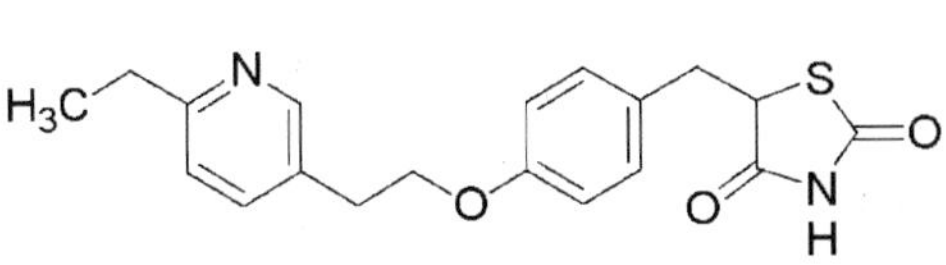

pioglitazone (Actos)

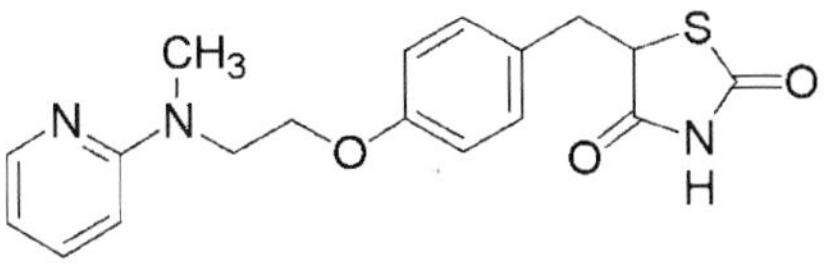

rosiglitazone (Avandia)

Chapter 7. Anesthetics

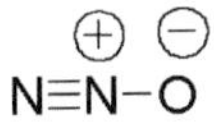

$N \equiv N^{+}-O^{-}$

nitrous oxide

$H_3C-CH_2-O-CH_2-CH_3$

diethyl ether

$H_3C-CH_2-OH \xrightarrow{H_2SO_4} H_3C-CH_2-O-CH_2-CH_3 + H_2O$

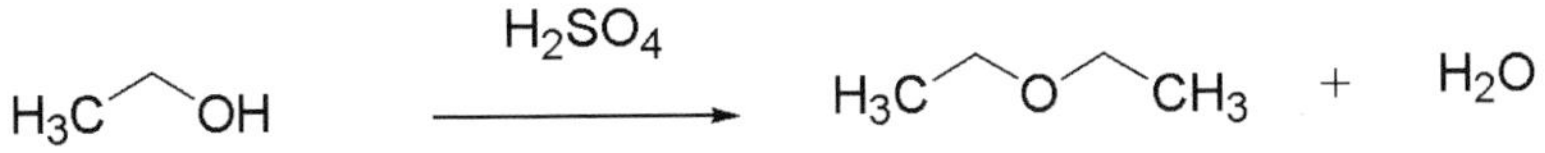

ethanol — sulfuric acid — diethyl ether — water

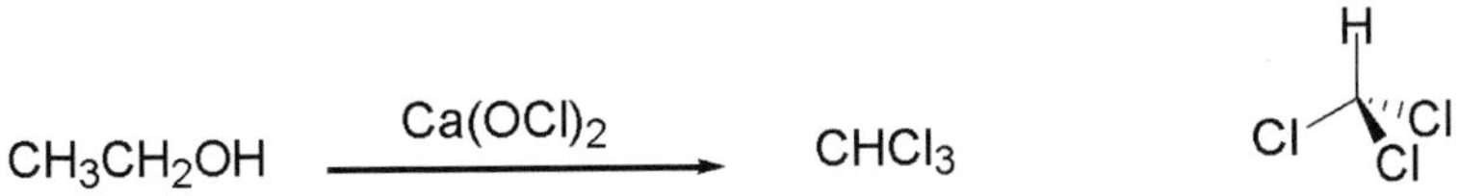

$CH_3CH_2OH \xrightarrow{Ca(OCl)_2} CHCl_3$

ethanol — chloroform — chlorform

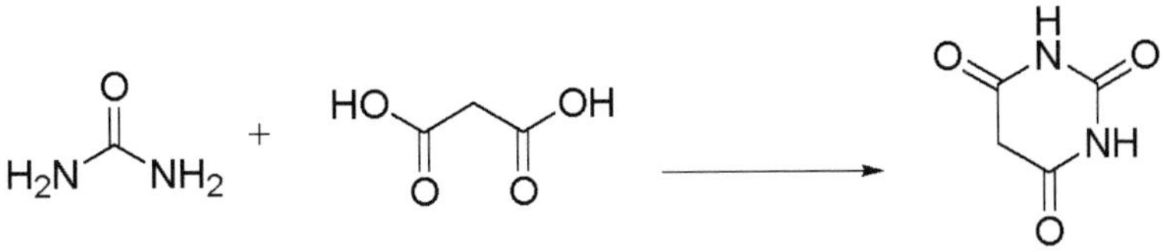

urea + malonic acid → barbituric acid

Barbital

Pentobarbital

Thiopentothal sodium

amobarbital (Sodium Amytal)

scopolamine

cocaine

procaine

benzocaine (Americaine)

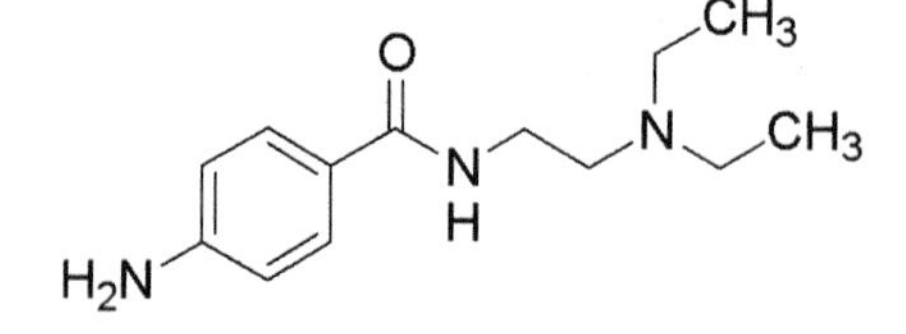

procainamide (Procamide)

Chapter 8. Anti-Inflammatory Drugs

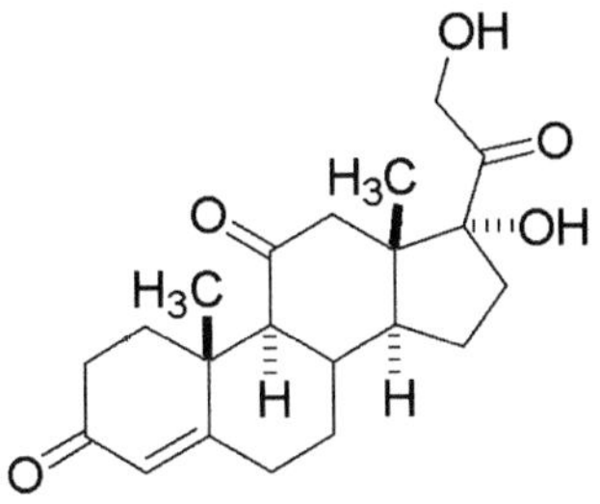

cortisone

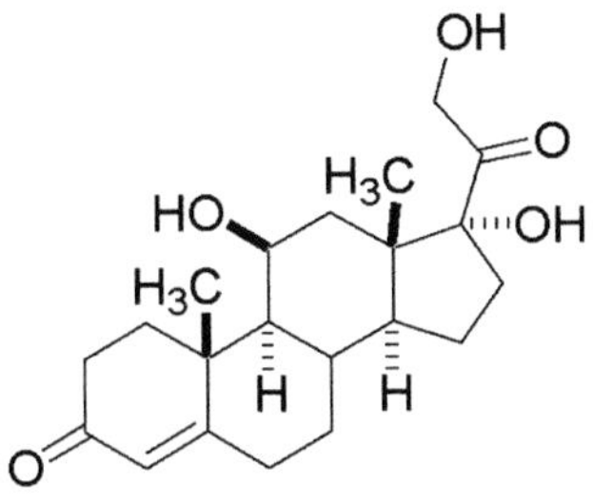

hydrocortisone (cortisol)

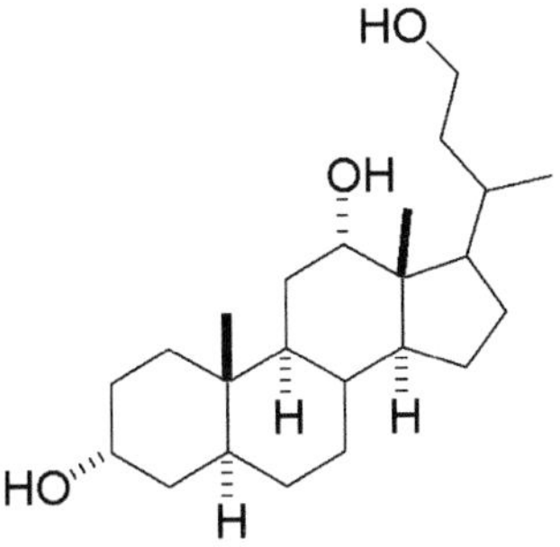

bile acid

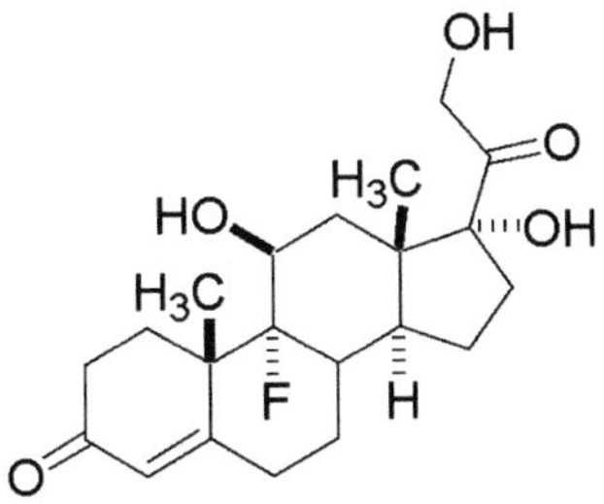

9-fluoro-cortisol

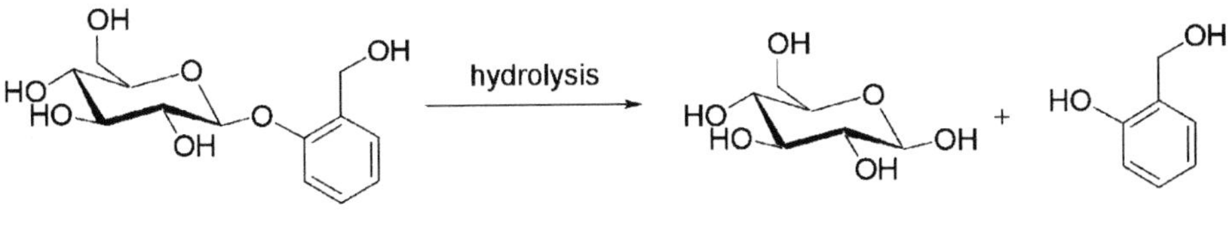

salicin — hydrolysis → glucose + salicyl alcohol

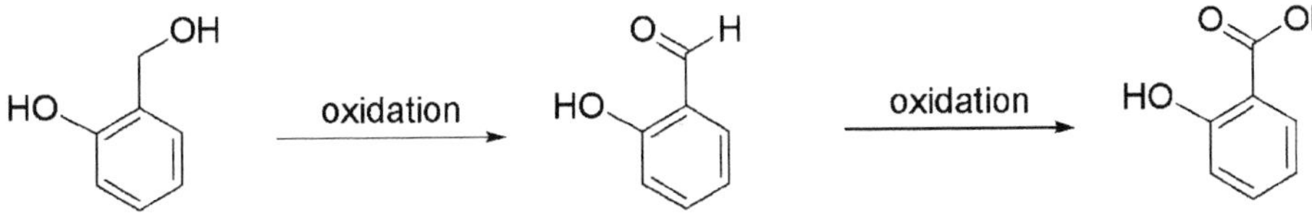

salicyl alcohol — oxidation → salicyl aldehyde — oxidation → salicylic acid

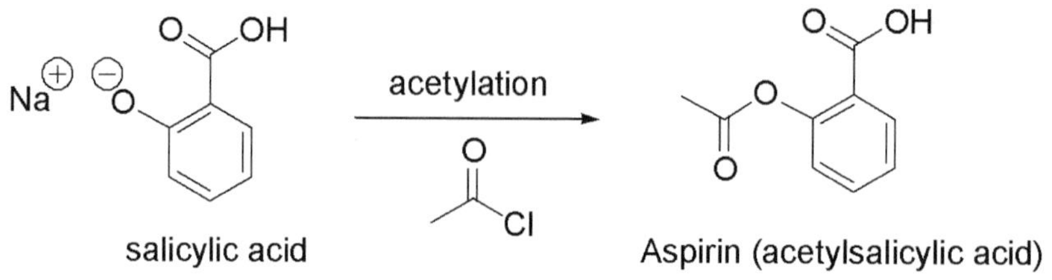

Na^{+} $^{-}$O O OH

acetylation

O Cl

O OH O O

salicylic acid

Aspirin (acetylsalicylic acid)

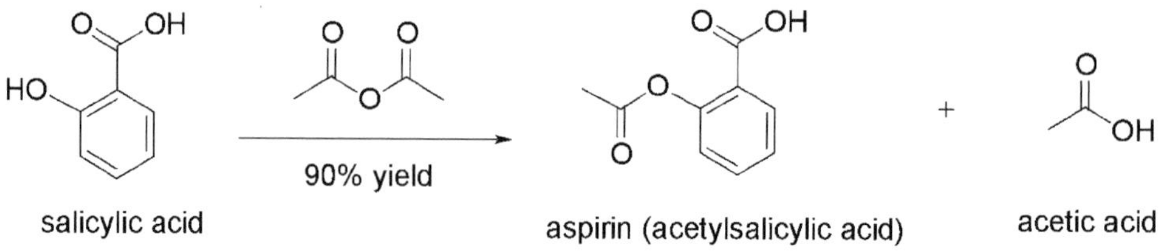

O OH HO

O O O

90% yield

O OH O O

\+

O OH

salicylic acid

aspirin (acetylsalicylic acid)

acetic acid

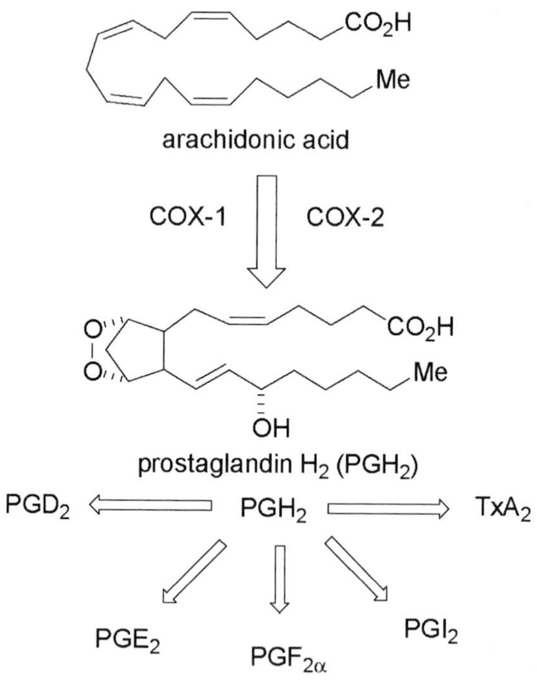

CO_2H

Me

arachidonic acid

COX-1

COX-2

O O

CO_2H

Me

OH

prostaglandin H_2 (PGH_2)

PGD_2 PGH_2 TxA_2

PGE_2 $PGF_{2\alpha}$ PGI_2

in vivo

de-ethylation

phenacetin

acetaminophen

in vivo

oxidation

acetanilide

indomethacin

ibuprofen

naproxen

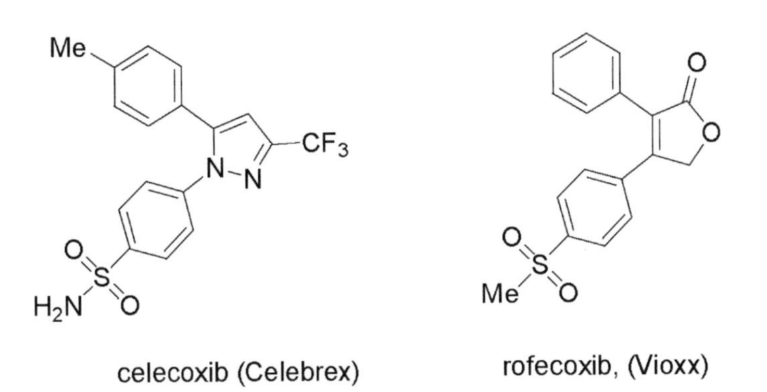

celecoxib (Celebrex)

rofecoxib, (Vioxx)

Fluticasone propionate (Flovent, Flunase)

Salmeterol xinafoate (Serevent)

Montelukast sodium (Singulair)

APPENDIX B

Trademarks of the Drugs

Trademarks	Generic Names	Companies
Abilify	aripiprazole	Bristol-Myers Squibb
Abilify	aripiprazole	Bristol-Myers Squibb
Accupril	quinapril	Parke-Davis
Actos	pioglitazone	Lilly
Adalat	nifedipine	Bayer
Ambien	zolpidem	Pharmacia
Antabuse	disulfiram	American Home Products
Arimidex	anastrozole	AstraZeneca
Aromasin	exemestane	Pfizer
Atacand	candesartan	Astra
Avandia	rosiglitazone	GlaxoSmithKline
Avapro	irbesartan	Sanofi
Aureomycin	chlortetracycline	Lederle
Avastin	bevacizumab	Genentech
Benadryl	diphenhydramine	Parke-Davis
Benicar	medoxomil	Sankyo
Briplatin	cisplatin	Bristol-Myers Squibb
Buspar	buspirone	Bristol-Myers Squibb
Caduet	atorvastatin calcium/ amlodipine	Pfizer
Camptosar	irinotecan	Pfizer
Capoten	captopril	Bristol-Myers Squibb

Trademarks	Generic Names	Companies
Caverject	prostaglandin E1/ alprostadil	Pharmacia & Upjohn
Celebrex	celecoxib	Pfizer
Cialis	tadalafil	Lilly
Cipro	ciprofloxacin	Bayer
Clarinex	desloratadine	Schering-Plough
Claritin	loratadine	Schering-Plough
Clozaril	clozapine	Novartis
Cozarr	losartan	DuPont
Crestor	rosuvastatin	AstraZeneca
Crixivan	indinavir	Merck
Cytoxan	cyclophosphamide	Bristol-Myers Squibb
Dalzic	practolol	Zeneca
Diovan	valsartan	Novartis
Disulfiram	acamprostate calcium	Lipha
Effexor	velafaxine	Wyeth
Emtrexate	methotrexate	Nordic
Enalaprilat	enalapril	Merck
Epivir	lamivudine	Glaxo Wellcome
Erbitux	cetuximab	ImClone/Bristol-Myers Squibb
Evista	raloxifene	Lilly
Femara	etrozole	Novartis
Geodon	ziprasidone	Pfizer
Gleevec	imatinib mesylate	Novartis
Glucophage	metformin	Bristol-Myers Squibb
Halcion	triazolam	Upjohn
Haldol	haloperidol	McNeil
Herceptin	trastuzumab	Genentech
Hycamtin	topotecan	SmithKline Beecham
Inderal	propranolol	Wyeth-Ayerst
Invirase	saquinavir	Roche
Iressa	gefitinib	AstraZeneca
Lentaron	formestane	Novartis
Lescol	fluvastatin	Novartis

Trademarks	Generic Names	Companies
Levitra	vardenafil hydrochloride	Bayer/GlaxoSmithKline
Librium	chlordiazepoxide	Roche
Lipitor	atorvastatin calcium	Pfizer
Marsilid	iproniazid	Roche
Mevacor	lovastatin	Merck
Micardis	telmisartan	Boehringer Ingelheim
Miltown	meprobamate	Wallace Laboratories
Navelbine	vinorelbine	Fabre
Noflo	norfloxacin	Kyorin
Nolvadex	tamoxifen	Zeneca
Norlutin	norethindrone	Parke-Davis
Norvasc	amlodipine	Pfizer
Norvir	ritonavir	Abbott
Oncovin	vincristine	Lilly
Paraplatin	carboplatin	Bristol-Myers Squibb
Parlodel	bromocriptine	Novartis
Paxil	paroxetine hydrochloride	SmithKline Beecham
Pravacol	pravastatin	Bristol-Myers Squibb
Prozac	fluoxetine hydrochloride	Lilly
Purinethol	6-mercaptopurine	Burroughs Wellcome
Risperdal	risperidone	Janssen
Ritalin	methylphenidate	Novartis
Rituxan	rituximab	Genentech
Seroquel	quetiapine fumarate	AstraZeneca
Tagamet	cimetidine	SmithKline Beecham
Tarceva	erlotinib	ISI/Genentech/Roche
Taxol	paclitaxel	Bristol-Myers Squibb
Taxotere	docetaxel	Aventis
Tenormin	atenolol	Zeneca
Terramycin	oxytetracycline	Pfizer
Teveten	eprosartan	Solvay
Thorazine	chlorpromazine	SmithKline Beecham
Tofranil	imipramine	Novartis
Valium	diazepam	Roche

Trademarks	Generic Names	Companies
Vasotec	enalapril	Merck
Velban	vinblastine	Lilly
Velcade	bortezomib	Millennium
Viagra	sildenafil citrate	Pfizer
Viramune	nevirapine	Boehringer Ingelheim
Wellbutrin	bupropion	Glaxo Wellcome
Zetia	ezetimibe	Schering-Plough
Ziagen	abacavir	Glaxo Wellcome
Zocor	simvastatin	Merck
Zoloft	sertraline hydrochloride	Pfizer
Zyprexa	olanzapine	Lilly
Zyrtec	cetirizine dihydrochloride	Pfizer
Zyvox	inezolid	Pfizer

BIBLIOGRAPHY

PREFACE

1. Bronowski, Jacob. *Ascent of Man* Little Brown & Co: London, UK, 1976, 14.

CHAPTER I. CANCER DRUGS

1. Gaines, Ann; Whiting, Jim. *Robert A. Winsberg and the Search for the Cause of Cancer.* Mitchell Lane Publishers: Bear, DE, 2001.
2. *Nobel Lectures, Physiology or Medicine, 1922–1941.* Elsevier: Amsterdam, 1965.
3. Proctor, Robert N. *Nazi War on Cancer.* Princeton University Press: Princeton, NJ, 1991.
4. Cornwell, John. *Hitler's Scientists, Science, War, and Devil's Pact.* Penguin Books: New York, 2002.
5. *Nobel Lectures, Physiology or Medicine, 1963–1970.* Elsevier: Amsterdam, 1972.
6. Varmus, Harold; Weinberg, Robert A. *Genes and Biology of Cancer.* Scientific American Library: New York, 1993.
7. Weinberg, Robert A. *Racing to the Beginning of the Road: The Search for the Origin of Cancer.* W. H. Freeman: New York, 1998.
8. Crotty, Shane. *Ahead of the Cure: David Baltimore's Life in Science.* University of California Press: Berkeley, 2001.
9. Infield, Glenn B. *Disaster at Bari: The True Story of World War II's Worst Chemical Warfare Disaster!* Ace Books: New York, 1971.
10. Rosenberg, Barnett. Platinum compounds: A new class of potential anti-tumor agents. *Nature* 1969, *222*, 385–386.
11. Rosenberg, Barnett. Cisplatin: Its history and possible mechanisms of action. In: Prestayko, Archie W.; Crooke, Stanley T.; Carter, Stephen K., eds., *Cisplatin: Current Status and New Development.* Academic Press: New York, 1980.
12. Eustace, P. History and development of Cisplatin in the management of malignant disease. *Cancer Nursing* 1980, *3(5)*, 373–378.
13. Lippard, Stephen J., ed. *Platinum, Gold, and Other Metal Chemotherapeutic Agents: Chemistry and Biochemistry.* American Chemical Society: Washington, DC, 1983.
14. Summer, Judith. *The Natural History of Medicinal Plants.* Timber Press: Portland, OR, 2000.

15. Goodman, Jordan; Walsh, Vivien. *The Story of Taxol: Nature and Politics in the Pursuit of an Anti-Cancer Drug.* Cambridge University Press: Oxford, UK, 2001.
16. Oberlies, Nicholas H.; Kroll, David J. Camptothecin and Taxol: Historic achievement in natural products research. *Journal of Natural Products* 2004, *67*, 129–135.
17. Howitz, Susan B. Personal recollections on the early development of Taxol. *Journal of Natural Products* 2004, *67*, 136–138.
18. Kingston, David G. I. Taxol: A molecule for all seasons. *Chemical Communications* 2001, 867–880.
19. Stille, Darlene R. *Extraordinary Women Scientists.* Children's Press: Chicago, 1995.
20. Bowden, Mary Ellen; Crow, Amy Beth; Sullivan, Tracy. *Pharmaceutical Achievers: The Human Faces of Pharmaceutical Research.* Chemical Heritage Press: Philadelphia, 2003, 170.
21. Stille, Darlene R. *Extraordinary Women Scientists.* Children's Press: Chicago, 1995.
22. Jordan, V. Craig. Tamoxifen: A most unlikely pioneering medicine. *Nature Review, Drug Discovery* 2005, *2*, 205–213.
23. Yaish, Pnina; Gilon, Chaim; Levitzki, Alexander. Blocking of EGF-dependent cell proliferation by EGF receptor kinase inhibitors. *Science* 1988, *242*, 933–935.
24. Racker, Efraim; Spector, Mark. Warburg effect revisited: merger of biochemistry and molecular biology. *Science* 1981, *213*, 303–307.
25. Spector, Mark; Pepinsky, Robert B.; Vogt, Volker, M; Racker, Efraim. A mouse homolog to the avian sarcoma virus src protein is a member of a protein kinase cascade. *Cell* 1981, *25*, 9–21.
26. Stein, Michael D. Spectacular cancer mechanism doubted. *Nature* 1981, *293*, 93.
27. Wade, Nicholas. The rise and fall of a scientific superstar. *New Scientist* 1981, *91*, 781–782.
28. McKean, Kevin. A scandal in the laboratory. *Discover* 1981, November, 18–23.
29. Newmark, Peter. Spectacular cancer mechanism doubted. *Nature* 1981, *293*, 93.
30. Newmark, Peter. Biochemical cascades and carcinogenesis. *Nature* 1981, *293*, 93.
31. Bazell, Robert. *HER-2, The Making of Herceptin: A Revolutionary Treatment for Breast Cancer.* Random House: New York, 1998.
32. Odelberg, Wilhelm. Ed. *Les Prix Nobel. The Nobel Prizes 1984.* Nobel Foundation: Stockholm, 1985.
33. Boehm, T.; Folkman, J.; Browder, T.; O'Reilly, M. S. Antiangiogenic therapy of experimental cancer does not induce acquired drug resistance. *Nature* 1997, 390(6658), 404–407.
34. Anand, Geeta. How drug's rebirth as treatment for cancer fueled price rises. *The Wall Street Journal,* November 15, 2004.
35. Cooke, Robert. *Dr. Folkman's War: Angiogenesis and the Struggle to Defeat Cancer.* Random House: New York, 2001, 282–285.
36. Ferrara, Napoleone; Hillan, Kenneth J.; Gerber, Hans-Peter; Novotny, William. Discovery and development of Bevacizumab: An anti-VEGF antibody for treating cancer. *Nature* 2004, *3*, 391–400.

37. Goodman, Laurie. Persistence, luck, Avastin. *Journal of Clinical Investigation* 2004, *113(7)*, 934.
38. Zimmermann, Jürg. Glivec: A new treatment modality for CML: A case history. *Chimia* 2002, *5/6*, 428–431.
39. Vasella, Daniel. *The Magic Cancer Bullet: How a Tiny Orange Pill Is Rewriting Medical History*. HarperBusiness: New York, 2003.

CHAPTER 2. DRUGS TO KILL GERMS

1. Bankston, John. *Joseph Lister and the Story of Antiseptics (Uncharted, Unexplored, and Unexplained)*. Mitchell Lane Publishers: Hockessin, DE, 2004.
2. Pasteur, Louis; Lister, Joseph. *Germ Theory and Its Applications to Medicine and on the Antiseptic Principle of the Practice of Surgery*. Prometheus Books: Amherst, NY, 1996.
3. Quétel, Claude. *History of Syphilis*. Johns Hopkins University Press: Baltimore, 1990.
4. Bäumler, Ernst. *Paul Ehrlich: Scientist for Life*. Grant Edwards, trans., Holmes & Merie: New York, 1984.
5. Bankston, John. *Gerhard Domagk and the Discovery of Sulfa*. Mitchell Lane: Bear, DE, 2003.
6. Oesper, R. Gerhard Domagk and chemotherapy. *Journal of Chemical Education* 1954, *31*, 188–191.
7. Ward, P. S. The American reception of Salvarsan. *Journal of History of Medicine* 1981, 44–62.
8. Otfinoski, Steven. *Alexander Fleming, Conquering Disease with Penicillin*. Facts on File: New York, 1993.
9. Macfarlance, Gwyn. *Howard Florey: The Making of a Great Scientist*. Oxford University Press: New York, 1979.
10. Lax, Eric. *The Mold in Dr. Florey's Coat*. John Macrae: New York, 2004.
11. Clark, Ronald W. *The Life of Ernst Chain: Penicillin and Beyond*. St. Martin's Press: New York, 1985.
12. Jacobs, Francine. *Breakthrough: The True Story of Penicillin*. Dodd, Mead: New York, 1985.
13. Hobby, Gladys L. *Penicillin, Meeting the Challenge*. Yale University Press: New Haven, CT, 1985.
14. Tanner, Ogden. *25 Years of Innovation: The Story of Pfizer Central Research*. Greenwich Publishing Group: Lyme, CT, 1996.
15. Cornforth, J. W.; Clarke, H. T. *et al.* Oxazoles and oxazolones. In: *The Chemistry of Penicillin*. Princeton University Press: New Jersey, 1949, 688–848.
16. Sheehan, John C. *Enchanted Ring: The Untold Story of Penicillin*. MIT Press: Cambridge, MA, 1982.
17. Sheehan, J. C.; Johnson, D. A. The synthesis of substituted penicillins and simpler structural analogs. VIII. Phthalimidomalonaldehydic esters: synthesis and condensation with penicillamine. *Journal of the American Chemical Society* 1954, *76*, 158.

18. Ryan, Frank. *The Forgotten Plague: How the Battle against Tuberculosis was Won—and Lost.* Little, Brown: New York, 1992.
19. Gordon, Karen. *Selman Waksman and the Discovery of Streptomycin.* Mitchell Lane Publishers: Bear, DE, 2003.
20. Schatz, A.; Bugie, E.; Waksman, S. A. Streptomycin, a substance exhibiting antibiotic activity against gram-positive and gram-negative bacteria. *Proceedings for Experimental Biology and Medicine* 1944, *55*, 66–9.
21. Waksman, Selman A. *My Life with the Microbes.* Simon & Schuster: New York, 1954.
22. Waksman, Selman A. *The Conquest of Tuberculosis.* University of California Press: Berkeley, 1964.
23. Waksman, S. A. Streptomycin, isolation, properties, and utilization. *Journal of History of Medicine* 1951, *6*, 318–347.
24. Reichman, Lee B.; Tanne, Janice Hopkins. *Time Bomb: The Global Epidemic of Multi-Drug-Resistant Tuberculosis.* McGraw-Hill: New York, 2002.
25. Stephens, C. R.; Conover, L. H.; Hochstein, F. A.; Regna, P. P.; Pilgrim, F. J.; Brunings, K. J.; Woodward, R. B. Degradation of Aureomycin. VIII. Structure of Aureomycin and Terramycin. *Journal of the American Chemical Society* 1952, *74*, 4976–4977.
26. Wentland, M. P. In Memoriam: George Y. Lesher, Ph.D. In: Hooper, D. C.; Wolfson, J. S., eds. *Quinolone Antimicrobial Agents,* 2nd ed. American Society for Microbiology: Washington, DC, 1993.
27. Hooper, David C.; Wolfson, John S., eds. *Quinolone Antimicrobial Agents,* 2nd ed. American Society Microbiology: Washington, DC, 1993.
28. Batts, Donald H.; Kollef, Marin H.; Lipsky, Benjamin A.; Nicolau, David P.; Weigelt, John A., eds. *Creation of a Novel Class: The Oxazalidinone Antibiotics.* Innova Institute for Medical Education: Tampa, FL, 2004.

CHAPTER 3. CARDIOVASCULAR DRUGS

1. Fox, Ruth. *Milestones of Medicine.* Random House: New York, 1950.
2. Gregory, Andrew. *Harvey's Heart: The Discovery of Blood Circulation.* Icon Books: Cambridge, UK, 2001.
3. Holmes, L. C.; DiCarlo, F. Nitroglycerin: The explosive drug. *Journal of Chemical Education* 1971, *48*, 573–576.
4. Brock, William H. *Justus von Liebig, The Chemical Gatekeeper.* Cambridge University Press: Cambridge, UK, 2002.
5. Butler, Anthony; Nicholson, Rosslyn. *Life, Death and Nitric Oxide.* Royal Society of Chemistry: Cambridge, UK, 2003.
6. Anon. *The Nobel Century.* Chapmans: London, UK, 1991.
7. Ackernecht, E. Aspects of the history of therapeutics, II. Digitalis and some other panaceas. *Bulletin of History of Medicine* 1962, *36*, 389–419.

8. Moore, Thomas J. *Heart Failure.* Random House: New York, 1989.
9. Friend, D. G. Digitalis after two centuries. *Archives of Surgery* 1976, *111*, 14–19.
10. Roddis, Louis H. *William Withering.* Paul B. Hoeber: New York, 1936.
11. Moorman, L. J. William Withering, his work, his health, his friends. *Bulletin History Medicine* 1942, *12*, 355–366.
12. Paterson, G. R.; Locock, M. J. The discovery of cardiac glucosides. *Applied Therapeutics* 1967, *9*, 60–65.
13. Vogl, A. The discovery of the organic mercurial diuretics. *American Heart Journal* 1950, *39*, 881–883.
14. Black, James. A personal view of pharmacology. *Annual Review of Pharmacology and Toxicology* 1996, *36*, 1–33.
15. Beyer, K. H. Discovery of the thiazide: Where biology and chemistry meet. *Perspectives in Biology and Medicine* 1977, *20*, 410–420.
16. deStevens, George. My odyssey in drug discovery. *Journal of Medicinal Chemistry* 1991, *34(9)*, 2665–2670.
17. Vane, John. The history of inhibitors of angiotensin-converting enzyme. In: D'Oléans-Juste, P.; Plante, G. E., eds. *ACE Inhibitors.* Birkhäuser Verlag: Switzerland, 2001, 1–10.
18. Smith, Charles G.; Vane, John. The discovery of Captopril. *FASEB Journal* 2003, *17(8)*, 788–789.
19. Patchett, Arthur A. Enalapril and Lisinopril. In: Lednicer, Daniel, ed. *Chronicles of Drug Discovery.* ACS: Washington, DC, 1993, 125–162.
20. Black, James. Drugs from emasculated hormones: The principle of syntopic antagonism. *In Vitro Cellular & Developmental Biology* 1989, *25(4)*, 311–320.
21. Ahlquist, Raymond P. Agents which block adrenergic β-receptors. *Annual Review of Pharmacology* 1968, *8*, 259–272.
22. Nayak, P. Ranganath; Ketteringham, John M. *Breakthrough!* Rawson Associates: New York, 1986.
23. Crowther, A. F. The discovery of the first β-adrenergic blocking agents. *Drug Design and Delivery* 1990, *6(2)*, 149–156.
24. Harrison, Donald Carey. Clinical pharmacology and history of beta-adrenergic blockade. *Circ. Eff. Clin. Uses Beta-Adrenergic Blocking Drugs* 1971, 1–19.
25. Laurence, Desmond R. History of the development of beta-adrenoceptor blocking drugs. *International Congress Series* 1974, 341
26. Laurence, Desmond R. Advances of beta-adrenergic blocking therapy-sotalol. *Proceedings of International Symposium* 1974, *1(1)*, 1–5.
27. Meridy, Howard W. Beta-adrenergic receptor blocking drugs and their use in anesthesia: a review. *Hartford Hospital Bulletin* 1969, *24(3)*, 182–195.
28. Mossberg, Kurt A.; Peel, Claire. Beta-adrenergic antagonists. *Performance-Enhancing Substances in Sport and Exercise,* 2002, 149–156.
29. Shanks, R. G. The history of beta-adrenoceptor blocking drugs. *Trends in Pharmaceutical Sciences* 1984, *5*, 405–409.

30. Vos, R.; Bodewitz, H. Pharmacological and therapeutic profiling in drug innovation: The early history of the beta-blockers. *Perspectives in Biology and Medicine* 1988, *31(4)*, 469–480.
31. Shanks, R. G. The history of beta-adrenoceptor blocking drugs. *Trends in Pharmaceutical Sciences* 1984, *5*, 405–409.
32. Triggle, David J. Pharmacology of Ca_v1 (L-type) channels. In: McDonough, Stefan I., ed. *Calcium Channel Pharmacology.* Kluwer Academic: New York, 2004.
33. Fleckenstein-Grün, Gisa. Historical development of calcium antagonists—In memoriam: Albrecht Fleckenstein. *High Blood Pressure* 1994, *3*, 284–290.
34. Nayler, Winifred G. *Calcium Antagonists.* Academic Press: London, UK, 1988.
35. Chew, Soon-Keong. History and development of beta-blockers. *Berita Farmasi* 1977, *4(8)*, 4, 7–12.
36. Bellosta, Stefano; Paoletti, Rodolfo; Corsini, Alberto. History and development of HMG-CoA reductase inhibitors. In: Schmitz, Gerd; Torzewski, Michael, eds., *HMG-CoA Reductase Inhibitors.* Birkhaeuser Verlag: Basel, Switzerland, 2002, 1–17.
37. Gaw, Allan; Packard, Christopher J.; Shepherd, James, eds. *Statins: The HMG-CoA Reductase Inhibitors in Perspective,* 2nd ed. Martin Dunitz: London, 2004.
38. Vagelos, P. Roy; Galambo, Louis. *Medicine, Science, and Merck.* Cambridge University Press: Cambridge, UK, 2004.

CHAPTER 4. SEX AND DRUGS

1. Sandroni, Paola. Aphrodisiacs past and present: A historical review. *Clinical Autonomic Research* 2001, *11(5)*, 303–7.
2. Taylor, N. *Plant Drugs That Changed the World.* George Allen & Unwin: London, UK, 1966.
3. Muller, J. Witch ointments and aphrodisiacs: A contribution to the cultural history of nightshade plants. *Gesnerus* 1998, *55(3–4)*, 205–220.
4. Nickell, Nancy L. *Nature's Aphrodisiacs.* Cross Press: Freedom, CA, 1999.
5. Taberner, Peter V. *Aphrodisiacs: The Science and the Myth.* University of Pennsylvania Press: Philadelphia, 1985.
6. Taberner, Peter V. Sex and Drugs: Aphrodite's Legacy. *Trends in Pharmacological Sciences* 1985, *6*, 49–54.
7. Waddell, Thomas G.; Jones, Hal; Keith, A. Lane. Legendary chemical aphrodisiacs. *Journal of Chemical Education* 1980, *57(5)*, 341–342.
8. Waddell, Thomas G.; Ibach, Daniel M. Modern chemical aphrodisiacs. *Indian Journal of Pharmaceutical Sciences* 1989, *51(3)*, 79–82.
9. Bose, A. Aphrodisiacs: A psychosocial perspective. *Indian Journal of History of Science* 1981, *16(1)*, 100–103.
10. Dubos, René. *Pasteur and Modern Science.* Scientific Revolutionaries: A Biographical Series. Science Tech Publishers: Madison, WI, 1960.

11. Bowden, Mary Ellen; Crow, Amy Beth; Sullivan, Tracy. *Pharmaceutical Achievers: The Human Faces of Pharmaceutical Research.* Chemical Heritage Press: Philadelphia, 2003.
12. Choudhary, Mohammad Iqbal; Atta-ur-Rahman. Elixirs of love. *Chemistry in Britain* 1997, October, 25–27.
13. New drugs may relieve sexual disorder. *Chemical and Engineering News* 1977, August 1, 17–18.
14. Jarow, Jonathan; Kloner, Robert A.; Holmes, Ann M. *Viagra.* M. Evans: New York, 1998.
15. Palmer, Elizabeth. Making of the love drug (Viagra). *Chemistry in Britain* 1999, *35*, 24–26.
16. Butenandt, A. The history of oestrone. *Trends in Biological Sciences* 1979, *4*, 215–216.
17. Pramik, Mary Jean, ed. *Norethindrone: The First Three Decades.* Syntex Laboratories: Palo Alto, CA, 1978.
18. Ringold, Howard. Norethindrone history and synthesis. In: Pramik, Mary Jean, ed. *Norethindrone: The First Three Decades.* Syntex Laboratories: Palo Alto, CA, 1978.
19. Birch, A. J. Steroid hormones and the Luftwaffe. *Steroid* 1992, *57*, 363–377.
20. Birch, Arthur J. *To See the Obvious.* American Chemical Society: Washington, DC, 1995.
21. Colton, Frank B. Steroids and "the pill": Early steroid research at Searle. *Steroids* 1992, *57(12)*, 624–630.
22. Djerassi, C. *From the Lab into the World.* American Chemical Society: Washington, DC, 1994.
23. Djerassi, C. *The Pill, Pygmy Chimps, and Degas' Horse.* Basic Books: New York, 1992.
24. Djerassi, C. *The Politics of Contraception: The Present and the Future.* W. H. Freeman: San Francisco, 1979.
25. Djerassi, C. *Steroids Made It Possible.* American Chemical Society: Washington, DC, 1990.
26. Djerassi, C. Steroid research at Syntex: The pill and cortisone. *Steroids* 1992, *57*, 631–641.
27. Djerassi, C. The mother of the pill. *Recent Progress in Hormone Research* 1995, *50*, 1–18.
28. Dodds, C. Oral contraceptives: The past and the future. *Clinical Pharmacology and Therapy* 1969, *10*, 147–162.
29. Drill, V. A. History of the first oral contraceptive. *Journal of Toxicology and Environmental Health* 1997, *3*,133–138.
30. McGinty, Daniel A.; Djerassi, Carl. Some chemical and biological properties of 19-nor-17α-ethylnyltestosterone. *Annals of the New York Academy of Sciences* 1958, *71*, 500–515.
31. Marks, Lara V. *Sexual Chemistry: A History of the Contraceptive Pill.* Yale University Press: New Haven, CT, 2001.
32. Asbell, Bernard. *The Pill: A Biography of the Drug That Changed the World.* Random House: New York, 1995.

33. Gallo, Robert C.; Montagnier, Luc. The AIDS epidemic. In: *The Science of AIDS: Readings from Scientific American.* W. H. Freeman: New York, 1989, 1–11.
34. Weiss, R. A.; Jaffe, H. W.; Duesberg, P. HIV and AIDS. *Nature* 1990, *345*, 659–700.
35. Gallo, Robert C. *Virus Hunting, AIDS, Cancer, and the Human Retrovirus: A Story of Scientific Discovery.* Basic Books: New York, 1991.
36. Levine, Arnold J. *Viruses.* Scientific American Library: New York, 1992.
37. Crawford, Dorothy H. *The Invisible Enemy: A Natural History of Viruses.* Oxford University Press: London, 2000.
38. Montagnier, Luc. *Virus: The Codiscoverer of HIV Tracks Its Rampage and Charts the Future.* W. W. Norton: New York, 2000.
39. Montagnier, Luc. A history of HIV discovery. *Science* 2002, *298*, 1727–1728.
40. Crewdson, John. *Science Fictions: A Scientific Mystery, a Massive Cover-Up and the Dark Legacy of Robert Gallo.* Back Bay Books: Boston, MA, 2003.
41. Fauci, Anthony S. HIV and AIDS: 20 years of science. *Nature Medicine* 2003, *9(7)*, 839–843.
42. Chu, Chung K.; Beach, J. Warren; Kosugi, Yoshiyuki. An efficient total synthesis of 3'-azido-3'-deoxythymidine (AZT) and 3'-azido-2',3'-dideoxyuridine (AZDDU, CS-87) from *D*-mannitol. *Tetrahedron Letters* 1988, *29*, 5349–5352.
43. Horwitz, Jerome P.; Chua, Jonathan; Noel, Michael. The mononesylates of 1-(2'-deoxy-beta-D-lyxofuranyosyl) thymidine. *Journal of Organic Chemistry* 1964, *29*, 2076–2078.
44. Duncan, Ian B.; Redshaw, Sally. Discovery and early development of saquinavir. *Infectious Diseases Therapy* 2002, *25*, 27–47.
45. Kempf, Dale J. Discovery and early development of Ritonavir and ABT-378. *Infectious Diseases Therapy* 2002, *25*, 49–64.
46. Dorsey, Bruce D.; Vacca, Joseph P. Discovery and early development of indinavir. *Infectious Diseases Therapy* 2002, *25*, 65–83.
47. Marshall, Eliot. Universities, NIH hear the price isn't right on essential drugs. *Science* 2001, *292*, 614–615.

CHAPTER 5. DRUGS OF THE MIND

1. McCoy, Elin. *Coffee and Tea*, 3rd ed. Thern Raines, Raines & Raines: New York, 1991.
2. Reid, T. R. Caffeine. *National Geographic* 2005, January, 3–31.
3. Sternbach, L. H. The discovery of Librium. *Agents and Actions* 1972, *2(4)*, 193–196.
4. Baenninger, Alex; Costa e Siva, Jorge Alberto; Moeller, Hans-Juergen; Rickles, Karl. *Good Chemistry: The Life and Legacy of Valium Inventor, Leo Sternbach.* McGraw-Hill: New York, 2004, 43.
5. Garattini, S.; Mussini, E.; Randall, L. O. *The Benzodiazepines.* Raven Press: New York, 1973.
6. Costa, Erminio, ed. *The Benzodiazepines: From Molecular Biology to Clinical Practice.* Raven Press: New York, 1983.

7. American Psychiatric Association. *Diagnostic and Statistical Manual or Mental Disorders (DSM-IV-TR).* American Psychiatric Association, Washington, DC, 2000.
8. Baumeister, Alan A.; Hawkins, Mike F.; Uzelac, Sarah M. The myth of reserpine-induced depression: Role in the historical development of the monoamine hypothesis. *Journal of the History of the Neurosciences* 2003, *12(2)*, 207–220.
9. Kulkarni, S. K.; Verma, Anita. Reserpine: Forty years and more. *Eastern Pharmacist* 1993, *36(428)*, 29–33.
10. Ayd, Frank J., Jr.; Blackwell, Barry, eds. *Discoveries in Biological Psychiatry.* J. B. Lippincott: Philadelphia, 1970.
11. Cade, John F. J.; Lithium salts in the treatment of psychotic excitement. *Medical Journal of Australia* 1949, *2*, 349–352.
12. Webb, John. The lithium story. *Journal of Chemical Education* 1976, *53(5)*, 291–292.
13. Cade, Jack F. John Frederick Joseph Cade: Family memories on the occasion of the 50th anniversary of his discovery of the use of lithium in mania. *The Australian and New Zealand Journal of Psychiatry* 1999, *33(5)*, 615–618.
14. Bliss, Michael. *The Discovery of Insulin,* 2nd ed. University of Toronto Press: Toronto, 1982.
15. Bosworth, D. M. Iproniazid: A brief review of its introduction and clinical use. *Annals of the New York Academy of Sciences* 1959, *80*, 809–819.
16. Davis, William A. History of marsilid. *Journal of Clinical and Experimental Psychopathology* 1958, *19(Suppl. 1)*, 1–10.
17. Kauffman, G. B. The history of iproniazid and its role in antidepressant therapy. *Journal of Chemical Education* 1979, *56*, 35–36.
18. Kramer, Peter D. *Listening to Prozac: A Psychiatrist Explores Antidepressant Drugs and the Making of the Self.* Penguin Books: New York, 1993.
19. Molloy, Bryan B.; Wong, David T.; Fuller, Ray W. The discovery of fluoxetine. *Pharmaceutical News* 1994, *1(2)*, 6–10.
20. Goodson, A. C. Frankenstein in the age of Prozac. *Literature and Medicine* 1996, *15(1)*, 16–32.
21. Snyder, Solomon H. *Drugs and the Brain.* Scientific American Library: New York, 1996.
22. Bender, George A.; Thom, Robert A. *Parke-Davis Pictorial Annals of Medicine and Pharmacy.* Warner Lambert: Morris Plain, NJ, 1999, 96.
23. Curzon, Gerald. How reserpine and chlorpromazine act: The impact of key discoveries on the history of psychopharmacology. *Trends in Pharmacological Sciences* 1990, *11(2)*, 61–63.
24. Thuillier, Jean. *Ten Years That Changed the Face of Mental Illness.* Martin Dunitz: London, 1999.
25. Granger, B. The discovery of haloperidol. *Encephale* 1999, *25(1)*, 59–66.
26. Black, James. A personal perspective on Paul Janssen. *Journal of Medicinal Chemistry* 2005, *48*, 1687–1688.

27. Galemmo, Robert A., Jr.; Janssen, Frans E.; Lewi, Paul J.; Maryanoff, Bruce E. In memorium: Dr. Paul A. J. Janssen (1926–2003). *Journal of Medicinal Chemistry* 2005, *48*, 1686.
28. Nunes, Fatima. LSD: A historical reevaluation. *Journal of Chemical Education* 1968, *45(11)*, 688–691.
29. Hofmann, A. Notes and documents concerning the discovery of LSD. *Agents Actions* 1970, *1*, 148–150.
30. Halford, Bethany. Clandestine chemistry: Legislators, drug companies try to keep cough syrup out of meth production. *Chemical and Engineering News* 2005, March 28, 26.
31. Man, Charles C.; Plummer, Mark L. *The Aspirin War: Money, Medicine, and 100 Years of Rampant Competition.* Alfred A. Knopf: New York, 1991.

CHAPTER 6. DIABETES DRUGS

1. Sanders, Lee J. *The Philatelic History of Diabetes: In Search of a Cure.* American Diabetes Association: Alexandria, VA, 2001.
2. Porter, Roy. *The Greatest Benefit to Mankind: A Medical History of Humanity.* W. W. Norton: New York, 1997.
3. Simmons, John Galbraith. *Doctors and Discoveries: Lives That Created Today's Medicine, from Hippocrates to the Present.* Houghton Mifflin: Boston, 2002.
4. Kyle, R. A.; Shampo, M. A. *Medicine and Stamps.* American Medical Association: Chicago, 1970.
5. Shafrir, Eleazar. *History and Perspective of Diabetes Illustrated by Postage Stamps.* Freund Publishing: London, UK, 1999.
6. Kornberg, Arthur. Remembering our teachers. *Journal of Biological Chemistry* 2001, *276(1)*, 3–11.
7. Raber, Linda. ACS honors work of Carl and Gerty Cori. *Chemical and Engineering News* 2004, October 25, 69–70.
8. Banting, F. G. Early work on insulin. *Science* 1937, *85*, 594–596.
9. Levine, I. E. *The Discovery of Insulin: Dr. Frederick Banting.* Julian Messner: New York, 1959.
10. Rowland, John. *The Insulin Man: The Story of Sir Frederick Banting.* Roy Publishers: New York, 1966.
11. Mayer, Ann Margaret. *Sir Frederick Banting: Doctor against Diabetes.* Chicago Creative Education: Mankato, MN, 1974.
12. McCarthy, Tom. *Frederick Banting and Charles Best: Discovers of Insulin.* Novalis: Ottawa, 1982.
13. Bliss, Michael. The discovery of insulin: How it really happened. *Receptors and Ligands in Intercellular Communication* 1985, *6*, 7–19.
14. Bliss, Michael. *Banting: A Biography.* University of Toronto Press: Toronto, Ontario, Canada, 1992.

15. Bankston, John. *Frederick Banting and the Discovery of Insulin.* Mitchell Lane Publishers: Bear, DE, 2002.
16. Rosenfeld, Louis. Insulin: Discovery and controversy. *Clinical Chemistry* 2002, *48(42),* 2270–2288.
17. Tamblyn, Diana. *Duty Must Be Done: The Story of Frederick Banting,* 2003. Available at: http://www.speedlines.com/MyComics/mycomics.htm, accessed in December 2004.
18. Webb, Michael, ed. *Frederick Banting: Discoverer of Insulin.* Copp Clark Pittman: Mississauga, Ontario, Canada, 1991.
19. Bliss, Michael. Who discovered insulin? *News in Physiological Sciences* 1986, *1,* 31–36.
20. Bliss, Michael. *The Discovery of Insulin,* 2nd ed. University of Toronto Press: Toronto, Ontario, Canada, 1982.
21. Bliss, Michael. J. J. R. Macleod and the discovery of insulin. *Quarterly Journal of Experimental Physiology* 1989, *74(2),* 87–96.
22. Best, C. H. Nineteen hundred twenty-one in Toronto. *Diabetes* 1972, *21,* (*Suppl. 2*), 385–395.
23. Bliss, M. Rewriting medical history: Charles Best and the Banting and Best myth. *Journal of the History of Medicine and Allied Sciences* 1993, *48(3),* 253–274.
24. Macleod, J. J. R. History of the research leading to the discovery of insulin [1922]: Letter of Macleod to Col. Albert Gooderham, Toronto, Sept. 20, 1922. *Bulletin of History of Medicine* 1978, *52,* 295–312.
25. Collip, J. B. Reminiscences on the discovery of insulin. *Canadian Medical Association Journal* 1962, *87,* 1055–1061.
26. Hargittai, István. *The Road to Stockholm: Nobel Prizes, Science, and Scientists.* Oxford University Press: Oxford, UK, 2002.
27. Murnaghan, J. H.; Talalay, P. John Jacob Abel and the crystallization of insulin. *Perspectives of Biology and Medicine* 1967, *11,* 334–380.
28. McElhenry, Victor K. Total synthesis of insulin in Red China. *Science* 1966, *153,* 281–283.
29. Kung, Y. T., Du, Y. C.; Huang, W. T.; Chen, C. C.; Ke, L. T. et al. The Total Synthesis of Crystalline Bovine Insulin *Scientia Sinica* 1965, *14,* 1710.
30. Stille, Darlene R. *Extraordinary Women Scientists.* Children's Press: Chicago, 1995, 85–87.

CHAPTER 7. ANESTHETICS

1. Fülöp-Miller, René. *Triumph over Pain.* Bobbs-Merrill: Indianapolis, IN, 1938, 91, 107, 169, 181.
2. Bishop, W. J. *The Early History of Surgery.* Barnes and Noble Books: New York, 1995.
3. Friedman, Meyer; Friedman, Gerald W. *Medicine's Greatest Discoveries.* Yale University Press: New Haven, CT, 1998.
4. Porter, Roy. *The Greatest Benefit to Mankind: A Medical History of Humanity.* W. W. Norton: New York, 1997.

5. Rushman, G. B.; Davies, N. J. H.; Atkinson, R. S. *A Short History of Anaesthesia: The First 150 Years.* Butterworth-Heinemann: Oxford, UK, 1996.
6. McKenzie, Alistair G. *A History of Anaesthesia through Postage Stamps.* Maclean Dubois: Edinburgh, 2000.
7. Fenster, Julie M. *Mavericks, Miracles, and Medicine: The Pioneers Who Risked Their Lives to Bring Medicine into the Modern Age.* Carroll & Graf: New York, 2003.
8. Wolfe, Richard J.; Menczer, Leonard F., eds. *I Awaken to Glory.* Boston Medical Library: Boston, 1994.
9. Fenster, Julie M. *Ether Day: The Strange Tale of America's Greatest Medical Discovery and the Haunted Men Who Made It.* Perennial: New York, 2002.
10. Taylor, F. L. Crawford W. Long. *Annals of Medical History* 1925, *7*, 267–296.
11. Williams, Guy. *The Age of Miracle: Medicine and Surgery in the Nineteenth Century.* Academy Chicago Publishers: Chicago, 1982.
12. Wolfe, Richard J. *Tarnished Idol: William Thomas Green Morton and the Introduction of Surgical Anesthesia, a Chronicle of the Ether Controversy.* Norman Publishing: San Anselmo, CA, 2000.
13. Read, John. *Humor and Humanism in Chemistry.* G. Bell & Sons: London, 1947.
14. Stratmann, Linda. *Chloroform: The Quest for Oblivion.* Sutton: Phoenix Mill, UK, 2003.
15. Vinten-Johansen, Peter; Brody, Howard; Paneth, Nigel; Rachman, Stephen; Rip, Michael. *Cholera, Chloroform, and the Science of Medicine: A Life of John Snow.* Oxford University Press: Oxford, 2003.
16. Kauffman, George B. Adolf von Baeyer and the naming of barbituric acid. *Journal of Chemical Education* 1980, *57*, 222–223.
17. Laylin, James K., ed. *Nobel Laureates in Chemistry, 1901–1992.* American Chemical Society: Washington, DC, 1993.
18. Bett, W. R. Discovery of veronal. *Chemist and Druggist* 1954, *161*, 168–169.
19. Kogan, Herman. *The Long White Line: The Story of Abbott Laboratories.* Random House: New York, 1963.
20. Hunter, A. R. Twenty years ago: Sodium Pentothal anaesthesia. *Anaesthesia* 1968, *23(3)*, 450–458.
21. Mohoney, John Thomas. *The Merchants of Life: An Account of the American Pharmaceutical Industry.* Harper & Brothers: New York, 1959.
22. Kanigel, Robert. *Apprentice to Genius: The Making of a Scientific Dynasty.* Macmillan: New York, 1986.
23. Andrews, George; Solomon, David, eds. *The Coca Leaf and Cocaine Papers.* Harcourt Brace Jovanovich: New York, 1975.
24. Ruetsch, Y. A.; Böni, T.; Borgeat, A. From cocaine to ropivacaine: The history of local anesthetic drugs. *Current Topics in Medicinal Chemistry* 2001, *1*, 175–182.

1. Shen, T. Y. Inflammatory mediators and anti-inflammatory agents. In: Krogsgaard-Larsen, Povl; Bundgaard, Hans, eds., *A Textbook of Drug Design And Development.* Harwood Academic Publishers: Chur, Switzerland, 1994, 434–463.
2. Dubos, René. *Pasteur and Modern Science.* Science Tech Publishers: Madison, WI, 1988.
3. Bryan, Charles S. *Osler, Inspiration from a Great Physician.* Oxford University Press: New York, NY, 1997.
4. Hench, P. S. A reminiscence of certain events before, during, and after the discovery of cortisone. *Minnesota Medicine* 1953, *July*, 705–711.
5. Lloyd, M. Philip Showalter Hench, 1896–1965. *Rheumatology* 2002, *41*, 582–584.
6. Plotz, C. M. Fifty years of cortisone. *The Journal of Rheumatology* 2002, *29(4)*, 853–854.
7. Neeck, G. Fifty years of experience with cortisone therapy in the study and treatment of rheumatoid arthritis. *Annals of the New York Academy of Science* 2002, *966*, 28–38.
8. Marks, H. M. Cortisone, 1949: a year in the political life of a drug. *Bulletin of History of Medicine* 1992, *66*, 419–439.
9. Chemerda, J. M.; Chamberlin, E. M.; Wilson, E. H.; Tishler, M. The synthesis of cortisone acetate. *Journal of the American Chemical Society* 1951, *73*, 4052.
10. Hirschmann, Ralph. The cortisone era; aspects of its impact. Some contributions of the Merck Laboratories. *Steroids* 1992, *57*, 579–592.
11. Glyn, John. The discovery and early use of cortisone. *Journal of Royal Society of Medicine* 1998, *91*, 513–517.
12. Djerassi, C. *Steroid Made It Possible*, American Chemical Society: Washington, DC, 1990.
13. Peterson, D. H.; Murray, H. C. Microbiological oxygenation of steroids at carbon 11. *Journal of the American Chemical Society* 1952, *74*, 1871.
14. Peterson, D. H. Autobiography of Durey H. Peterson. *Steroids* 1985, *45(1)*, 1–17.
15. Corey, E. J.; Gregoriou, G. A.; Peterson, D. H. Stereochemistry of 11α -hydroxylation of steroids. *Journal of the American Chemical Society* 1958, *80*, 2338.
16. Jeffreys, Diarmuid. *Aspirin: The Remarkable Story of a Wonder Drug.* Bloomsbury: New York, NY. 2004.
17. Mann, Charles C.; Plummer, Mark L. *The Aspirin War, Money, Medicine, and 100 Years of Rampant Competition.* Alfred A. Knopf: New York, 1991.
18. Wasson, Richard L. Aspirin. In: McKetta, John J.; Cunningham, William A., eds. *Encyclopedia of Chemical Process and Development.* Dekker: New York, 1977, 24–30.
19. Piper, P. J.; Vane, John R. Release of additional factors in anaphylaxis and its antagonism by anti-inflammatory drugs. *Nature* 1969, *223*, 29–35.
20. Vane, John R. Inhibition of prostaglandin synthesis as a mechanism of action for aspirin-like drugs *Nature* 1971, *231(25)*, 232–235.
21. Adam, Stewart S. Ibuprofen, the propionics and NSAIDS: personal reflections over four decades. *Inflammopharmacology* 1999, *7(3)*, 191–197.

22. Brune, K.; Hinz, B. The discovery and development of anti-inflammatory drugs. *Arthritis and Rheumatism* 2004, *50(8)*, 2391–2399.
23. Flower, R. J. Non-steroidal anti-inflammatory drugs: back to the future *Rheumatology* 1999, *38(8)*, 693–696.
24. Botting, Regina M.; Botting, Jack H. The discovery of COX-2. In: Pairet, M.; van Ryn, J., eds. *COX-2 Inhibitors.* Birkhäser Verlag: Basel, 2004.
25. Simmons, Daniel L.; Botting, Regina M.; Hla, Timothy Cyclooxygenase isozymes: the biology of prostaglandin synthesis and inhibition. *Pharmacological Reviews* 2004, *56(3)*, 387–437.
26. Brody, Anne Y. *Rochester vs. Searle:* complying with the written description and enablement requirements in early-stage drug discovery. *Biotechnology Law Report* 2003, *22(5)*, 472–475.
27. Vane, John R. Back to an aspirin a day? *Science* 2002, *296*, 474–475.
28. Mpofu, S.; Fatima, F.; Moots, R. J. Anti-TNF-α therapies: they are all the same (aren't they?). *Rheumatology* 2005, *44*, 271–273.
29. Old, Lloyd J. Tumor necrosis factor. *Scientific American* 1988, *258*, 69–75, 59–60.
30. Vilèel, Jan; Feldmann, Marc. Historical review: cytokines as therapeutics and targets of therapeutics. *Trends in Pharmacological Sciences* 2004, *25(4)*, 201–209.
31. Toussirot, Éric; Wending, Daniel. The use of THF-α blocking agents in rheumatoid arthritis: an overview. *Expert Opinion in Pharmacotherapy* 2004, *5(3)*, 581–594.
32. Mikuls, Ted R.; Moreland, Larry W. TNF blockade in the treatment of rheumatoid arthritis: infliximab versus etanercept. *Expert Opinion in Pharmacotherapy* 2001, *2(1)*, 75–84.
33. Zashin, Scott J.; Hesser, M. Laurette. *Arthritis without Pain: The Miracle of the TNF Blockers.* Sarah Allison: Dallas, TX, 2004.
34. Lovinger, Sarah P. Use of biologics for rheumatoid arthritis tempered by concerns over safety, cost. *JAMA* 2003, *289(24)*, 3229–3230.

CHAPTER 9. REFLECTIONS

1. Lynn, Matthew. *The Billion-Dollar Battle: Merck v. Glaxo.* Heinemann: London, UK, 1991, 125.

INDEX